Emerging Drugs and Targets for Alzheimer's Disease
Volume 1: Beta-Amyloid, Tau Protein and Glucose Metabolism

RSC Drug Discovery Series

Editor-in-cheif:
Professor David Thurston, *London School of Pharmacy, UK*

Series Editor:
Dr David Fox, *Pfizer Global Research and Development, Sandwich, UK*
Professor Salvatore Guccione, *University of Catania, Italy*
Professor Ana Martinez, *Instituto de Quimica Medica-CSIC, Spain*
Dr David Rotella, *Wyeth Research, USA*

Advisor to the Board:
Professor Robin Ganellin, *University College London, UK*

Titles in the Series:
  1: Metabolism, Pharmacokinetics and Toxicity of Functional Groups: Impact of the Building Blocks of Medicinal Chemistry on ADMET
  2: Emerging Drugs and Targets for Alzheimer's Disease; Volume 1: Beta-Amyloid, Tau Protein and Glucose Metabolism
  3: Emerging Drugs and Targets for Alzheimer's Disease; Volume 2: Neuronal Plasticity, Neuronal Protection and Other Miscellaneous Strategies

How to obtain future titles on publication:
A standing order plan is available for this series. A standing order will bring delivery of each new volume immediately on publication.

For further information please contact:
Book Sales Department, Royal Society of Chemistry,
Thomas Graham House, Science Park, Milton Road, Cambridge, CB4 0WF, UK
Telephone: +44 (0)1223 420066, Fax: +44 (0)1223 420247, Email: books@rsc.org
Visit our website at http://www.rsc.org/Shop/Books/

Emerging Drugs and Targets for Alzheimer's Disease

Volume 1: Beta-Amyloid, Tau Protein and Glucose Metabolism

Edited by

Ana Martinez
Instituto de Quimica Medica-CSIC, Madrid, Spain

RSCPublishing

RSC Drug Discovery Series No. 2

ISBN: 978-1-84973-063-1
ISSN: 2041-3203

A catalogue record for this book is available from the British Library

© Royal Society of Chemistry 2010

Published by The Royal Society of Chemistry,
Thomas Graham House, Science Park, Milton Road,
Cambridge CB4 0WF, UK

Registered Charity Number 207890

For further information see our web site at www.rsc.org

Foreword

It is a great honour to have been asked to write the foreword to this book in which several distinguished experts in the field have commented on possible therapeutic approaches for the treatment of Alzheimer's disease (AD) and, in some cases, for the possible side effects of those treatments.

One hundred years ago, the first patient with this disease was described,[1] but during this time we have been unable not only to cure but even to prevent the disease. However, during the last 25 years knowledge of several molecular aspects of the disease has dramatically increased. In familial AD, we know that the origin of the neurodegeneration is a consequence of mutations in one of the three following genes: app, ps-1 and ps-2.[2] APP (amyloid precursor protein) is the product of app,[3] and ps-1 and ps-2 code for the proteins, presenilin-1 and presenilin-2.[4] From the pioneering work of Glenner and Wang,[5] we know that the main component of senile plaques, found in the brain of Alzheimer patients is a fragment (the beta amyloid peptide) of APP. It is well known that the main component of the other aberrant structure found in the brain of those patients, neurofibrillary tangles, is the cytoskeletal protein tau, modified by hyperphosphorylation.[6] The kinases involved in that aberrant phosphorylation and the aggregation of tau have been also studied.[6–8] Less is known about the origin of sporadic Alzheimer disease, but there are some risk factors, like aging (that could promote oxidative damage), or the presence of a specific isoform of the protein ApoE, involved in cholesterol transport in the nervous system.[9]

On the other hand, mutations in ps-1 gene could affect glial cells, like microglia, inducing an aberrant activation of those cells.[10] That activation could have some consequences, such as aberrant induction of proliferation of neuronal precursors and inflammation. Indeed, AD can be defined as a CNS disorder with an inflammatory component.

Other studies have suggested that AD is an insulin-resistance disease or that factors related to insulin could act as putative therapeutic agents for the disorder.[11] Other changes in the amount of growth factors, or changes in brain metabolism, have been associated with other aspects of the disease, and, in addition, from the pioneer work of David and Maloney,[12] changes in the level of different neurotransmitters have been found in AD patients.

Inspection of the web site, clinicaltrials.gov, reveals that more than two hundred compounds are currently in clinical trials for AD treatment, and currently there are no broadly effective agents for AD treatment. Thus, the aim of this book is to comment on emerging drugs and targets for Alzheimer disease. Among those targets are those proteases (β and γ secretase) that are involved in APP cleavage, yielding beta amyloid peptide (Aβ), although inhibition of these proteins (mainly γ-secretase) could result in dangerous secondary effects.[13] Also, since dimers, dodecamers, other oligomers and higher polymers of Aβ could be toxic for neurons,[14] Aβ anti-aggregates have been tested.

On the other hand, some kinases or phosphatases involved in the aberrant phosphorylation of tau protein, the main components of NFT, have been described as possible targets to prevent tau pathology, together with antitau aggregates that could prevent the formation of filamentous tau polymers.

In the case of sporadic Alzheimer disease, alterations in the metabolism of cholesterol have been studied,[15] and modulation of cholesterol levels has been suggested for the treatment of the disease.

The 27 chapters of this book provide a broad based and deep review on some of the above-mentioned targets for Alzheimer disease. In Volume 1, the initial section considers strategies aimed at lowering Aβ levels and blocking Aβ aggregation by targetting Aβ secretases, or by other approaches, including immunotherapy. This part includes six chapters, from Drs. Wolfe, Xia, Dotti and Ledesma, Soreq, McLaurin and Solomon. Therapeutic targets focused on tau protein follows, focusing on the regulation of its dephosphorylation, mainly by tau kinase I,[16,17] (or GSK3), or its dephosphorylation by phosphatases like PP2A.[18] This section includes a discussion of the rationale for tau-aggregation inhibitor therapy in the last chapter of this part. The authors for these aspects related to tau pathology are Drs. Iqbal, Jope, Martínez, Hovens and Wischik.

The third part of the book is related to neuron glucose metabolism as a possible target, discussing the possibility of AD being an insulin-resistance disease, about the possible use of insulin-like growth factor 1 as an AD therapeutic or the use of ketone bodies in the potential treatment of AD. The authors for these three chapters are Drs. De la Monte, Torres and Henderson.

In Volume II the first part is related to aspects of neuronal plasticity. The possibility for regeneration of degenerated brain and the use of growth factors like NGF is discussed. The use of induced pluripotent cells from patients could facilitate the search for specific drugs in the treatment of the disease or how to promote synaptic resilience in AD patients. The authors for these chapters are Drs Grundke-Iqbal, Cattaneo, Sugaya and Menniti.

The second part of Volume II is related to different ways for neuronal protection, trying to prevent oxidative damage, microtubule breakdown or inflammatory processes. Four chapters are dedicated to those subjects from Drs. Smith, Ghozes, Vallano and Heneka.

The final section of Volume II is entitled "Miscellaneous targets and strategies" and, as indicated by the title, tries to look for additional solutions beyond the targets previously described. Five chapters are dedicated to this topic from Drs. Lezoualc'h, Ridell, Melchiorre, Nawrot and Kennedy.

In summary, a wide variety of conceptually distinct approaches are presented in an attempt to treat or prevent Alzheimer disease, a disease discovered 100 years ago whose incidence has dramatically increased since that time. Hopefully, memories of this dreaded disease will disappear before the 200th anniversary of its description.

Jesús Avila
Centro de Biología Molecular "Severo Ochoa" (CSIC-UAM), Madrid, Spain

References

1. Alzheimer, *Psych. genchtl. Med.*, 1907, **64**, 146.
2. D. L. Price and S. S. Sisodia, *Ann. Rev. Neurosci.*, 1998, **21**, 479.
3. C. L. Masters, G. Simms, N. A. Weinman, G. Multhaup, B. L. McDonald and K. Beyreuther, *Proc. Nat. Acad. Sci. USA.*, 1985, **82**, 4245.
4. R. Sherrington, E. I. Rogaev, Y. Liang, E. A. Rogaeva, G. Levesque, M. Ikeda, H. Chi, C. Lin, G. Li and K. Holman, *et al., Nature*, 1995, **375**, 754.
5. G. G. Glenner and C. W. Wong, *Biochem. Biophys. Res. Commun.*, 1984, **122**, 1131.
6. I. Grundke-Iqbal, K. Iqbal, M. Quinlan, Y. C. Tung, M. S. Zaidi and H. M. Wisniewski, *J. Biol. Chem.*, 1986, **261**, 6084.
7. I. Grundke-Iqbal, K. Iqbal, Y. C. Tung, M. Quinlan, H. M. Wisniewski and L. I. Binder, *Proc. Nat. Acad. Sci. USA.*, 1986, **83**, 4913.
8. E. Montejo de Garcini, L. Serrano and J. Avila, *Biochem. Biophys. Res. Commun.*, 1986, **141**, 790–796.
9. E. H. Corder, A. M. Saunders, W. J. Strittmatter, D. E. Schmechel, P. C. Gaskell, G. W. Small, A. D. Roses, J. L. Haines and M. A. Pericak-Vance, *Science*, 1993, **261**, 921.
10. S. H. Choi, K. Veeraraghavalu, O. Lazarov, S. Marler, R. M. Ransohoff, J. M. Ramirez and S. S. Sisodia, *Neuron*, 2008, **59**, 568.
11. E. Carro, J. L. Trejo, T. Gomez-Isla, D. LeRoith and I. Torres-Aleman, *Nature Med.*, 2002, **8**, 1390.
12. P. Davies and A. J. Maloney, *Lancet*, 1976, **2**, 1403.
13. E. Inoue, M. Deguchi-Tawarada, A. Togawa, C. Matsui, K. Arita, S. Katahira-Tayama, T. Sato, E. Yamauchi, Y. Oda and Y. Takai, *J. Cell Biol.*, 2009, **185**, 551.
14. C. Haass and D. J. Selkoe, *Nature Rev.*, 2007, **8**, 101.
15. M. D. Ledesma, J. Abad-Rodriguez, C. Galvan, E. Biondi, P. Navarro, A. Delacourte, C. Dingwall and C. G. Dotti, *EMBO Rep.*, 2003, **4**, 1190.
16. K. Ishiguro, A. Shiratsuchi, S. Sato, A. Omori, M. Arioka, S. Kobayashi, T. Uchida and K. Imahori, *FEBS Lett.*, 1993, **325**, 167.
17. J. J. Lucas, F. Hernandez, P. Gomez-Ramos, M. A. Moran, R. Hen and J. Avila, *EMBO J.*, 2001, **20**, 27.
18. C. X. Gong, T. Lidsky, J. Wegiel, L. Zuck, I. Grundke-Iqbal and K. Iqbal, *The J. Biol. Chem.*, 2000, **275**, 5535.

Preface

"Over the next century, experts estimate that Alzheimer's disease will be more prevalent than AIDS, cancer and all cardiovascular diseases"

World Health Organization

Alzheimer's disease, the most common form of dementia, is a progressive and neurodegenerative brain disorder that results in symptoms such as loss of memory, impaired reasoning and changes in mood and behavior. The risk of suffering this pathology increases sharply with age and poses one of the major public health problems, both emotional and financial, not only to the present society, but even more to the coming generation. The lack of a cure and an effective form of treatment means that our current society must find effective ways to slow the progression of Alzheimer's disease and develop effective and efficient strategies to meet the needs of those with this challenging disease.

Recent research advances in molecular biology and technology have provided multiple credible hypotheses around which therapeutic agents can be developed. This book collects some of the most outstanding examples of new drugs currently under pharmaceutical development or new targets in the validation process that will reach the Alzheimer's drugs market over the next few years as disease modifying drugs.

I wish to thank all the contributors to the chapters in this book for their efforts to summarize the most relevant data on their topics and mainly, for their faith in the project. I also extend special thanks to my colleagues Guadalupe Mengod, Carmen Gil, Ana Perez-Castillo, Manuel Sarasa, Jesus Avila, Khalid Iqbal and Daniel Perez for their support in reviewing some of the chapters. My special thanks and undying gratitude to Valle Palomo for her help with the

RSC Drug Discovery Series No. 2
Emerging Drugs and Targets for Alzheimer's Disease
Volume 1: Beta-Amyloid, Tau Protein and Glucose Metabolism
Edited by Ana Martinez
© Royal Society of Chemistry 2010
Published by the Royal Society of Chemistry, www.rsc.org

cover design, to the staff at RSC, especially Gwen Jones, for their patience and support in bringing this book to completion and to my family, especially my husband and my seven children for their patience during my endless days.

Finally, I hope that this complete book, split in two different volumes and written by distinguished experts on the field, can serve as an essential resource of information for scientists in the pharmaceutical and biotechnology industries and academics working in the neurodegenerative field. I would hope that all of them find new clues in this book to accelerate their own research successfully. My last words are to express the hope that support of research aimed at finding an effective therapy to this disease, which devastates the brain of the patients and the heart of their caregivers, should have the highest priority both in government and private funding.

Ana Martinez, PhD
Professor of Research
Instituto de Quimica Médica-CSIC
Madrid, Spain

Contents

RSC Drug Discovery Series No. 2
Emerging Drugs and Targets for Alzheimer's Disease
Volume 1: Beta-Amyloid, Tau Protein and Glucose Metabolism
Edited by Ana Martinez
© Royal Society of Chemistry 2010
Published by the Royal Society of Chemistry, www.rsc.org

Tau Protein as Target

Glucose Metabolism as Target

VOLUME 2

Neuronal Plasticity as Target

Chapter 15   Regeneration of Degenerated Brain: A Promising Therapeutic Target   3
Inge Grundke-Iqbal and Khalid Iqbal

Chapter 16   Promoting Synaptic Resilience in Alzheimer's Disease Patients Through Phosphodiesterase Inhibition   21
Kelly R. Bales and Frank S. Menniti

Neuronal Protection as Target

Chapter 19   Targeting Oxidative Mechanisms in Alzheimer Disease   **97**
Hyun P. Lee, Raj K. Rolston, Xiongwei Zhu,
Michael W. Marlatt, Rudy J. Castellani, Akihiko Nunomura,
Hyoung-gon Lee, Gemma Casadesus, George Perry
and Mark A. Smith

**Chapter 20   Davunetide (NAP) Pharmacology: Neuroprotection and
Tau**   **108**
Illana Gozes

Miscellaneous Targets and Strategies

Chapter 23 Serotonin 5-HT$_4$ Receptors as Pharmacological Targets for the Treatment of Alzheimer's Disease **169**

Isabelle Berque-Bestel and Frank Lezoualc'h

Chapter 24 Targeting ApoE in Alzheimer's Disease: Liver X Receptor Agonists as Potential Therapeutics **191**

David R. Riddell and David J. O'Neill

Chapter 25 Discovery of Memoquin, a Multitarget-Directed Ligand (MTDL) for the Treatment of Alzheimer's Disease 213
M. Laura Bolognesi, Anna Minarini and Carlo Melchiorre

Chapter 26 RNA Interference of Genes Related to Alzheimer's Disease 228
Barbara Nawrot, Malgorzata Sierant and Alina Paduszynska

Chapter 27 Medicinal Plants, Phytochemicals and Alzheimer's Disease 269

David O. Kennedy, Emma L. Wightman and Edward J. Okello

BETA-AMYLOID AS TARGET

The Amyloid Hypothesis of Alzheimer's Disease and Prospects for Therapeutics

MICHAEL S. WOLFE

Center for Neurologic Diseases, 77 Avenue Louis Pasteur, H.I.M. 754, Brigham and Women's Hospital and Harvard Medical School, Boston, MA, USA

1.1 A Brief History of Alzheimer's Disease

In 1906, the Bavarian physician Alois Alzheimer described the form of progressive senile dementia that bears his name. This discovery ultimately led to the recognition that age-dependent deficits in learning and memory are not simply "senility" but are due to specific pathological processes. Alzheimer described the two principal types of brain pathology that are still considered the definitive diagnosis of the disease: cerebral plaques and neurofibrillary tangles. The plaques are dense proteinaceous cores, containing the amyloid-β peptide (Aβ) as the primary component, surrounded by destroyed and damaged neurons. Neurofibrillary tangles are found inside neurons and neuronal projections, and these are primarily composed of a filamentous hyperphosphorylated form of the microtubule-associated protein tau.[1]

RSC Drug Discovery Series No. 2
Emerging Drugs and Targets for Alzheimer's Disease
Volume 1: Beta-Amyloid, Tau Protein and Glucose Metabolism
Edited by Ana Martinez

Published by the Royal Society of Chemistry, www.rsc.org

Alzheimer's first description of this brain pathology raised two major questions. Are plaques and tangles causative or merely disease markers? If causative, which of these two features is closer to the initial pathogenic event? The debate over these issues continued for most of the century, but a general consensus eventually emerged that both Aβ aggregates and tau filaments are fundamental to the disease process, but the former is upstream of the latter.[2] Strong evidence, gathered using a variety of approaches, favours the "amyloid hypothesis" of Alzheimer's disease (AD). However, understanding this evidence requires an appreciation of how Aβ is produced.

1.2   APP Processing and the Amyloid Hypothesis

The amyloid-β precursor protein (APP) is a single-pass integral membrane protein that is processed by several proteases called secretases.[3] APP is first cut by either α- or β-secretase (Figure 1.1). α-Secretase cleaves close to the transmembrane domain, in the middle of the Aβ region of APP, to release a soluble ectodomain (α-APP$_s$) and produce a membrane-bound 83 residue C-terminal fragment (C83). Alternatively, β-secretase sheds the APP ectodomain (β-APP$_s$), cleaving further from the membrane to produce a 99-residue C-terminal fragment with the N-terminus of Aβ. Both C99 and C83 are substrates for a third protease, γ-secretase, which catalyses an unusual hydrolysis within the

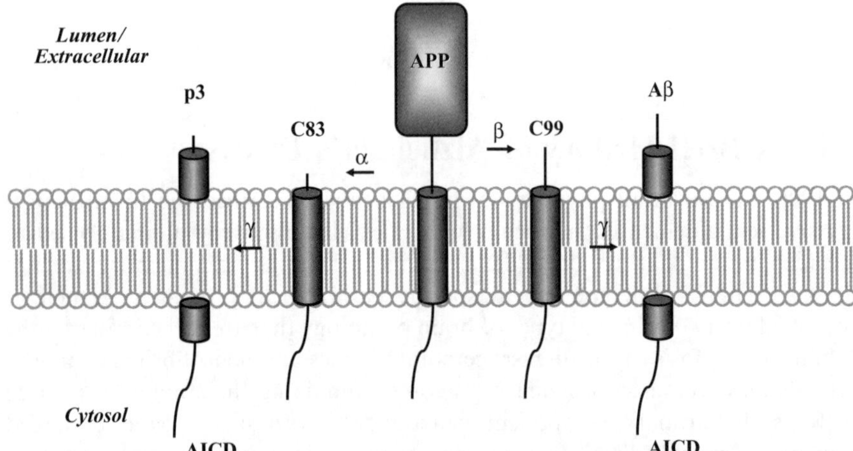

Figure 1.1   APP processing. The amyloid-β precursor protein (APP) is proteolytically processed either by α-secretase or β-secretase. α-Secretase processing releases the ectodomain, α-APP$_s$, leaving an 83-residue C-terminal fragment (C83) in the membrane. Alternatively, cleavage by β-secretase produces β-APP$_s$ and a 99-residue fragment (C99). C99 is hydrolysed within its transmembrane domain by γ-secretase to form the amyloid-β peptide (Aβ) implicated in the pathogenesis of Alzheimer's disease. C83 is likewise cleaved by γ-secretase to produce p3, an N-terminally truncated variant of Aβ. Cleavage of APP substrates by γ-secretase to release the APP intracellular domain (AICD).

transmembrane region. Proteolysis of C99 by γ-secretase produces the 4 kDa Aβ, while C83 is cut into p3, an N-terminally truncated form of Aβ. γ-Secretase cleavage primarily results in a 40 amino acid Aβ peptide (Aβ40). However, other shorter and longer forms of Aβ are also produced, in particular a 42 amino acid version (Aβ42) that has a very high propensity to aggregate and that is the major Aβ species found in plaques.

Compelling genetic evidence for the amyloid hypothesis came in 1991, when missense mutations in the gene encoding APP were found to cause AD in certain families.[4–6] These familial forms of AD are dominant and have their age of onset quite early, in the sixth decade of life. Otherwise, the disease progresses the same and shows identical pathology (*i.e.* plaques and tangles) to sporadic late-onset AD. These AD-causing missense mutations occur in and around the Aβ region of APP,[7] altering the production or properties of Aβ. Most intriguing are several different AD-causing mutations that occur near the γ-secretase cleavage site. These mutations lead to specific increases in Aβ42, implicating this Aβ variant in particular as the molecular culprit in AD. One family has been identified with a duplication of the APP gene, demonstrating that overexpression of wild-type APP can also lead to AD.[8] Other families with dominant, early onset AD were found in 1995 to carry mutations in two homologous genes called presenilin-1 (PS1) and presenilin-2 (PS2).[9–11] Remarkably, over 100 different missense mutations in the presenilins have been identified so far.[12] These mutations are also associated with increases in Aβ42 production,[7] indicating that presenilins somehow modulate γ-secretase activity to determine the specificity of this protease. Subsequent biochemical studies demonstrated that presenilin is part of a larger γ-secretase complex and that this protein is the catalytic component.[13] Thus, the major sites of genetic mutations that cause AD are found in γ-secretase and its substrate APP.

1.3   Present and Future Alzheimer Therapeutics

The current therapeutics for AD are primarily cholinergic agents, specifically inhibitors of acetylcholinesterase.[14] The basis for this approach is the fact that AD causes dramatic losses in cholinergic neurons. Thus, these agents increase levels of acetylcholine to keep the remaining cholinergic neurons firing. Another drug called memantine dampens the action of glutamate at NMDA receptors,[14] which is thought to reduce excitotoxicity. Unfortunately, neither type of therapy stops the progressive loss of neurons, and these drugs eventually becomes ineffective. Moreover, multiple neurotransmitter systems are altered in AD. Agents that affect the molecules responsible for the degeneration (Aβ or tau) and downstream neurotoxic events would be better. Major efforts to find such agents have been bearing fruit in recent years, with promising therapeutics at various stages of development. This chapter will highlight new anti-amyloid strategies that hold the promise of delaying or reversing the devastation of AD, a need that will only increase as the human population continues to become more aged.

1.3.1 β-Secretase and Inhibitors

Because Aβ is so closely associated with AD, the proteases that produce this peptide, β- and γ-secretases, have been top targets for therapeutic development.[15] β-Secretase (also called BACE1) is a membrane-anchored aspartyl protease in the pepsin family,[16] and a soluble recombinant form of this protease has been cocrystallised with an active site-directed inhibitor.[17] The search for β-secretase inhibitors has been greatly facilitated by its resemblance to other aspartyl protease targets, particularly renin and HIV protease, and the ability to do structure-based design. Despite these advantages, the development of β-secretase inhibitors as practical drug candidates has been challenging. One reason is the nature of β-secretase active site, which is relatively long, shallow and hydrophilic, making difficult the discovery of small, potent inhibitors that penetrate biological membranes. The other problem is the tendency of β-secretase inhibitors to be substrates for P-glycoprotein,[18] which transports the compounds out of the brain. These problems are gradually being overcome, however, and the first β-secretase inhibitor has entered into clinical trials.

This clinical candidate is a compound from CoMentis. Although the structure of this compound has not yet been made public, it is apparently related to the hydroxylamine compound **1** (Figure 1.2), which is a highly potent inhibitor of β-secretase (K_i ~ 1.8 nM) that shows some selectivity *versus* other human aspartyl proteases (*e.g.*, cathepsin D and BACE2).[19] Unlike many previously reported β-secretase inhibitors, compound **1** displayed potency for blocking Aβ production at the level of β-secretase in cells (IC_{50} ~ 1 nM). The compound was also shown to acutely lower Aβ levels in APP transgenic mice after intraperitoneal administration, although its ability to penetrate the brain is not clear. Another promising compound has recently been reported from GlaxoSmithKline, hydroxylamine compound **2**.[20] Although closely similar to **1**, the subtle changes (*e.g.*, cyclic sulfonamide, ethylamine, and trifluoromethyl group) resulted in a compound that is orally bioavailable and able to lower brain Aβ levels upon chronic administration in an AD mouse model. Compound **3** from Merck was taken a step further, demonstrating an ability to lower Aβ in the cerebrospinal fluid (CSF) of rhesus monkeys, albeit transiently due to metabolism by the cytochrome P450 CYP3A4.[21] However, coadministration with ritonavir, a potent CYP3A4 inhibitor, led to sustained reduction in CSF Aβ levels.

Membrane secretases typically process a number of different substrates, and β-secretase is no exception. The protease also cuts the transmembrane protein ST6Gal, a sialyltransferase particularly important to the function of B cells in the immune system.[22] Neuregulin-1 is also a substrate, and β-secretase processing of neuregulin-1 is apparently critical for myelination in the peripheral nervous system.[23] While these findings might raise concerns about β-secretase as a therapeutic target, the knockout mice show dramatic reductions in brain Aβ levels, while being fertile and showing only modest behavioural alterations.[24,25] Further investigation of these mice, however, revealed a schizophrenia-like phenotype mediated through decreased neuregulin-1 signalling.[26]

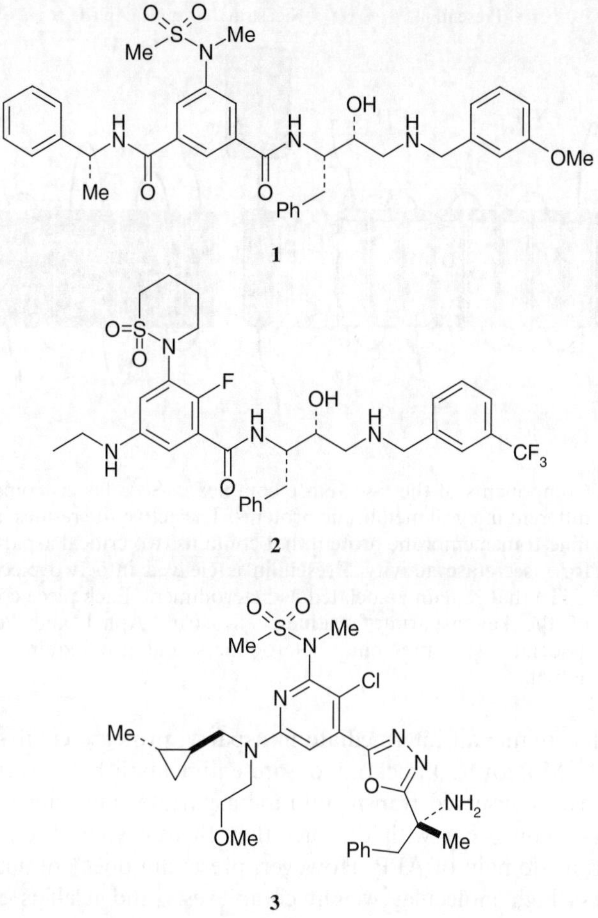

Figure 1.2 β-Secretase inhibitors.

The effect of conditional knockout or knockdown of β-secretase in adult mice is unknown, but pharmacological inhibition of β-secretase (with the compound delivered by intracerebroventricular administration) significantly lowered Aβ without affecting neuregulin-1 processing.[27] In short, although β-secretase appears to be a promising target, the effect of long-term inhibition of this protease in aging adult humans remains to be seen.

1.3.2 γ-Secretase, Inhibitors and Modulators

γ-Secretase is considerably more complicated than β-secretase. While a single protein is responsible for β-secretase activity, γ-secretase is a complex of integral membrane proteins that includes presenilin (Figure 1.3).[13] Pharmacological evidence suggested an aspartyl protease mechanism for this enzyme: peptide

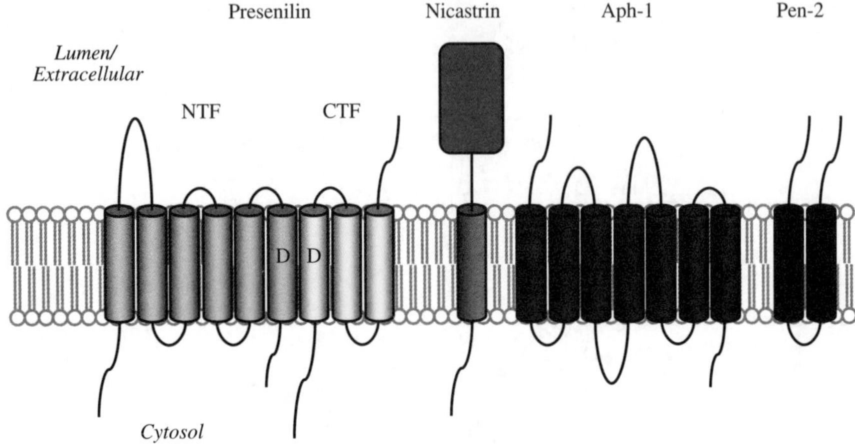

Figure 1.3 Components of the γ-secretase complex. γ-Secretase is composed of four different integral membrane proteins. The active site resides in presenilin, a nine-transmembrane protein that contains two critical aspartates required for γ-secretase activity. Presenilin is cleaved into two pieces (NTF and CTF) that remain associated as heterodimers. Each piece contributes one of the key aspartate residues. Nicastrin, Aph-1 and Pen-2 proteins associate with presenilin heterodimers and are required for protease activity.

analogues that mimic an intermediate in aspartyl protease catalysis inhibit γ-secretase.[28,29] Moreover, knockout of presenilin abolishes γ-secretase activity,[30,31] and two conserved transmembrane aspartates are critical for γ-secretase activity,[32] consistent with the fact that this enzyme cleaves within the transmembrane domain of APP. However, presenilin does not act alone. This protein enters high molecular weight complexes[33] and itself is cut into two pieces that remain associated.[34,35] These heterodimers are metabolically stable,[36,37] and their formation is tightly regulated,[38] suggesting that they are the bioactive form of presenilin. Indeed, active site-directed inhibitors of γ-secretase bind directly to presenilin heterodimers,[39,40] strong evidence that the catalytic centre of this protease resides at the heterodimeric interface. It is now clear that the γ-secretase complex consists of presenilin heterodimers,[41,42] a single-pass membrane protein called nicastrin,[42,43] and two other integral membrane proteins called aph-1 and pen-2.[44,45]

Despite the complexity of γ-secretase, identifying cell-permeable inhibitors has been easier for this protease than for β-secretase. The likely reason is that the active site is hydrophobic, reflected by the fact that the enzyme recognises and cleaves transmembrane domains. Thus, inhibitors should also have hydrophobic character to interact with the enzyme, and this same characteristic aids membrane permeability. The most serious concern is mechanism-based toxicity: γ-secretase also cleaves the Notch receptor as part of an essential signalling event,[46] and blocking γ-secretase with inhibitors causes gastro-intestinal toxicity and immunosuppression in mice.[47,48] Nevertheless, one

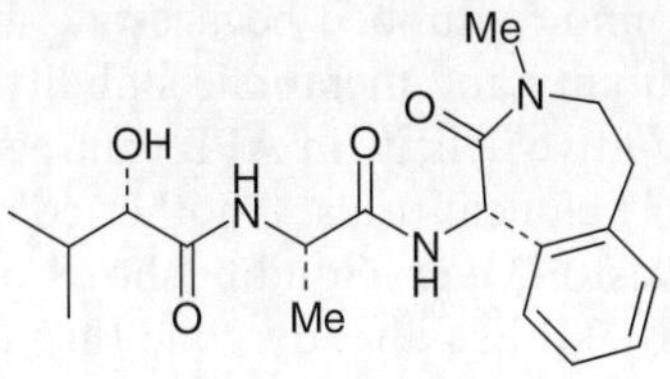

4

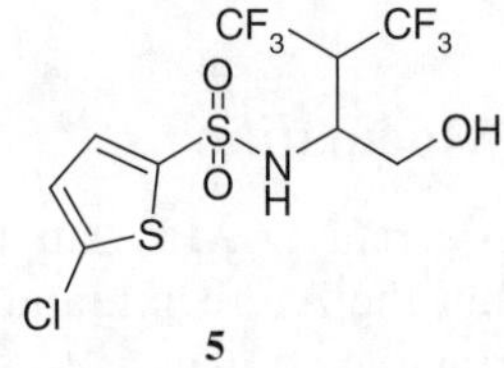

5

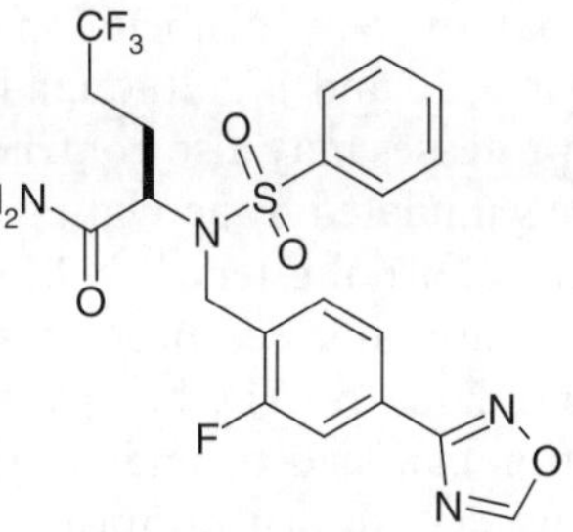

6

Figure 1.4 γ-Secretase inhibitors and modulators.

inhibitor has advanced into late clinical trials: Eli Lilly's benzodiazepine LY450139 (compound **4** in Figure 1.4). LY-450139 was chronically administered for five months to young APP transgenic mice, leading to reduced total brain Aβ and slower formation of Aβ plaques and without overt toxicity. In initial human trials in healthy volunteers,[49,50] single doses of LY-450139 up to 140 mg were apparently safe and reduced plasma Aβ levels by up to 72.6%. However, steady-state Aβ in the cerebrospinal fluid was not affected, and it is unclear if higher doses can lower brain Aβ and without Notch-related side effects.

In contrast to γ-secretase inhibitors, with their potential to cause Notch-related toxicities, compounds that can modulate the enzyme to alter or block Aβ production with little or no effect on Notch would bypass this potential roadblock to therapeutics. One agent has recently been reported to be a Notch-sparing γ-secretase modulator: the sulfonamide inhibitor **5** developed by researchers at Wyeth.[51] Although said to be Notch-sparing, the compound does inhibit Notch proteolysis by γ-secretase, but with a 14-fold higher IC$_{50}$ than for

APP proteolysis. Compound **5**, dubbed begacestat, showed good pharmaceutical properties (*e.g.*, solubility and metabolic stability), lowered brain Aβ and purportedly reversed cognitive deficits in APP transgenic mice.[51] Compound **5** has now entered into early clinical trials. Another compound, the sulfonamide BMS-708163 (**6**) from Bristol-Myers-Squibb, shows better selectivity for APP *vis-à-vis* Notch (200-fold) and has already gone through Phase-I clinical trials with promising results, including the demonstration of Aβ lowering in the cerebrospinal fluid.[52]

1.3.3   α-Secretase and Modulators

Instead of processing by β-secretase, APP can be alternatively shed by α-secretases, which cleave within the Aβ region and therefore preclude Aβ production. Thus, another therapeutic strategy is to shunt APP processing down the α-secretase pathway. Indeed, stimulation of α-secretase-mediated shedding of APP reduces Aβ formation. Several metalloproteases display α-secretase activity, including ADAMs 9, 10 and 17, although knockout of any can lead to compensation and other proteases may also contribute.[53] However, ADAM 17 (also called TACE) can be stimulated by activating protein kinase C (PKC), for instance by treatment with phorbol esters.[54,55] More practically, PKC can be activated by M1 and M3 muscarinic receptor agonists,[56] and such agents are considered reasonable candidates for AD therapy. M1 activation appears more promising, as M3 activation can lead to undesirable side effects, and an M1-selective muscarinic agonist has shown promising results in an APP/PS1/tau triple transgenic AD mouse model, decreasing both Aβ and tau pathology in these mice.[57]

1.3.4   Inhibitors of Cholesterol Biosynthesis

Another gene associated with AD is *apoE*, which encodes for a lipid-binding protein. This gene has three allelic variants: *apoE2*, *apoE3*, and *apoE4*. Those with one or two copies of *apoE4* are at higher risk for AD, while carriers of the *apoE2* allele have a lower risk compared to the general population.[58,59] How could alterations in a lipid-binding protein lead to AD? One clue comes from epidemiological studies that suggest a connection between high cholesterol levels during midlife and an increased risk of AD later in life.[60] Moreover, people taking drugs called statins that inhibit cholesterol biosynthesis have a reduced risk of developing AD.[61,62] Adding a further connection to the amyloid hypothesis, rabbits fed cholesterol had elevated brain Aβ levels and developed amyloid plaques.[63] Further, cells treated with statins produced much less Aβ.[64,65] and similar results were observed in the plasma and cerebrospinal fluid of guinea pigs given high doses of statins.[65] These drugs are widely used and quite safe for long-term treatment, so the possibility that they may prevent or delay the onset of AD has generated keen interest. However, studies designed to test this idea have led to equivocal results so far.

1.3.5 Amyloid Aggregation Inhibitors

Because plaques containing Aβ fibrils are a defining pathological feature of AD and mutations that alter Aβ production or aggregation properties cause AD, blocking the assembling of Aβ into oligomers and fibrils is considered a potential strategy for preventing the toxic effects of this peptide. A key question, however, is which of these assembled states of Aβ is responsible for the devastation of AD? The answer is not entirely clear, but evidence suggests that oligomeric Aβ may be the primary culprit,[66] with Aβ plaques being less deleterious but a potential reservoir for soluble oligomers. Aβ oligomers formed inside cells in culture were secreted into medium, and these oligomers markedly inhibited hippocampal long-term potentiation (LTP) in rats *in vivo*. LTP is a measure of the formation and strength of synaptic connections and is thought to be essential for learning and memory. In a follow-up study, soluble Aβ oligomers isolated from the cerebral cortex of AD brains likewise inhibited LTP and also reduced the dendritic spine density in normal rodent hippocampus and disrupted memory.[67] These effects were specifically attributed to Aβ dimers. However, another report suggests higher-order Aβ oligomers may be responsible for disrupted memory, at least in mice.[68] Recently, the cellular prion protein has been identified as a candidate receptor for soluble Aβ oligomers that mediates synaptic dysfunction.[69]

The above findings suggest that compounds that prevent the formation of Aβ fibrils but not oligomers may have effects *in vivo* that are the opposite of what is intended: Blocking fibril formation may lead to an elevation in levels of toxic oligomers. Thus, it is critical to determine the points along the aggregation pathway in which a compound works. Among investigational compounds reported to block Aβ oligomer formation, *scyllo*-inositol (compound **7**, Figure 1.5) is perhaps the most advanced (see Chapter 5 for more information). This orally bioavailable compound was shown to inhibit high-molecular weight Aβ oligomer formation in the brain and ameliorate several AD-like phenotypes in transgenic mice, including impaired cognition, altered synaptic physiology, cerebral Aβ pathology and accelerated mortality.[70] *Scyllo*-inositol, under development by Transition Therapeutics in partnership with Elan, is currently in advanced clinical trials for the treatment of AD. Another strategy for inhibiting Aβ aggregation involves lowering copper and zinc levels, as these two metals can apparently form complexes with Aβ and zinc in particular can

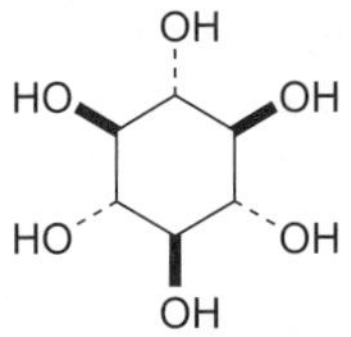

Figure 1.5 Aβ aggregation inhibitor scyllo-inositol.

induce aggregation at physiologically relevant concentrations.[71] The antibiotic clioquinol chelates zinc and copper, and oral administration of this drug for only nine weeks in transgenic mice led to a substantial reduction in brain Aβ deposition.[72] A clioquinol analogue, PBT2 from Prana Biotechnology, is advancing in clinical trials, with some promising results so far.[73]

1.3.6 Immunological Approaches

Another general strategy for preventing Aβ build-up is to enhance the clearance of this peptide. A novel approach toward this end is immunisation, either actively with the Aβ peptide itself or passively with anti-Aβ antibodies. In 1999, Shenk and colleagues at Elan Pharmaceuticals reported the effects of immunising with Aβ42 in transgenic mice harboring an AD-causing mutant human APP that normally develop Aβ-containing plaques within 3–6 months.[74] Immunisation of young mice prevented the formation of Aβ-containing plaques and subsequent AD-like neuropathological changes, while immunisation of older mice led to substantial reduction of plaque burden and pathology. Active immunisation was also shown to reduce cognitive dysfunction in transgenic mice.[75,76]

Passive immunisation with anti-Aβ antibodies can also reduce plaque burdens in transgenic mice[77] and reverse memory deficits.[78] These antibodies can apparently enter the brain since they can be found attached to plaques and can even clear pre-existing plaques through Fc-mediated microglial phagocytosis. A related study by researchers at Eli Lilly showed that administration of anti-Aβ antibodies in transgenic mice leads to a rapid and large increase in peripheral Aβ levels[79] and can reverse memory deficits.[78] In this case, the antibodies apparently did not bind to brain Aβ deposits, and evidence suggested that the sequestration of plasma Aβ causes a marked reduction in Aβ deposition in the brain. Despite the uncertainty of the mechanism by which anti-Aβ antibodies exert their effects in mice, the overall results are promising, because they suggest that active immunisation may not be necessary. Such findings are of immediate practical importance because early clinical trials for active immunisation with Aβ42 caused serious, even lethal, encephalitis in a small but unacceptable fraction of the subjects. One antibody from Elan and Wyeth has advanced into late-stage clinical trials, although the mid-phase trial gave equivocal results. Passive immunisation may not be the only solution to this problem though. Other strategies, such as using the N-terminal region of Aβ as the immunogen or hapten may avoid the cytotoxic T-cell response elicited by full-length Aβ.[80]

1.4 Perspective

Our understanding of the molecular basis of AD has advanced tremendously in the past 25 years. Evidence from pathology, genetics, biochemistry, molecular biology, cell biology and animal models strongly support the amyloid hypothesis of AD pathogenesis, in which aggregated forms of Aβ initiate a cascade leading to neurotoxicity and neurodegeneration. Although there are still gaps in our understanding (*e.g.*, how does Aβ cause neurotoxicity?), the sense is that enough is known to develop effective therapeutics for prevention or treatment. This translational research, however, has proven challenging for a variety of reasons, depending on the target and the strategy. For instance, β-secretase is a key target that appears to be relatively safe; however, the development of potent inhibitors that access the brain has been difficult. In contrast, many γ-secretase inhibitors readily enter the brain, but these agents should lower Aβ production while leaving critical functions of γ-secretase (*e.g.*, Notch signalling) intact. Immunotherapy seems promising, but questions remain about the mechanism of action, and clinical trials have so far been equivocal.

The main difficulty in general may be the advanced stage of the AD patients in the various trials run to date. Giving anti-Aβ agents to an Alzheimer patient may be akin to giving cholesterol-lowering statins to a patient having a heart attack. Clearly in the latter case, the ability to identify people at high risk of cardiovascular disease has been invaluable, saving many lives. Early diagnosis or assessment of risk of AD will be critical for designing appropriate clinical trials to give experimental therapeutics the best chance of showing positive outcomes. In this regard, the identification of biomarkers is essential and will also provide a means of discerning if a therapeutic candidate is effective at the molecular level. Despite the many difficulties, there is the palpable feeling that disease-modifying therapies for AD that target Aβ will soon be at hand: although each target and strategy has its issues, the many agents entering or already in the pipeline raise the hope that effective therapeutics will emerge at the other end.

References

1. M. Goedert and M. G. Spillantini, *Science*, 2006, **314**, 777–781.
2. J. Hardy and D. J. Selkoe, *Science*, 2002, **297**, 353–356.
3. W. P. Esler and M. S. Wolfe, *Science*, 2001, **293**, 1449–1454.
4. A. Goate, M. C. Chartier-Harlin, M. Mullan, J. Brown, F. Crawford, L. Fidani, L. Giuffra, A. Haynes, N. Irving and L. James, *et al.*, *Nature*, 1991, **349**, 704–706.
5. M. C. Chartier-Harlin, F. Crawford, H. Houlden, A. Warren, D. Hughes, L. Fidani, A. Goate, M. Rossor, P. Roques and J. Hardy, *et al.*, *Nature*, 1991, **353**, 844–846.
6. J. Murrell, M. Farlow, B. Ghetti and M. D. Benson, *Science*, 1991, **254**, 97–99.
7. D. J. Selkoe, *Physiol. Rev.*, 2001, **81**, 741–766.
8. A. Rovelet-Lecrux, D. Hannequin, G. Raux, N. Le Meur, A. Laquerriere, A. Vital, C. Dumanchin, S. Feuillette, A. Brice, M. Vercelletto, F. Dubas, T. Frebourg and D. Campion, *Nature Genet.*, 2006, **38**, 24–26.

9. R. Sherrington, E. I. Rogaev, Y. Liang, E. A. Rogaeva, G. Levesque, M. Ikeda, H. Chi, C. Lin, G. Li and K. Holman, *et al.*, *Nature*, 1995, **375**, 754–760.

10. E. Levy-Lahad, W. Wasco, P. Poorkaj, D. M. Romano, J. Oshima, W. H. Pettingell, C. E. Yu, P. D. Jondro, S. D. Schmidt and K. Wang, *et al.*, *Science*, 1995, **269**, 973–977.

11. E. I. Rogaev, R. Sherrington, E. A. Rogaeva, G. Levesque, M. Ikeda, Y. Liang, H. Chi, C. Lin, K. Holman and T. Tsuda, *et al.*, *Nature*, 1995, **376**, 775–778.

12. http://www.alzforum.org/res/com/mut/pre/default.asp.

13. M. S. Wolfe, *Biochemistry*, 2006, **45**, 7931–7939.

14. G. K. Wilcock, in *Pharmacological Mechanisms in Alzheimer Therapeutics*, ed. A. C. Cuello, Springer, New York NY, 2008, pp. 36–49.

15. M. S. Wolfe, *J. Med. Chem.*, 2001, **44**, 2039–2060.

16. S. L. Cole and R. Vassar, *J. Biol. Chem.*, 2008, **23**, 23.

17. L. Hong, G. Koelsch, X. Lin, S. Wu, S. Terzyan, A. K. Ghosh, X. C. Zhang and J. Tang, *Science*, 2000, **290**, 150–153.

18. J. E. Meredith Jr, L. A. Thompson, J. H. Toyn, L. Marcin, D. M. Barten, J. Marcinkeviciene, L. Kopcho, Y. Kim, A. Lin, V. Guss, C. Burton, L. Iben, C. Polson, J. Cantone, M. Ford, D. Drexler, T. Fiedler, K. A. Lentz, J. E. Grace Jr, J. Kolb, J. Corsa, M. Pierdomenico, K. Jones, R. E. Olson, J. E. Macor and C. F. Albright, *J. Pharmacol. Exp. Ther.*, 2008, **326**, 502–513.

19. A. K. Ghosh, *J. Med. Chem.*, 2009, **52**, 2163–2176.

20. I. Hussain, J. Hawkins, D. Harrison, C. Hille, G. Wayne, L. Cutler, T. Buck, D. Walter, E. Demont, C. Howes, A. Naylor, P. Jeffrey, M. I. Gonzalez, C. Dingwall, A. Michel, S. Redshaw and J. B. Davis, *J. Neurochem.*, 2007, **100**, 802–809.

21. S. Sankaranarayanan, M. A. Holahan, D. Colussi, M. C. Crouthamel, V. Devanarayan, J. Ellis, A. Espeseth, A. T. Gates, S. L. Graham, A. R. Gregro, D. Hazuda, J. H. Hochman, K. Holloway, L. Jin, J. Kahana, M. T. Lai, J. Lineberger, G. McGaughey, K. P. Moore, P. Nantermet, B. Pietrak, E. A. Price, H. Rajapakse, S. Stauffer, M. A. Steinbeiser, G. Seabrook, H. G. Selnick, X. P. Shi, M. G. Stanton, J. Swestock, K. Tugusheva, K. X. Tyler, J. P. Vacca, J. Wong, G. Wu, M. Xu, J. J. Cook and A. J. Simon, *J. Pharmacol. Exp. Ther.*, 2009, **328**, 131–140.

22. S. Kitazume, Y. Tachida, R. Oka, K. Shirotani, T. C. Saido and Y. Hashimoto, *Proc. Natl. Acad. Sci. USA*, 2001, **98**, 13554–13559.

23. M. Willem, A. N. Garratt, B. Novak, M. Citron, S. Kaufmann, A. Rittger, B. DeStrooper, P. Saftig, C. Birchmeier and C. Haass, *Science*, 2006, **314**, 664–666.

24. Y. Luo, B. Bolon, S. Kahn, B. D. Bennett, S. Babu-Khan, P. Denis, W. Fan, H. Kha, J. Zhang, Y. Gong, L. Martin, J. C. Louis, Q. Yan, W. G. Richards, M. Citron and R. Vassar, *Nature Neurosci.*, 2001, **4**, 231–232.

25. H. Cai, Y. Wang, D. McCarthy, H. Wen, D. R. Borchelt, D. L. Price and P. C. Wong, *Nature Neurosci.*, 2001, **4**, 233–234.

26. A. V. Savonenko, T. Melnikova, F. M. Laird, K. A. Stewart, D. L. Price and P. C. Wong, *Proc. Natl. Acad. Sci. USA*, 2008, **105**, 5585–5590.
27. S. Sankaranarayanan, E. A. Price, G. Wu, M. C. Crouthamel, X. P. Shi, K. Tugusheva, K. X. Tyler, J. Kahana, J. Ellis, L. Jin, T. Steele, S. Stachel, C. Coburn and A. J. Simon, *J. Pharmacol. Exp. Ther.*, 2008, **324**, 957–969.
28. M. S. Wolfe, W. Xia, C. L. Moore, D. D. Leatherwood, B. Ostaszewski, I. O. Donkor and D. J. Selkoe, *Biochemistry*, 1999, **38**, 4720–4727.
29. M. S. Shearman, D. Beher, E. E. Clarke, H. D. Lewis, T. Harrison, P. Hunt, A. Nadin, A. L. Smith, G. Stevenson and J. L. Castro, *Biochemistry*, 2000, **39**, 8698–8704.
30. A. Herreman, L. Serneels, W. Annaert, D. Collen, L. Schoonjans and B. De Strooper, *Nature Cell. Biol.*, 2000, **2**, 461–462.
31. Z. Zhang, P. Nadeau, W. Song, D. Donoviel, M. Yuan, A. Bernstein and B. A. Yankner, *Nature Cell. Biol.*, 2000, **2**, 463–465.
32. M. S. Wolfe, W. Xia, B. L. Ostaszewski, T. S. Diehl, W. T. Kimberly and D. J. Selkoe, *Nature*, 1999, **398**, 513–517.
33. G. Yu, F. Chen, G. Levesque, M. Nishimura, D. M. Zhang, L. Levesque, E. Rogaeva, D. Xu, Y. Liang, M. Duthie, P. H. St George-Hyslop and P. E. Fraser, *J. Biol. Chem.*, 1998, **273**, 16470–16475.
34. G. Thinakaran, D. R. Borchelt, M. K. Lee, H. H. Slunt, L. Spitzer, G. Kim, T. Ratovitsky, F. Davenport, C. Nordstedt, M. Seeger, J. Hardy, A. I. Levey, S. E. Gandy, N. A. Jenkins, N. G. Copeland, D. L. Price and S. S. Sisodia, *Neuron*, 1996, **17**, 181–190.
35. A. Capell, J. Grunberg, B. Pesold, A. Diehlmann, M. Citron, R. Nixon, K. Beyreuther, D. J. Selkoe and C. Haass, *J. Biol. Chem.*, 1998, **273**, 3205–3211.
36. T. Ratovitski, H. H. Slunt, G. Thinakaran, D. L. Price, S. S. Sisodia and D. R. Borchelt, *J. Biol. Chem.*, 1997, **272**, 24536–24541.
37. H. Steiner, A. Capell, B. Pesold, M. Citron, P. M. Kloetzel, D. J. Selkoe, H. Romig, K. Mendla and C. Haass, *J. Biol. Chem.*, 1998, **273**, 32322–32331.
38. G. Thinakaran, C. L. Harris, T. Ratovitski, F. Davenport, H. H. Slunt, D. L. Price, D. R. Borchelt and S. S. Sisodia, *J. Biol. Chem.*, 1997, **272**, 28415–28422.
39. Y. M. Li, M. Xu, M. T. Lai, Q. Huang, J. L. Castro, J. DiMuzio-Mower, T. Harrison, C. Lellis, A. Nadin, J. G. Neduvelil, R. B. Register, M. K. Sardana, M. S. Shearman, A. L. Smith, X. P. Shi, K. C. Yin, J. A. Shafer and S. J. Gardell, *Nature*, 2000, **405**, 689–694.
40. W. P. Esler, W. T. Kimberly, B. L. Ostaszewski, T. S. Diehl, C. L. Moore, J.-Y. Tsai, T. Rahmati, W. Xia, D. J. Selkoe and M. S. Wolfe, *Nature Cell. Biol.*, 2000, **2**, 428–434.
41. Y. M. Li, M. T. Lai, M. Xu, Q. Huang, J. DiMuzio-Mower, M. K. Sardana, X. P. Shi, K. C. Yin, J. A. Shafer and S. J. Gardell, *Proc. Natl. Acad. Sci. USA.*, 2000, **97**, 6138–6143.
42. W. P. Esler, W. T. Kimberly, B. L. Ostaszewski, W. Ye, T. S. Diehl, D. J. Selkoe and M. S. Wolfe, *Proc. Natl. Acad. Sci. USA.*, 2002, **99**, 2720–2725.

43. G. Yu, M. Nishimura, S. Arawaka, D. Levitan, L. Zhang, A. Tandon, Y. Q. Song, E. Rogaeva, F. Chen, T. Kawarai, A. Supala, L. Levesque, H. Yu, D. S. Yang, E. Holmes, P. Milman, Y. Liang, D. M. Zhang, D. H. Xu, C. Sato, E. Rogaev, M. Smith, C. Janus, Y. Zhang, R. Aebersold, L. S. Farrer, S. Sorbi, A. Bruni, P. Fraser and P. St George-Hyslop, *Nature*, 2000, **407**, 48–54.

44. C. Goutte, M. Tsunozaki, V. A. Hale and J. R. Priess, *Proc. Natl. Acad. Sci. USA*, 2002, **99**, 775–779.

45. R. Francis, G. McGrath, J. Zhang, D. A. Ruddy, M. Sym, J. Apfeld, M. Nicoll, M. Maxwell, B. Hai, M. C. Ellis, A. L. Parks, W. Xu, J. Li, M. Gurney, R. L. Myers, C. S. Himes, R. Hiebsch, C. Ruble, J. S. Nye and D. Curtis, *Dev. Cell.*, 2002, **3**, 85–97.

46. B. De Strooper, W. Annaert, P. Cupers, P. Saftig, K. Craessaerts, J. S. Mumm, E. H. Schroeter, V. Schrijvers, M. S. Wolfe, W. J. Ray, A. Goate and R. Kopan, *Nature*, 1999, **398**, 518–522.

47. G. H. Searfoss, W. H. Jordan, D. O. Calligaro, E. J. Galbreath, L. M. Schirtzinger, B. R. Berridge, H. Gao, M. A. Higgins, P. C. May and T. P. Ryan, *J. Biol. Chem.*, 2003, **278**, 46107–46116.

48. G. T. Wong, D. Manfra, F. M. Poulet, Q. Zhang, H. Josien, T. Bara, L. Engstrom, M. Pinzon-Ortiz, J. S. Fine, H. J. Lee, L. Zhang, G. A. Higgins and E. M. Parker, *J. Biol. Chem.*, 2004, **279**, 12876–12882.

49. E. R. Siemers, J. F. Quinn, J. Kaye, M. R. Farlow, A. Porsteinsson, P. Tariot, P. Zoulnouni, J. E. Galvin, D. M. Holtzman, D. S. Knopman, J. Satterwhite, C. Gonzales, R. A. Dean and P. C. May, *Neurology*, 2006, **66**, 602–604.

50. E. R. Siemers, R. A. Dean, S. Friedrich, L. Ferguson-Sells, C. Gonzales, M. R. Farlow and P. C. May, *Clin. Neuropharmacol.*, 2007, **30**, 317–325.

51. S. C. Mayer, A. F. Kreft, B. Harrison, M. Abou-Gharbia, M. Antane, S. Aschmies, K. Atchison, M. Chlenov, D. C. Cole, T. Comery, G. Diamantidis, J. Ellingboe, K. Fan, R. Galante, C. Gonzales, D. M. Ho, M. E. Hoke, Y. Hu, D. Huryn, U. Jain, M. Jin, K. Kremer, D. Kubrak, M. Lin, P. Lu, R. Magolda, R. Martone, W. Moore, A. Oganesian, M. N. Pangalos, A. Porte, P. Reinhart, L. Resnick, D. R. Riddell, J. Sonnenberg-Reines, J. R. Stock, S. C. Sun, E. Wagner, T. Wang, K. Woller, Z. Xu, M. M. Zaleska, J. Zeldis, M. Zhang, H. Zhou and J. S. Jacobsen, *J. Med. Chem.*, 2008, **14**, 14.

52. C. Drahl, *Chem. Eng. News*, 2009, **87**, 31.

53. R. Postina, *Curr. Alzheimer. Res.*, 2008, **5**, 179–186.

54. A. Y. Hung, C. Haass, R. M. Nitsch, W. Q. Qiu, M. Citron, R. J. Wurtman, J. H. Growdon and D. J. Selkoe, *J. Biol. Chem.*, 1993, **268**, 22959–22962.

55. J. D. Buxbaum, E. H. Koo and P. Greengard, *Proc. Natl. Acad. Sci. USA.*, 1993, **90**, 9195–9198.

56. R. M. Nitsch, B. E. Slack, R. J. Wurtman and J. H. Growdon, *Science*, 1992, **258**, 304–307.

57. A. Caccamo, S. Oddo, L. M. Billings, K. N. Green, H. Martinez-Coria, A. Fisher and F. M. LaFerla, *Neuron*, 2006, **49**, 671–682.
58. E. H. Corder, A. M. Saunders, W. J. Strittmatter, D. E. Schmechel, P. C. Gaskell, G. W. Small, A. D. Roses, J. L. Haines and M. A. Pericak-Vance, *Science*, 1993, **261**, 921–923.
59. E. H. Corder, A. M. Saunders, N. J. Risch, W. J. Strittmatter, D. E. Schmechel, P. C. Gaskell Jr, J. B. Rimmler, P. A. Locke, P. M. Conneally and K. E. Schmader, *et al., Nature Genet.*, 1994, **7**, 180–184.
60. M. Kivipelto, E. L. Helkala, M. P. Laakso, T. Hanninen, M. Hallikainen, K. Alhainen, H. Soininen, J. Tuomilehto and A. Nissinen, *BMJ*, 2001, **322**, 1447–1451.
61. B. Wolozin, W. Kellman, P. Ruosseau, G. G. Celesia and G. Siegel, *Arch. Neurol.*, 2000, **57**, 1439–1443.
62. H. Jick, G. L. Zornberg, S. S. Jick, S. Seshadri and D. A. Drachman, *Lancet*, 2000, **356**, 1627–1631.
63. D. L. Sparks, S. W. Scheff, J. C. Hunsaker 3rd, H. Liu, T. Landers and D. R. Gross, *Exp. Neurol.*, 1994, **126**, 88–94.
64. M. Simons, P. Keller, B. De Strooper, K. Beyreuther, C. G. Dotti and K. Simons, *Proc. Natl. Acad. Sci. USA*, 1998, **95**, 6460–6464.
65. K. Fassbender, M. Simons, C. Bergmann, M. Stroick, D. Lutjohann, P. Keller, H. Runz, S. Kuhl, T. Bertsch, K. von Bergmann, M. Hennerici, K. Beyreuther and T. Hartmann, *Proc. Natl. Acad. Sci. USA*, 2001, **98**, 5856–5861.
66. D. M. Walsh, I. Klyubin, J. V. Fadeeva, W. K. Cullen, R. Anwyl, M. S. Wolfe, M. J. Rowan and D. J. Selkoe, *Nature*, 2002, **416**, 535–539.
67. G. M. Shankar, S. Li, T. H. Mehta, A. Garcia-Munoz, N. E. Shepardson, I. Smith, F. M. Brett, M. A. Farrell, M. J. Rowan, C. A. Lemere, C. M. Regan, D. M. Walsh, B. L. Sabatini and D. J. Selkoe, *Nature. Med.*, 2008, **14**, 837–842.
68. S. Lesne, M. T. Koh, L. Kotilinek, R. Kayed, C. G. Glabe, A. Yang, M. Gallagher and K. H. Ashe, *Nature*, 2006, **440**, 352–357.
69. J. Lauren, D. A. Gimbel, H. B. Nygaard, J. W. Gilbert and S. M. Strittmatter, *Nature*, 2009, **457**, 1128–1132.
70. D. Fenili, M. Brown, R. Rappaport and J. McLaurin, *J. Mol. Med.*, 2007, **85**, 603–611.
71. A. I. Bush and R. E. Tanzi, *Proc. Natl. Acad. Sci. USA*, 2002, **99**, 7317–7319.
72. R. A. Cherny, C. S. Atwood, M. E. Xilinas, D. N. Gray, W. D. Jones, C. A. McLean, K. J. Barnham, I. Volitakis, F. W. Fraser, Y. Kim, X. Huang, L. E. Goldstein, R. D. Moir, J. T. Lim, K. Beyreuther, H. Zheng, R. E. Tanzi, C. L. Masters and A. I. Bush, *Neuron*, 2001, **30**, 665–676.
73. L. Lannfelt, K. Blennow, H. Zetterberg, S. Batsman, D. Ames, J. Harrison, C. L. Masters, S. Targum, A. I. Bush, R. Murdoch, J. Wilson and C. W. Ritchie, *Lancet Neurol.*, 2008, **7**, 779–786.
74. D. Schenk, R. Barbour, W. Dunn, G. Gordon, H. Grajeda, T. Guido, K. Hu, J. Huang, K. Johnson-Wood, K. Khan, D. Kholodenko, M. Lee,

Z. Liao, I. Lieberburg, R. Motter, L. Mutter, F. Soriano, G. Shopp, N. Vasquez, C. Vandevert, S. Walker, M. Wogulis, T. Yednock, D. Games and P. Seubert, *Nature*, 1999, **400**, 173–177.

75. D. Morgan, D. M. Diamond, P. E. Gottschall, K. E. Ugen, C. Dickey, J. Hardy, K. Duff, P. Jantzen, G. DiCarlo, D. Wilcock, K. Connor, J. Hatcher, C. Hope, M. Gordon and G. W. Arendash, *Nature*, 2000, **408**, 982–985.

76. C. Janus, J. Pearson, J. McLaurin, P. M. Mathews, Y. Jiang, S. D. Schmidt, M. A. Chishti, P. Horne, D. Heslin, J. French, H. T. Mount, R. A. Nixon, M. Mercken, C. Bergeron, P. E. Fraser, P. St George-Hyslop and D. Westaway, *Nature*, 2000, **408**, 979–982.

77. F. Bard, C. Cannon, R. Barbour, R. L. Burke, D. Games, H. Grajeda, T. Guido, K. Hu, J. Huang, K. Johnson-Wood, K. Khan, D. Kholodenko, M. Lee, I. Lieberburg, R. Motter, M. Nguyen, F. Soriano, N. Vasquez, K. Weiss, B. Welch, P. Seubert, D. Schenk and T. Yednock, *Nature Med.*, 2000, **6**, 916–919.

78. J. C. Dodart, K. R. Bales, K. S. Gannon, S. J. Greene, R. B. DeMattos, C. Mathis, C. A. DeLong, S. Wu, X. Wu, D. M. Holtzman and S. M. Paul, *Nature Neurosci.*, 2002, **5**, 452–457.

79. R. B. DeMattos, K. R. Bales, D. J. Cummins, J. C. Dodart, S. M. Paul and D. M. Holtzman, *Proc. Natl. Acad. Sci. USA.*, 2001, **98**, 8850–8855.

80. C. A. Lemere, M. Maier, Y. Peng, L. Jiang and T. J. Seabrook, *Curr. Alzheimer. Res.*, 2007, **4**, 427–436.

Targeting Alzheimer's γ-Secretase: Genetic and Chemical Modulation

WEIMING XIA

Center for Neurologic Diseases, Department of Neurology, Brigham and Women's Hospital, Harvard Medical School, Harvard Institute of Medicine, Boston, MA 02115, USA

2.1 γ-Secretase: The Key Enzyme for the Generation of Amyloid β Protein

Alzheimer's disease (AD) is characterised by the presence of extracellular neuritic plaques and intracellular neurofibrillary tangles.[1] The neuritic plaques are formed by the accumulation of amyloid β protein (Aβ), and the neurofibrillary tangles are composed of hyperphosphorylated Tau protein.[2] Many species of Aβ peptides with different lengths have been identified, and among various Aβ isoforms, the most common ones are 40-residue Aβ (Aβ40) and 42-residue Aβ (Aβ42). All species of Aβ peptides are produced by sequential cleavage of amyloid precursor protein (APP) by β-secretase and followed by γ-secretase.[1,2] Once APP is cleaved by β-secretase, it generates a 12-kDa C-terminal stub of APP (C99), which can be cleaved by γ-secretase to yield Aβ peptides. The other product derived from γ-secretase cleavage is amyloid intracellular domain, AICD, which might be involved in the regulation of

RSC Drug Discovery Series No. 2
Emerging Drugs and Targets for Alzheimer's Disease
Volume 1: Beta-Amyloid, Tau Protein and Glucose Metabolism
Edited by Ana Martinez
© Royal Society of Chemistry 2010
Published by the Royal Society of Chemistry, www.rsc.org

downstream gene expression.[3] Alternative to β-secretase cleavage of APP, α-secretase cleaves the majority of APP and generates APP$_s$-α and C83, followed by the γ-secretase cleavage to generate p3, precluding Aβ formation ("non-amyloidogenic").[4,5] The γ-secretase is responsible for the final cleavage to generate different lengths of Aβ peptides.[6,7] The increased generation of shorter Aβ peptides, like 37, 38 and 39 residue-Aβ species, usually accompanies a selective decrease of longer Aβ peptide, like Aβ42 and Aβ43. These longer Aβ peptides are more toxic than the shorter Aβ peptide.

Genetic studies have identified many autosomal dominant mutations in presenilin (PS) and APP genes, and missense mutations in PS and APP genes account for the majority of early onset familial AD (FAD) cases. Because APP is the substrate of the γ-secretase, much effort has been devoted to study PS, the γ-secretase and its substrate APP.

2.2   PS1 Carries the Active Site of the γ-Secretase

Presenilin 1 (PS1) on chromosome 14[8] and its homologue presenilin 2 (PS2) on chromosome 1,[9,10] are the major genes linked to FAD. Spanning the whole PS1 gene, over a hundred point mutations linked to FAD have been identified. PS1 and PS2 are 467 and 448 amino acid polypeptides with ~60% homology, and more mutations have been found in PS1 gene than in PS2 gene.

The functional form of PS1 existing in cells are the N- and C-terminal fragments (NTF/CTF) derived from full-length (FL) PS1, and low levels of the holoprotein are detected in cells and tissues.[11,12] While the levels of PS1 fragments can be pharmacologically reduced by applying specific γ-secretase inhibitors,[13,14] endogenous PS1 NTF/CTF are very stable.[15–17]

Earlier studies have found that PS1 and γ-secretase were closely associated. In neurons derived from PS1 knockout embryos, Aβ production was decreased.[18] In PS1/PS2 double knockout neurons[19] and in adult brains of conditional PS1 knockout mice,[20] a clear reduction of Aβ was found. Furthermore, PS1 and PS2 bind to APP[21,22] and C99/C83, the immediate γ-secretase substrates, at the sites of Aβ generation, *i.e.* Golgi/trans-Golgi network (TGN)-type vesicles.[23]

The key discovery of the active site of γ-secretase came from the identification of two aspartate residues that locate at the transmembrane domain 6 and 7 of PS1. Mutations in either of two aspartates abolish γ-secretase cleavage of APP.[24] The analogous aspartate residues in PS2 are similarly required for APP processing by γ-secretase.[25,26] Binding assays with transition-state analogue affinity reagents for γ-secretase provided convincing evidence that these γ-secretase inhibitors not only prevent Aβ generation but also bind directly to PS1 and PS2,[27,28] demonstrating that PS1 carries the active site of the γ-secretase.

2.3   γ-Secretase Complex Is Composed of Four Components

Besides PS1, additional components were believed to participate in the formation of γ-secretase complex. A number of studies have shown very stable high molecular weight complexes that contain PS NTF and CTF.[29,30] The high molecular weight complexes showed enrichment of NTF/CTF but not full-length PS1,[29–32] and expression of FAD mutations in PS1 did not change the size revealed by blue native gel electrophoresis.[33] Nicastrin was first found to associate with the CTF fragments of APP and Notch[34,35] and is required for γ-secretase activity.[36–41] Entry of Nicastrin into the active γ-secretase complex occurs after N-glycosylation in the Golgi apparatus, and the glycosylated Nicastrin directly interacts with the PS1 heterodimers.[42–47]

Genetic screens in *C. elegans* revealed two additional components of the γ-secretase components, APH-1 (anterior pharynx defective), a new protein with seven TM domains,[48] and PEN-2 (presenilin enhancer), a small protein of 101 amino acids with two TM domains.[36,49]

The functions of these four components have been extensively examined in knockout and knockdown animals. Since all four proteins play a central role in γ-secretase complex, animals deficient in individual components show certain phenotypic similarities. Knockout of PS1/PS2 in mice leads to phenotypes indistinguishable from Nicastrin knockouts, including embryonic lethality, several patterning defects, and abnormal somite segmentation.[41] Three APH-1 homologues have been identified in mice, and knockout of each of the three *Aph-1* genes leads to different and tissue-specific phenotypes.[50] Similar to Nicastrin or PS null embryos, knockout of APH-1A is lethal and causes defects in angiogenesis, neural tube formation, and somitogenesis;[50,51] however, knockout of APH-1B or APH-1C causes very mild phenotypes and the mice survive into adulthood.[50] APH-1B has recently been found to play a major role in the γ-secretase cleavage of APP to generate Aβ.[52] PEN-2 knockout mice have not been reported, and zebrafish deficient in PEN-2 showed enhanced apoptosis.[53,54]

In vitro studies have clearly documented the essential function of these four components in the γ-secretase cleavage of APP and other substrates. APH-1 forms a subcomplex with Nicastrin, and the stabilised proteins enter the γ-secretase complex.[55,56] Downregulation of APH-1[36,37] or PEN-2[36,57] by RNAi is associated with reduced levels of PS NTF/CTF heterodimers and subsequent deficient γ-secretase activity. Although reducing PEN-2 decreases endo-proteolytic processing of PS1, reducing APH-1 stabilises the FL PS1.[55,58,59] In mammalian cells overexpressing all four components, increased γ-secretase activity was observed.[55,58–62] In yeast, enhanced γ-secretase activity was observed after the complex was reconstituted.[63]

2.4　γ-Secretase Cleaves Multiple Substrates, Including APP and Notch

Besides APP, γ-secretase cleaves dozens of other Type-I transmembrane proteins, and many of them are critically involved in essential metabolic pathways like cell–cell adhesion, transcriptional regulation, and regulation of ion conductance. A number of substrates have been examined in the context of its relationship to APP processing and AD pathogenesis, *e.g.*, Notch,[64] ErbB-4,[65,66] E- and N-cadherin,[67,68] DCC,[69] and APLP1 and APLP2.[70–72] All known γ-secretase substrates are Type-I transmembrane proteins and the cleavage of all these proteins can be blocked by γ-secretase inhibitors.

Notch is a key γ-secretase substrate.[64,73] During embryonic development and in adulthoods, Notch signalling is critical to a wide variety of cell-fate determinations. The full-length Notch protein undergoes ectodomain shedding, and the remaining membrane-bound C-terminal stub is cleaved by γ-secretase to release the notch intracellular domain (NICD).[74] NICD is subsequently translocated to the nucleus to regulate downstream gene expression.[73,75,76]

In mice, Notch deficiency produces embryonic-lethal phenotypes, demonstrating the essential nature of Notch signalling in embryonic development. Knocking out PS1 resulted in decreased Notch signalling like Notch1 conditional knockout mice.[77] When both PS1 and PS2 are knocked out, the phenotype of mice is indistinguishable from the Notch knockout mice, and no Aβ is produced.[78] Proper Notch signalling is equally important in adulthood, and perturbed Notch signalling has been found in different types of tumor cell lines. By reducing Notch signalling, γ-secretase inhibitors were shown to suppress tumorigenesis pathways.[79–82]

2.5   Chemical Modulation of γ-Secretase Activity

Different classes of γ-secretase inhibitors have been explored to block the cleavage of APP and the generation of Aβ. Designing inhibitors of γ-secretase complex and quantifying the reduction of Aβ production is no longer a challenge, and a number of potent γ-secretase inhibitors have been published. Among these γ-secretase inhibitors, some of them selectively block the cleavage of APP with minimum or no inhibition of the γ-secretase cleavage of Notch to generate NICD.

2.5.1   Targetting γ-Secretase in Zebrafish

A variety of γ-secretase inhibitors were tested in animals (*e.g.*, zebrafish and mice) and phenotypes related to Notch signalling have been examined to determine the selectivity of APP *versus* Notch. In zebrafish, abnormal Notch signalling leads to altered phenotypes that are among the most studied molecular events, and a number of Notch mutants have been characterised. These mutants display defective anteroposterior polarity and increased neurogenin 1 (ngn1) positive cells. Defective somitogenesis leads to the classic phenotype of curved tails, and abnormal neurogenesis causes reduction in neural crest cell migration and subsequently a loss of pigmentation.[83,84] Some of these phenotypes can be replicated in zebrafish treated with a potent γ-secretase inhibitor, DAPT, at the late blastula stage.[85] The DAPT-treated embryos show high similarity to those Notch mutants like bea, des, aei, and wit.[83,86] This is in contrast to control vehicle-treated embryos that show a wild-type phenotype and developed normally. Those control embryos develop normal eyes, straight trunk and tail, V-shaped somites, and normal pigmentation at 6 days postfertilisation.[85,87] The reduction of Notch signalling in zebrafish could be rescued in part by microinjection of NICD mRNA, which suppresses the increased number of primary neurons in DAPT-treated embryos.[85] The Aβ-lowering JLK nonpeptidic isocoumarin inhibitors[88] and compound E[87] were also tested for their effect on the Notch pathway responsible for somitogenesis in zebrafish embryos.

2.5.2 Targeting γ-Secretase in Rodents

In mice, a potent γ-secretase inhibitor LY411,575 (Figure 2.1A) was used for 15 days, and pathological analysis of drug-treated animals revealed many unwanted side effects, including impaired lymphocyte development, and increased goblet cell number in intestine with abnormal tissue morphology.[89] Similar phenotypes were observed in rats treated with a different γ-secretase inhibitor, compound X (Figure 2.1B),[90] and most of these side effects were caused by perturbed Notch signalling downstream of deficient γ-secretase cleavage of Notch.

Apparently, inhibiting the production of Aβ by targeting γ-secretase is attractive for developing new treatments of AD, but has potential toxic side effects, *i.e.* inhibiting γ-secretase not only prevents APP cleavage and Aβ production, but also blocks the cleavage of other important substrates like Notch. Efforts have been made to develop γ-secretase inhibitors that block

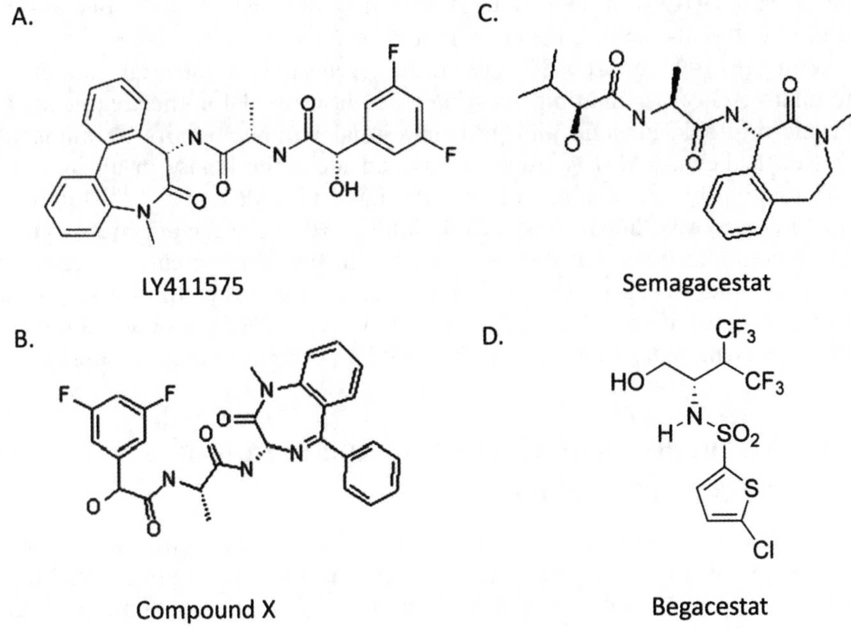

Figure 2.1   Structure of γ-secretase inhibitors.

APP cleavage while sparing other substrates, and these are considered as γ-secretase modulators.

One of the best-characterised γ-secretase modulators belongs to the class of NSAIDs. In cultured cells, NSAIDs (including ibuprofen, indomethacin, and sulindac sulfide) specifically decrease Aβ42 while increasing shorter Aβ peptides like Aβ38.[91] Treatment of transgenic mice overexpressing APP with the NSAID ibuprofen for 4 months does not induce major side effects. Quantification of CD45 and CD11b indicates that microglial activation is suppressed. Ibuprofen reduces SDS-soluble Aβ42 with less effect on Aβ40. Histopathological analysis has shown that amyloid plaque load was reduced in the cortex of the treated animals.[92]

2.5.3 Targeting γ-Secretase in Humans

Although NSAID treatment of a transgenic animal can efficiently reduce Aβ levels, amyloid plaque burden, and microglial activation in mouse brains, human studies have yet shown clear outcomes. Among a number of NSAIDs that significantly reduces Aβ42 in APP transgenic mice,[93] the best candidate (R-flurbiprofen) failed Phase-III clinical trials for the treatment of AD.

Currently, several γ-secretase inhibitors and modulators are under clinical trials for the treatment of AD, such as Semagacestat (Figure 2.1C) from Eli Lilly[94–97] and Begacestat (Figure 1.1D) from Wyeth.[98] Phase-II trial of Semagacestat showed a reduction of plasma Aβ.[96,97] Begacestat, a γ-secretase modulator, has 14-fold selectivity of APP over Notch, with an inhibition concentration (IC_{50}) at 15 nM. In APP transgenic Tg2576 mice, Begacestat reduced both Aβ40 and Aβ42 in brain and reversed cognitive deficits.[98]

Among the FDA-approved drugs, Gleevec seems to be a potential γ-secretase modulator. Gleevec (imatinib mesylate) has been used for the treatment of chronic myeloid leukemia and gastrointestinal stromal tumors. It binds to Abelson leukemia (Abl) tyrosine kinase and locks the kinase in an inactive conformation by interacting with aspartate and phenylalanine.[99–101] Interestingly, Gleevec was shown to selectively inhibit APP cleavage and Aβ production at concentrations that do not affect Notch signalling in cultured cells.[102] This reduction was also observed in rat primary neuronal cultures and guinea pig brain. In addition, Gleevec has been shown to reduce Aβ production in a cell-free system, indicating a direct effect on the γ-secretase complex/substrates.

2.6 Mechanisms of Chemical Modulation of γ-Secretase Activity

The γ-secretase complex is comprised of four components with many transmembrane domains, therefore, elucidating the crystal structure of such a large membrane protein complex is not an easy task. A docking site, along with an active site, has been proposed for γ-secretase inhibitors/substrates to interact with the complex, and a direct contact of γ-secretase modulator with the substrate has been explored. Some representative inhibitors are depicted in Figure 2.1.

2.6.1 Binding of γ-Secretase Inhibitors to the Active Site

A number of potent γ-secretase inhibitors directly bind to PS1. An aspartyl protease transition state mimic, L-685,458, has an IC_{50} of 17 nM for inhibiting Aβ generation. Photoactivated L-685,458 can covalently label the PS1 NTF and CTF.[103] Similarly, a biotinylated and bromoacetylated transition-state analogue affinity compound was shown to bind to PS1.[27] Even a nontransition-state mimic benzophenone analogue specifically photocrosslinks several major membrane polypeptides, including PS1 NTF and CTF. In the absence of PS1, the binding of the compound to the γ-secretase complex was significantly reduced.[104]

A γ-secretase inhibitor that is under clinical trial, MK-0752, is also believed to bind to the active site of the γ-secretase. MK-0752 does not differentiate APP and Notch, and it was developed as a Notch inhibitor to treat acute lymphoblastic leukemia[105] and breast cancer.[106]

2.6.2 Binding of γ-Secretase Inhibitors to the Docking Site

A docking site within the γ-secretase complex has been proposed to interact with substrate, followed by cleavage at the active site and release of products.[107] Since FL PS1 was hypothesised to undergo autoproteolysis,[24] cleavage of FL PS1 into functional NTF/CTF may free the substrate docking site and convert the immature enzyme into the mature γ-secretase complex.[108] In addition, helical peptide inhibitors were developed to bind to the putative docking site and block γ-secretase complex. Those results suggest that a docking site for γ-secretase substrates might be located at the PS subunit interface and close to the active site.[109]

2.6.3 Binding of γ-Secretase Modulator to the Substrate

While the development of γ-secretase modulators becomes the trend to target the γ-secretase complex, few studies have provided a clear mechanism for γ-secretase modulators to regulate the protease activity. Since the discovery of NSAIDs as promising γ-secretase modulators,[91] multiple mechanisms have been proposed and explored for its action. First, NSAIDs were shown to induce a decrease of Aβ42 with a simultaneous increase of Aβ38, and NSAIDs were thought to modulate γ-secretase directly. This is supported by an *in vitro* study showing that a radiolabelled transition-state γ-secretase inhibitor was replaced by NSAIDs in a cell-free system. Furthermore, the NSAIDs were found to induce noncompetitive inhibition of γ-secretase activity, and it is likely to bind to the site other than the docking and active sites of the γ-secretase complex.

Indeed, recent studies have shown that NSAIDs directly bind to the substrate APP, which is distinguishable from other substrates like Notch.[110]

2.6.4 Pathways Independent of the γ-Secretase Complex

An interesting mechanism described by an early report suggests that NSAIDs might selectively reduce Aβ42 via the inhibition of the small GTPase Rho.[111] An inhibitor of Rho-associated kinase (ROCK) decreased the levels of Aβ42 in brains of APP transgenic mice.[111] The effect seemed to be specific, and among the NSAIDs examined, only those that are capable of inhibiting Rho reduced Aβ42. However, these ROCK inhibitors did not reduce Aβ production in an *in vitro* γ-secretase activity assay, suggesting an indirect effect on the γ-secretase complex.[112]

The mechanism of Gleevec for selective inhibition of Aβ generation without affecting Notch signalling is less clear. A combination of enhanced Notch signalling and Abl tyrosine kinase activity was shown to promote acute lymphocytic leukemia,[113] and it seems that Gleevec contains separate functionalities that cause inhibition of Abl tyrosine kinase and γ-secretase. The reduction of Aβ generation by Gleevec was also detected in fibroblasts cultured from Abl knockout mice.[102] These results demonstrated that inhibition of Aβ is not related to the inhibition of Abl kinase activity by Gleevec. It is also possible that the metabolite(s) of Gleevec (instead of Gleevec itself) may inhibit the γ-secretase cleavage of APP.[114] Further studies are needed to understand the unique mechanisms independent of its kinase inhibitory effect of Gleevec to block Aβ production.

2.7 Genetic Modulation of γ-Secretase Activity by Four Components

Accumulative evidence has shown that all four γ-secretase components play essential roles in the cleavage of multiple substrates. Recent studies have provided much insight information of individual components that play distinctive roles in modulating the γ-secretase activity.

2.7.1 Presenilin

Conversion of full-length PS1 into functional NTF/CTF is a critical step in forming the active γ-secretase complex. Endoproteolysis of PS is mediated by an unknown enzyme termed "presenilinase", which could be inhibited by γ-secretase inhibitors.[11–12,14,115] These two subunits of NTF and CTF may move closer to each other when FAD-linked mutations are introduced to PS1 protein.[107] The fluorescence resonance energy transfer (FRET) microscopy approach was used to determine spatial relationship between N and C termini of the PS1 molecule and any conformational changes.[107] In addition, FAD mutant PS1 induces an alteration in PS1-APP interaction and reduces proximity between the APP C-terminal fragment and PS1 loop domain.[107] Two critical Asp residues are essential for PS1 endoproteolysis and γ-secretase activity, and mutations in other residues linked to FAD have shown different effects on the γ-secretase activity.[116] Recent studies have shown that expression of FAD-linked mutant PS1 causes a change in the equilibrium of PS1 and PS2-containing γ-secretase complex. Both in cultured cells and in knock-in mice, expression of FAD mutant PS1 increases PS2-containing γ-secretase complex, and PS2-containing γ-secretase complex might be responsible for an increase of Aβ42/Aβ40 ratio.[117] A similar effect was observed in cells overexpressing PEN-2 (see below). Therefore, the participation of PS1 or PS2 into the active γ-secretase complex directly modulates the cleavage site of APP.

2.7.2 Nicastrin

Besides PS1, Nicastrin was first identified to interact with APP and Notch and is an essential γ-secretase component.[34] It is absolutely required for γ-secretase cleavage of APP and Notch.[36–39,118] Nicastrin-deficient mice showed phenotypes similar to PS1/PS2 double knockouts due to disrupted Notch signalling.[41,118] Biochemical analysis of Nicastrin revealed the interaction between its ectodomain and the γ-secretase substrates APP C-terminal fragment and the Notch.[119] Using recombinant γ-secretase components from transfected Sf9 cells, *in vitro* γ-secretase activity assay was carried out to compare the substrate recognition and binding followed by γ-secretase cleavage. The conserved DYIGS and *p*eptidase homologous region (DAP) of Nicastrin, and Glu333 in particular, was found to play a key role in modulating the γ-secretase cleavage of APP and Notch.[119] Although Glu333 might be critical for the maturation of the γ-secretase rather than for the recognition of the substrates and function as the actual binding site,[120] it is clear that Nicastrin holds a key position to select γ-secretase substrates such as APP and Notch. Therefore, regulating the function of Nicastrin will potentially lead to modulation of γ-secretase cleavage of APP and Notch.[119]

2.7.3 APH-1

There are two homologous *aph-1* genes, *aph-1A* and *aph-1B*. In mice, there is an additional duplicate of *aph-1B, aph-1C*. APH-1B is more important for the production of Aβ. In contrast to APH-1A-enriched γ-secretase complex, APH-1B-enriched γ-secretase complex precipitated from human brain homogenates promotes higher levels of Aβ and AICD generation in an *in vitro* γ-secretase activity assay. Consistent with this finding, removal of APH-1B by immune-depletion reduces Aβ and AICD generation. *In vivo*, deletion of APH-1B in mice decreases levels of Aβ in the brain and rescues learning and memory deficits.[52] Compared to APH-1A deficient mice that are embryonic lethal,[50,51] mice deficient in APH-1B survive into adulthood and show very mild phenotype.[50] Therefore, APH-1B-enriched γ-secretase is an ideal target for intervention, as it is critical for Aβ production but not for Notch signalling.

2.7.4 PEN-2

PEN-2 is the only γ-secretase component that a mouse knockout model has not been reported, but its unique function in the γ-secretase complex starts to emerge. In cultured cells[121] and zebrafish,[53] lacking PEN-2 caused enhanced apoptosis that might not be related to the γ-secretase.[54] In cultured cells and mice deficient in PS, PEN-2 is ubiquitinated and degraded by the proteasome.[122] PEN-2 facilitates the cleavage of full-length PS into active NTF and CTF.[36,57] PEN-2 stabilises PS1 NTF/CTF and closely associates with PS1 before participating into the active γ-secretase complex.[123] Molecular dissection of the PEN-2 protein indicates that the sequence and length of the C-terminus of PEN-2 are critical for intermolecular interactions and function of PS complexes.[124,125] Deletion of the last 12 amino acids of PEN-2 abrogates the ability of PEN-2 to stabilise PS fragments and fails to maintain γ-secretase activity when the endogenous PEN-2 is knocked down by RNAi.[124,125] Elongation at the N-terminus of PEN-2 causes increased generation of Aβ42, suggesting that PEN-2 is directly involved in modulating the γ-secretase cleavage site of APP.[126]

The involvement of PEN-2 in modulating γ-secretase was further supported by recent studies on PEN-2-mediated formation of different γ-secretase complexes.[117] Overexpression of PEN-2 alters the equilibrium of PS1- and PS2-containing active complexes and increases the PS2 containing active γ-secretase complex. Since the levels of PS2-containing γ-secretase complexes are positively correlated with increased Aβ42:Aβ40 ratios, high levels of PEN-2 lead to an increase of Aβ42 to Aβ40 ratio in cultured cells and in an *in vitro* system detected by γ-secretase activity assay, suggesting a specific modulation of the γ-secretase cleavage sites.[117]

2.7.5 The γ-Secretase Cofactor

In addition to the intrinsic role of γ-secretase components that modulate the γ-secretase activity, regulation of γ-secretase cleavage of APP and Notch could also be achieved by a cofactor TMP21.[127] Knocking down TMP21 by siRNA leads to an increase of γ-secretase activity and the generation of Aβ without changing the levels of all four γ-secretase components. TMP21 is not involved in altering the γ-secretase complex assembly and stability. It directly interacts with the protease complex and negatively suppresses the generation of Aβ. Importantly, it does not affect the cleavage of APP at the ε-site, leaving AICD unchanged.[127]

In general, chemical and genetic modulations of the γ-secretase complex are not two separate events. Chemical modulation of the γ-secretase is also dependent on the conformation of the complex. Earlier reports have demonstrated that FAD mutant PS1 could modulate the pharmacological properties of γ-secretase inhibitors, *e.g.*, the efficacy of a peptidomimetic γ-secretase inhibitor is significantly reduced in cells overexpressing FAD mutant PS1.[128] Furthermore, a NSAID sulindac sulfide was shown to reduce Aβ42 in the wild-type cells[91] and in cells expressing two FAD mutant PS1 (M146L and A285V), but not in cells that express other FAD mutant PS1 like L424R, L166P.[116] Interestingly, wild-type and all mutant PS1 expressing cells show an increase in Aβ38 in the presence of sulindac sulfide.[116] Apparently, chemical modulation of the γ-secretase activity is closely associated with individual γ-secretase components, and its effect correlates with the active conformation of the γ-secretase complex.

2.8 Conclusion

The γ-secretase complex, composed of PS1, APH-1, Nicastrin and PEN-2, mediates the proteolytic process of over 50 substrates, including APP and Notch. Identifying selective γ-secretase inhibitors that only block the cleavage of APP but not other substrates is ideal. Earlier studies have shown that a number of NSAIDs and Gleevec specifically block Aβ production without affecting the γ-secretase cleavage of Notch, and the NSAID directly binds to the substrate APP. Recent studies have shown that γ-secretase components function as modulators for distinguishing different substrates such as APP and Notch. Among them, Nicastrin is not only required for γ-secretase complex but also serves as an initial binding partner for γ-secretase substrates like APP and Notch.[119] PEN-2 has been found to modulate the formation of PS1 *versus* PS2 containing γ-secretase complex.[117] Finally, APH-1B plays a selective role in mediating the γ-secretase cleavage of APP.[52] These *in vitro* and *in vivo* studies have demonstrated that chemical and genetic modulation of γ-secretase for selective substrate recognition and Aβ production can be achieved by key components of the γ-secretase complex and potent γ-secretase modulators.

Acknowledgements

Some of the studies described in this chapter were carried out in the author's laboratory and were supported by NIH grant AG015379.

References

1. D. J. Selkoe, *Nature*, 1999, **399**(Supp 1), A23–A31.
2. D. J. Selkoe, *Ann. Intern. Med.*, 2004, **140**, 627–638.
3. X. Cao and T. C. Sudhof, *Science*, 2001, **293**, 115–120.
4. F. S. Esch, P. S. Keim, E. C. Beattie, R. W. Blacher, A. R. Culwell, T. Oltersdorf, D. McClure and P. J. Ward, *Science*, 1990, **248**, 1122–1124.
5. S. S. Sisodia, E. H. Koo, K. Beyreuther, A. Unterbeck and D. L. Price, *Science*, 1990, **248**, 492–495.

6. C. Haass, M. Schlossmacher, A. Y. Hung, C. Vigo-Pelfrey, A. Mellon, B. Ostaszewski, I. Lieberburg, E. H. Koo, D. Schenk, D. Teplow and D. Selkoe, *Nature*, 1992, **359**, 322–325.

7. M. Shoji, T. E. Golde, J. Ghiso, T. T. Cheung, S. Estus, L. M. Shaffer, X. Cai, D. M. McKay, R. Tintner, B. Frangione and S. G. Younkin, *Science*, 1992, **258**, 126–129.

8. R. Sherrington, E. I. Rogaev, Y. Liang, E. A. Rogaeva, G. Levesque, M. Ikeda, H. Chi, C. Lin, G. Li, K. Holman, T. Tsuda, L. Mar, J. -F. Foncin, A. C. Bruni, M. P. Montesi, S. Sorbi, I. Rainero, L. Pinessi, L. Nee, I. Chumakov, D. A. Pollen, A. D. Roses, P. E. Fraser, J. M. Rommens and P. H. St. George-Hyslop, *Nature*, 1995, **375**, 754–760.

9. E. I. Rogaev, R. Sherrington, E. A. Rogaeva, G. Levesque, M. Ikeda, Y. Liang, H. Chi, C. Lin, K. Holamn, T. Tsuda, L. Mar, S. Sorbi, B. Nacmias, S. Piacentini, L. Amaducci, I. Chumakov, D. Cohen, L. Lannfelt, P. E. Fraser, J. M. Rommens and P. H. St. George-Hyslop, *Nature*, 1995, **376**, 775–778.

10. E. Levy-Lahad, E. M. Wijsman, E. L. Nemens, L. Anderson, K. A. Goddard, J. L. Weber, T. D. Bird and G. D. Schellenberg, *Science*, 1995, **269**, 970–973.

11. G. Thinakaran, D. R. Borchelt, M. K. Lee, H. H. Slunt, L. Spitzer, G. Kim, T. Rotovitsky, F. Davenport, C. Nordstedt, M. Seeger, J. Hardy, A. I. Levey, S. E. Gandy, N. A. Jenkins, N. G. Copeland, D. L. Price and S. S. Sisodia, *Neuron*, 1996, **17**, 181–190.

12. D. R. Borchelt, G. Thinakaran, C. B. Eckman, M. K. Lee, F. Davenport, T. Ratovitsky, C.-M. Prada, G. Kim, S. Seekins, D. Yager, H. H. Slunt, R. Wang, M. Seeger, A. I. Levey, S. E. Gandy, N. G. Copeland, N. A. Jenkins, D. L. Price, S. G. Younkin and S. S. Sisodia, *Neuron*, 1996, **17**, 1005–1013.

13. D. Beher, J. D. Wrigley, A. Nadin, G. Evin, C. L. Masters, T. Harrison, J. L. Castro and M. S. Shearman, *J. Biol. Chem.*, 2001, **276**, 45394–45402.

14. W. Campbell, M.-K. Iskandar, M. Reed and W. Xia, *Biochem.*, 2002, **41**, 3372–3379.

15. G. Thinakaran, C. L. Harris, T. Ratovitski, F. Davenport, H. H. Slunt, D. L. Price, D. R. Borchelt and S. S. Sisodia, *J. Biol. Chem.*, 1997, **272**, 28415–28422.

16. T. Ratovitski, H. H. Slunt, G. Thinakaran, D. L. Price, S. S. Sisodia and D. R. Borchelt, *J. Biol. Chem.*, 1997, **272**, 24536–24541.

17. J. Zhang, D. E. Kang, W. Xia, M. Okochi, H. Mori, D. J. Selkoe and E. H. Koo, *J. Biol. Chem.*, 1998, **273**, 12436–12442.

18. B. De Strooper, P. Saftig, K. Craessaerts, H. Vanderstichele, G. Gundula, W. Annaert, K. Von Figura and F. Van Leuven, *Nature*, 1998, **391**, 387–390.

19. A. Herreman, L. Serneels, W. Annaert, D. Collen, L. Schoonjans and B. De Strooper, *Nature Cell. Biol.*, 2000, **2**, 461–462.

20. H. Yu, C. Saura, S. Choi, L. Sun, X. Yang, M. Handler, T. Kawarabayashi, L. Younkin, B. Fedeles, M. Wilson, S. Younkin, E. Kandel, A. Kirkwood and J. Shen, *Neuron*, 2001, **31**, 713–726.

21. W. Xia, J. Zhang, R. Perez, E. H. Koo and D. J. Selkoe, *Proc. Natl. Acad. Sci. USA.*, 1997, **94**, 8208–8213.

22. A. Weidemann, K. Paliga, U. Durrwang, C. Czech, G. Evin, C. L. Masters and K. Beyreuther, *Nature Med.*, 1997, **3**, 328–332.

23. W. Xia, W. J. Ray, B. L. Ostaszewski, T. Rahmati, W. T. Kimberly, M. S. Wolfe, J. Zhang, A. M. Goate and D. J. Selkoe, *Proc. Natl. Acad. Sci. USA.*, 2000, **97**, 9299–9304.

24. M. S. Wolfe, W. Xia, B. L. Ostaszewski, T. S. Diehl, W. T. Kimberly and D. J. Selkoe, *Nature*, 1999, **398**, 513–517.

25. H. Steiner, K. Duff, A. Capell, H. Romig, M. G. Grim, S. Lincoln, J. Hardy, X. Yu, M. Picciano, K. Fechteler, M. Citron, R. Kopan, B. Pesold, S. Keck, M. Baader, T. Tomita, T. Iwatsubo, R. Baumeister and C. Haass, *J. Biol. Chem.*, 1999, **274**, 28669–28673.

26. W. T. Kimberly, W. Xia, R. Rahmati, M. S. Wolfe and D. J. Selkoe, *J. Biol. Chem.*, 2000, **275**, 3173–3178.

27. W. P. Esler, W. T. Kimberly, B. L. Ostaszewski, T. S. Diehl, C. L. Moore, J. Y. Tsai, T. Rahmati, W. Xia, D. J. Selkoe and M. S. Wolfe, *Nature Cell. Biol.*, 2000, **2**, 428–434.

28. Y.-M. Li, M. Xu, M. -T. Lai, Q. Huang, J. L. Castro, J. DiMuzio-Mower, T. Harrison, C. Lellis, A. Nadin, J. G. Neduvelli, R. B. Register, M. K. Sardana, M. S. Shearman, A. L. Smith, X. -P. Shi, K. -C. Yin, J. A. Shafer and S. J. Gardell, *Nature*, 2000, **405**, 689–694.

29. A. Capell, J. Grunberg, B. Pesold, A. Diehlmann, M. Citron, R. Nixon, K. Beyreuther, D. J. Selkoe and C. Haass, *J. Biol. Chem.*, 1998, **273**, 3205–3211.

30. G. Yu, F. Chen, G. Levesque, M. Nishimura, D. M. Zhang, L. Levesque, E. Rogaeva, D. Xu, Y. Liang, M. Duthie, P. H. St George-Hyslop and P. E. Fraser, *J. Biol. Chem.*, 1998, **273**, 16470–16475.

31. G. Yu, F. Chen, M. Nishimura, H. Steiner, A. Tandon, T. Kawarai, S. Arawaka, A. Supala, Y. Q. Song, E. Rogaeva, E. Holmes, D. M. Zhang, P. Milman, P. E. Fraser, C. Haass and P. S. George-Hyslop, *J. Biol. Chem.*, 2000, **275**, 27348–27353.

32. H. Laudon, P. M. Mathews, H. Karlstrom, A. Bergman, M. R. Farmery, R. A. Nixon, B. Winblad, S. E. Gandy, U. Lendahl, J. Lundkvist and J. Naslund, *J. Neurochem.*, 2004, **89**, 44–53.

33. J. G. Culvenor, N. T. Ilaya, M. T. Ryan, L. Canterford, D. E. Hoke, N. A. Williamson, C. A. McLean, C. L. Masters and G. Evin, *Eur. J. Biochem.*, 2004, **271**, 375–385.

34. G. Yu, M. Nishimura, S. Arawaka, D. Levitan, L. Zhang, A. Tandon, Y. Q. Song, E. Rogaeva, F. Chen, T. Kawarai, A. Supala, L. Levesque, H. Yu, D. S. Yang, E. Holmes, P. Milman, Y. Liang, D. M. Zhang, D. H. Xu, C. Sato, E. Rogaev, M. Smith, C. Janus, Y. Zhang, R. Aebersold, L.

S. Farrer, S. Sorbi, A. Bruni, P. Fraser and P. St George-Hyslop, *Nature*, 2000, **407**, 48–54.

35. F. Chen, G. Yu, S. Arawaka, M. Nishimura, T. Kawarai, H. Yu, A. Tandon, A. Supala, Y. Song, E. Rogaeva, P. Milman, C. Sato, C. Yu, C. Janus, J. Lee, L. Song, L. Zhang, P. Fraser and P. St George-Hyslop, *Nature Cell. Biol.*, 2001, **3**, 751–754.

36. R. Francis, G. McGrath, J. Zhang, D. Ruddy, M. Sym, J. Apfeld, M. Nicoll, M. Maxwell, B. Hai, M. C. Ellis, A. L. Parks, W. Xu, J. Li, M. Gurney, R. L. Myers, C. S. Himes, R. Hiebsch, C. Ruble, J. S. Nye and D. Curtis, *Dev. Cell.*, 2002, **3**, 85–97.

37. S. Lee, S. Shah, H. Li, C. Yu, W. Han and G. Yu, *J. Biol. Chem.*, 2002, **277**, 45013–45019.

38. R. Siman and J. Velji, *J. Neurochem.*, 2003, **84**, 1143–1153.

39. K. Shirotani, D. Edbauer, A. Capell, J. Schmitz, H. Steiner and C. Haass, *J. Biol. Chem.*, 2003, **278**, 16474–16477.

40. C. Alves da Costa, M. P. Mattson, K. Ancolio and F. Checler, *J. Biol. Chem.*, 2003, **278**, 12064–12069.

41. T. Li, G. Ma, H. Cai, D. L. Price and P. C. Wong, *J. Neurosci.*, 2003, **23**, 3272–3277.

42. W. T. Kimberly, M. J. LaVoie, B. L. Ostaszewski, W. Ye, M. S. Wolfe and D. J. Selkoe, *J. Biol. Chem.*, 2002, **277**, 35113–35117.

43. J. Y. Leem, S. Vijayan, P. Han, D. Cai, M. Machura, K. O. Lopes, M. L. Veselits, H. Xu and G. Thinakaran, *J. Biol. Chem.*, 2002, **277**, 19236–19240.

44. T. Tomita, R. Katayama, R. Takikawa and T. Iwatsubo, *FEBS. Lett.*, 2002, **520**, 117–121.

45. S. Arawaka, H. Hasegawa, A. Tandon, C. Janus, F. Chen, G. Yu, K. Kikuchi, S. Koyama, T. Kato, P. E. Fraser and P. St George-Hyslop, *J. Neurochem.*, 2002, **83**, 1065–1071.

46. D.-S. Yang, A. Tandon, F. Chen, G. Yu, H. Yu, S. Arawaka, H. Hasegawa, M. Duthie, S. Schmidt, R. Nixon, T. Ramabhadran, P. Mathews, S. Gandy, H. Mount, P. St George-Hyslop and P. Fraser, *J. Biol. Chem.*, 2002, **277**, 28135–28142.

47. F. Chen, S. N. Tandon A, G. u. YJ, H. Hasegawa, S. Arawaka, F. J. Lee, X. Ruan, P. Mastrangelo, S. Erdebil, L. Wang, D. Westaway, H. T. Mount, B. Yankner, P. E. Fraser and P. S. George-Hyslop, *J. Biol. Chem.*, 2003, **278**, 19974–19979.

48. C. Goutte, M. Tsunozaki, V. A. Hale and J. R. Priess, *Proc. Natl. Acad. Sci.*, 2002, **99**, 775–779.

49. A. Crystal, V. A. Morais, T. C. Pierson, D. S. Pijak, D. Carlin, V. M. Lee and R. W. Doms, *J. Biol. Chem.*, 2003, **278**, 20117–20123.

50. L. Serneels, T. Dejaegere, K. Craessaerts, K. Horre, E. Jorissen, T. Tousseyn, S. Hebert, M. Coolen, G. Martens, A. Zwijsen, W. Annaert, D. Hartmann and B. De Strooper, *Proc. Natl. Acad. Sci. USA*, 2005, **102**, 1719–1724.

51. G. Ma, T. Li, D. L. Price and P. C. Wong, *J. Neurosci.*, 2005, **25**, 192–198.

52. L. Serneels, J. Van Biervliet, K. Craessaerts, T. Dejaegere, K. Horre, T. Van Houtvin, H. Esselmann, S. Paul, M. K. Schafer, O. Berezovska, B. T. Hyman, B. Sprangers, R. Sciot, L. Moons, M. Jucker, Z. Yang, P. C. May, E. Karran, J. Wiltfang, R. D'Hooge and B. De Strooper, *Science*, 2009, **324**, 639–642.
53. W. A. Campbell, H. W. Yang, H. Zetterberg, S. Baulac, J. A. Sears, T. Liu, S. T. C. Wong, T. P. Zhong and W. Xia, *J. Neurochem.*, 2006, **96**, 1423–1440.
54. H. Zetterberg, W. A. Campbell, H. W. Yang and W. Xia, *J. Biol. Chem.*, 2006, **281**, 11933–11939.
55. Y. Hu and M. Fortini, *J. Cell. Biol.*, 2003, **161**, 685–690.
56. M. J. LaVoie, P. C. Fraering, B. L. Ostaszewski, W. Ye, W. T. Kimberly, M. S. Wolfe and D. J. Selkoe, *J. Biol. Chem.*, 2003, **278**, 37213–37222. Epub 32003 Jul 37211.
57. H. Steiner, E. Winkler, D. Edbauer, S. Prokop, G. Basset, A. Yamasaki, M. Kostka and C. Haass, *J. Biol. Chem.*, 2002, **277**, 39062–39065.
58. N. Takasugi, T. Tomita, I. Hayashi, M. Tsuruoka, M. Niimura, Y. Takahashi, G. Thinakaran and T. Iwatsubo, *Nature*, 2003, **422**, 438–441.
59. W. J. Luo, H. Wang, H. Li, B. S. Kim, S. Shah, H. J. Lee, G. Thinakaran, T. W. Kim, G. Yu and H. Xu, *J. Biol. Chem.*, 2003, **278**, 7850–7854.
60. S. Baulac, M. J. LaVoie, W. T. Kimberly, J. Strahle, M. S. Wolfe, D. J. Selkoe and W. Xia, *Neurobiol. Dis.*, 2003, **14**, 194–204.
61. B. De Strooper, *Neuron*, 2003, **38**, 9–12.
62. W. Kimberly, M. LaVoie, B. L. Ostaszewski, W. Ye, M. S. Wolfe and D. J. Selkoe, *Proc. Natl. Acad. Sci. USA.*, 2003, **100**, 6382–6387.
63. D. Edbauer, E. Winkler, J. T. Regula, B. Pesold, H. Steiner and C. Haass, *Nature Cell. Biol.*, 2003, **5**, 486–488.
64. B. De Strooper, W. Annaert, P. Cupers, P. Saftig, K. Craessaerts, J. S. Mumm, E. H. Schroeter, V. Schrijvers, M. S. Wolfe, W. J. Ray, A. Goate and R. Kopan, *Nature*, 1999, **398**, 518–522.
65. C. Y. Ni, M. P. Murphy, T. E. Golde and G. Carpenter, *Science*, 2001, **294**, 2179–2181.
66. H. J. Lee, K. M. Jung, Y. Z. Huang, L. B. Bennett, J. S. Lee, L. Mei and T. W. Kim, *J. Biol. Chem.*, 2002, **277**, 6318–6323.
67. P. Marambaud, J. Shioi, G. Serban, A. Georgakopoulos, S. Sarner, V. Nagy, L. Baki, P. Wen, S. Efthimiopoulos, Z. Shao, T. Wisniewski and N. K. Robakis, *Embo. J.*, 2002, **21**, 1948–1956.
68. P. Marambaud, P. H. Wen, A. Dutt, J. Shioi, A. Takashima, R. Siman and N. K. Robakis, *Cell*, 2003, **114**, 635–645.
69. Y. Taniguchi, S. H. Kim and S. S. Sisodia, *J. Biol. Chem.*, 2003, **278**, 30425–30428. Epub 32003 Jul 30422.
70. M. H. Scheinfeld, E. Ghersi, K. Laky, B. J. Fowlkes, L. D'Adamio, R. Roncarati, N. Sestan, B. E. Berechid, P. A. Lopez, O. Meucci, J. C. McGlade and P. Rakic, *J. Biol. Chem.*, 2002, **277**, 44195–44201.
71. D. M. Walsh, J. V. Fadeeva, M. J. LaVoie, K. Paliga, S. Eggert, W. T. Kimberly, W. Wasco and D. J. Selkoe, *Biochemistry*, 2003, **42**, 6664–6673.

72. S. Eggert, K. Paliga, P. Soba, G. Evin, C. L. Masters, A. Weidemann and K. Beyreuther, *J. Biol. Chem.*, 2004, **279**, 18146–18156.
73. M. Okochi, H. Steiner, A. Fukumori, H. Tanii, T. Tomita, T. Tanaka, T. Iwatsubo, T. Kudo, M. Takeda and C. Haass, *Embo. J.*, 2002, **21**, 5408–5416.
74. J. S. Mumm, E. H. Schroeter, M. T. Saxena, A. Griesemer, X. Tian, D. J. Pan, W. J. Ray and R. Kopan, *Mol. Cell.*, 2000, **5**, 197–206.
75. M. E. Fortini, *Nature Rev. Mol. Cell. Biol.*, 2002, **3**, 673–684.
76. R. Kopan and A. Goate, *Neuron*, 2002, **33**, 321–324.
77. X. Yang, R. Klein, X. Tian, H. T. Cheng, R. Kopan and J. Shen, *Dev. Biol.*, 2004, **269**, 81–94.
78. D. Donoviel, A. Hadjantonakis, M. Ikeda, H. Zheng, P. St George Hyslop and A. Bernstein, *Genes Dev.*, 1999, **13**, 2801–2810.
79. C. Grabher, H. von Boehmer and A. T. Look, *Nature Rev. Cancer.*, 2006, **6**, 347–359.
80. T. Kindler, M. G. Cornejo, C. Scholl, J. Liu, D. S. Leeman, J. E. Haydu, S. Frohling, B. H. Lee and D. G. Gilliland, *Blood*, 2008, **28**, 28.
81. J. Yao, L. Duan, M. Fan and X. Wu, *Oral. Dis.*, 2007, **13**, 555–563.
82. S. Suwanjunee, W. Wongchana and T. Palaga, *Anticancer Drugs*, 2008, **19**, 477–486.
83. Y. J. Jiang, M. Brand, C. P. Heisenberg, D. Beuchle, M. Furutani-Seiki, R. N. Kelsh, R. M. Warga, M. Granato, P. Haffter, M. Hammerschmidt, D. A. Kane, M. C. Mullins, J. Odenthal, F. J. van Eeden and C. Nusslein-Volhard, *Development*, 1996, **123**, 205–216.
84. M. Granato, F. J. van Eeden, U. Schach, T. Trowe, M. Brand, M. Furutani-Seiki, P. Haffter, M. Hammerschmidt, C. P. Heisenberg, Y. J. Jiang, D. A. Kane, R. N. Kelsh, M. C. Mullins, J. Odenthal and C. Nusslein-Volhard, *Development*, 1996, **123**, 399–413.
85. A. Geling, H. Steiner, M. Willem, L. Bally Cuif and C. Haass, *EMBO. Rep.*, 2002, **3**, 688–694.
86. F. J. van Eeden, M. Granato, U. Schach, M. Brand, M. Furutani-Seiki, P. Haffter, M. Hammerschmidt, C. P. Heisenberg, Y. J. Jiang, D. A. Kane, R. N. Kelsh, M. C. Mullins, J. Odenthal, R. M. Warga, M. L. Allende, E. S. Weinberg and C. Nusslein-Volhard, *Development*, 1996, **123**, 153–164.
87. T. Yang, D. Arslanova, Y. Gu, C. Augelli-Szafran and W. Xia, *Mol. Brain.*, 2008, **1**, 15.
88. A. Petit, A. Pasini, C. Alves Da Costa, E. Ayral, J. F. Hernandez, C. Dumanchin-Njock, C. J. Phiel, P. Marambaud, S. Wilk, M. Farzan, P. Fulcrand, J. Martinez, D. Andrau and F. Checler, *J. Neurosci. Res.*, 2003, **74**, 370–377.
89. G. T. Wong, D. Manfra, F. M. Poulet, Q. Zhang, H. Josien, T. Bara, L. Engstrom, M. Pinzon-Ortiz, J. S. Fine, H. J. Lee, L. Zhang, G. A. Higgins and E. M. Parker, *J. Biol. Chem.*, 2004, **279**, 12876–12882. Epub 12004 Jan 12876.
90. G. H. Searfoss, W. H. Jordan, D. O. Calligaro, E. J. Galbreath, L. M. Schirtzinger, B. R. Berridge, H. Gao, M. A. Higgins, P. C. May and T. P. Ryan, *J. Biol. Chem.*, 2003, **278**, 46107–46116.

91. S. Weggen, J. L. Eriksen, P. Das, S. A. Sagi, R. Wang, C. U. Pietrzik, K. A. Findlay, T. E. Smith, M. P. Murphy, T. Bulter, D. E. Kang, N. Marquez-Sterling, T. E. Golde and E. H. Koo, *Nature*, 2001, **414**, 212–216.

92. Q. Yan, J. Zhang, H. Liu, S. Babu-Khan, R. Vassar, A. L. Biere, M. Citron and G. Landreth, *J. Neurosci.*, 2003, **23**, 7504–7509.

93. J. L. Eriksen, S. A. Sagi, T. E. Smith, S. Weggen, P. Das, D. C. McLendon, V. V. Ozols, K. W. Jessing, K. H. Zavitz, E. H. Koo and T. E. Golde, *J. Clin. Invest.*, 2003, **112**, 440–449.

94. E. R. Siemers, J. F. Quinn, J. Kaye, M. R. Farlow, A. Porsteinsson, P. Tariot, P. Zoulnouni, J. E. Galvin, D. M. Holtzman, D. S. Knopman, J. Satterwhite, C. Gonzales, R. A. Dean and P. C. May, *Neurology.*, 2006, **66**, 602–604.

95. E. R. Siemers, R. A. Dean, S. Friedrich, L. Ferguson-Sells, C. Gonzales, M. R. Farlow and P. C. May, *Clin. Neuropharmacol.*, 2007, **30**, 317–325.

96. A. S. Fleisher, R. Raman, E. R. Siemers, L. Becerra, C. M. Clark, R. A. Dean, M. R. Farlow, J. E. Galvin, E. R. Peskind, J. F. Quinn, A. Sherzai, B. B. Sowell, P. S. Aisen and L. J. Thal, *Arch. Neurol.*, 2008, **65**, 1031–1038.

97. D. B. Henley, P. C. May, R. A. Dean and E. R. Siemers, *Expert. Opin. Pharmacother.*, 2009, **10**, 1657–1664.

98. S. C. Mayer, A. F. Kreft, B. Harrison, M. Abou-Gharbia, M. Antane, S. Aschmies, K. Atchison, M. Chlenov, D. C. Cole, T. Comery, G. Diamantidis, J. Ellingboe, K. Fan, R. Galante, C. Gonzales, D. M. Ho, M. E. Hoke, Y. Hu, D. Huryn, U. Jain, M. Jin, K. Kremer, D. Kubrak, M. Lin, P. Lu, R. Magolda, R. Martone, W. Moore, A. Oganesian, M. N. Pangalos, A. Porte, P. Reinhart, L. Resnick, D. R. Riddell, J. Sonnenberg-Reines, J. R. Stock, S. C. Sun, E. Wagner, T. Wang, K. Woller, Z. Xu, M. M. Zaleska, J. Zeldis, M. Zhang, H. Zhou and J. S. Jacobsen, *J. Med. Chem.*, 2008, **51**, 7348–7351.

99. B. Nagar, W. G. Bornmann, P. Pellicena, T. Schindler, D. R. Veach, W. T. Miller, B. Clarkson and J. Kuriyan, *Cancer Res.*, 2002, **62**, 4236–4243.

100. T. Schindler, W. Bornmann, P. Pellicena, W. T. Miller, B. Clarkson and J. Kuriyan, *Science*, 2000, **289**, 1938–1942.

101. M. E. Noble, J. A. Endicott and L. N. Johnson, *Science*, 2004, **303**, 1800–1805.

102. W. J. Netzer, F. Dou, D. Cai, D. Veach, S. Jean, Y. Li, W. G. Bornmann, B. Clarkson, H. Xu and P. Greengard, *Proc. Natl. Acad. Sci. USA*, 2003, **100**, 12444–12449. Epub 12003 Oct 12441.

103. Y. M. Li, M. Xu, M. T. Lai, Q. Huang, J. L. Castro, J. DiMuzio-Mower, T. Harrison, C. Lellis, A. Nadin, J. G. Neduvelil, R. B. Register, M. K. Sardana, M. S. Shearman, A. L. Smith, X. P. Shi, K. C. Yin, J. A. Shafer and S. J. Gardell, *Nature*, 2000, **405**, 689–694.

104. D. Seiffert, J. D. Bradley, C. M. Rominger, D. H. Rominger, F. Yang, J. E. Meredith Jr, Q. Wang, A. H. Roach, L. A. Thompson, S. M. Spitz, J. N. Higaki, S. R. Prakash, A. P. Combs, R. A. Copeland, S. P. Arneric, P.

R. Hartig, D. W. Robertson, B. Cordell, A. M. Stern, R. E. Olson and R. Zaczek, *J. Biol. Chem.*, 2000, **275**, 34086–34091.

105. D. J. Deangelo, R. M. Stone, L. B. Silverman, W. Stock, E. C. Attar, I. Fearen, A. Dallob, C. Matthews, J. Stone, S. J. Freedman and J. Aster, *J. Clin. Oncol. (Meeting. Abstracts)*, 2006, **24**, 6585.

106. I. E. Krop, M. Kosh, I. Fearen, J. Savoie, A. Dallob, C. Matthews, J. Stone, E. Winer, S. J. Freedman and P. Lorusso, *J. Clin. Oncol. (Meeting Abstracts)*, 2006, **24**, 10574.

107. O. Berezovska, P. Ramdya, J. Skoch, M. S. Wolfe, B. J. Bacskai and B. T. Hyman, *J. Neurosci.*, 2003, **23**, 4560–4566.

108. K. S. Knappenberger, G. Tian, X. Ye, C. Sobotka-Briner, S. V. Ghanekar, B. D. Greenberg and C. W. Scott, *Biochemistry*, 2004, **43**, 6208–6218.

109. A. Y. Kornilova, F. Bihel, C. Das and M. S. Wolfe, *Proc. Natl. Acad. Sci. USA*, 2005. **102**, 3230–3235. Epub 2005 Feb 3218.

110. T. L. Kukar, T. B. Ladd, M. A. Bann, P. C. Fraering, R. Narlawar, G. M. Maharvi, B. Healy, R. Chapman, A. T. Welzel, R. W. Price, B. Moore, V. Rangachari, B. Cusack, J. Eriksen, K. Jansen-West, C. Verbeeck, D. Yager, C. Eckman, W. Ye, S. Sagi, B. A. Cottrell, J. Torpey, T. L. Rosenberry, A. Fauq, M. S. Wolfe, B. Schmidt, D. M. Walsh, E. H. Koo and T. E. Golde, *Nature*, 2008, **453**, 925–929.

111. Y. Zhou, Y. Su, B. Li, F. Liu, J. W. Ryder, X. Wu, P. A. Gonzalez-DeWhitt, V. Gelfanova, J. E. Hale, P. C. May, S. M. Paul and B. Ni, *Science*, 2003, **302**, 1215–1217.

112. S. Leuchtenberger, M. P. Kummer, T. Kukar, E. Czirr, N. Teusch, S. A. Sagi, R. Berdeaux, C. U. Pietrzik, T. B. Ladd, T. E. Golde, E. H. Koo and S. Weggen, *J. Neurochem.*, 2006, **96**, 355–365. Epub 2005 Nov 2021.

113. T. Mizuno, N. Yamasaki, K. Miyazaki, T. Tazaki, R. Koller, H. Oda, Z. I. Honda, M. Ochi, L. Wolff and H. Honda, *Oncogene*, 2008, **27**, 3465–3474.

114. P. C. Fraering, W. Ye, M. J. LaVoie, B. L. Ostaszewski, D. J. Selkoe and M. S. Wolfe, *J. Biol. Chem.*, 2005, **280**, 41987–41996. Epub 42005 Oct 41919.

115. W. Campbell, M. L. Reed, J. Strahle, M. S. Wolfe and W. Xia, *J. Neurochem.*, 2003, **85**, 1563–1574.

116. R. M. Page, K. Baumann, M. Tomioka, B. I. Perez-Revuelta, A. Fukumori, H. Jacobsen, A. Flohr, T. Luebbers, L. Ozmen, H. Steiner and C. Haass, *J. Biol. Chem.*, 2008, **283**, 677–683.

117. L. Placanica, L. Tarassishin, G. Yang, E. Peethumnongsin, S. H. Kim, H. Zheng, S. Sisodia and Y. M. Li, *J. Biol. Chem.*, 2009, **284**, 2967–2977.

118. J. Li, G. J. Fici, C. A. Mao, R. L. Myers, R. Shuang, G. P. Donoho, A. M. Pauley, C. S. Himes, W. Qin, I. Kola, K. M. Merchant and J. S. Nye, *J. Biol. Chem.*, 2003. 278, 33445–33449. Epub 32003 Jun 33418.

119. S. Shah, S. F. Lee, K. Tabuchi, Y. H. Hao, C. Yu, Q. LaPlant, H. Ball, C. E. Dann 3rd, T. Sudhof and G. Yu, *Cell*, 2005, **122**, 435–447.

120. L. Chavez-Gutierrez, A. Tolia, E. Maes, T. Li, P. C. Wong and B. de Strooper, *J. Biol. Chem.*, 2008, **283**, 20096–20105.

121. Z. Xie, D. M. Romano, D. M. Kovacs and R. E. Tanzi, *J. Biol. Chem.*, 2004. **279**, 34130–34137. Epub 32004 Jun 34137.
122. A. S. Crystal, V. A. Morais, R. R. Fortna, D. Carlin, T. C. Pierson, C. A. Wilson, V. M. Lee and R. W. Doms, *Biochemistry*, 2004, **43**, 3555–3563.
123. P. C. Fraering, M. J. LaVoie, W. Ye, B. L. Ostaszewski, W. T. Kimberly, D. J. Selkoe and M. S. Wolfe, *Biochemistry*, 2004, **43**, 323–333.
124. H. Hasegawa, N. Sanjo, F. Chen, Y. J. Gu, C. Shier, A. Petit, T. Kawarai, T. Katayama, S. D. Schmidt, P. M. Mathews, G. Schmitt-Ulms, P. E. Fraser and P. St George-Hyslop, *J. Biol. Chem.*, 2004, **279**, 46455–46463.
125. S. H. Kim and S. S. Sisodia, *J. Biol. Chem.*, 2005, **280**, 1992–2001.
126. N. Isoo, C. Sato, H. Miyashita, M. Shinohara, N. Takasugi, Y. Morohashi, S. Tsuji, T. Tomita and T. Iwatsubo, *J. Biol. Chem.*, 2007, **282**, 12388–12396.
127. F. Chen, H. Hasegawa, G. Schmitt-Ulms, T. Kawarai, C. Bohm, T. Katayama, Y. Gu, N. Sanjo, M. Glista, E. Rogaeva, Y. Wakutani, R. Pardossi-Piquard, X. Ruan, A. Tandon, F. Checler, P. Marambaud, K. Hansen, D. Westaway, P. St George-Hyslop and P. Fraser, *Nature*, 2006, **440**, 1208–1212.
128. W. Xia, B. L. Ostaszewski, W. T. Kimberly, T. Rahmati, C. L. Moore, M. S. Wolfe and D. J. Selkoe, *Neurobiol. Dis.*, 2000, **7**, 673–681.

CHAPTER 3

Cholesterol and Alzheimer's Disease: The Molecules, the Targets

M. DOLORES LEDESMA[1] AND CARLOS G. DOTTI[2,3]

[1] Centro de Biología Molecular Severo Ochoa, CSIC-UAM, Nicolás Cabrera 1, 28049, Madrid, Spain; [2] Department of Molecular and Developmental Genetics, VIB, Herestrasse 49, 3000 Leuven, Belgium; [3] Center for Human Genetics, K.U. Leuven, Herestrasse 49, 3000 Leuven, Belgium

3.1 Introduction

The sole mention of the term "cholesterol" brings to many the idea of deleteriousness, because of the association between high blood cholesterol with atherosclerosis, heart disease, obesity and other health problems. That high blood cholesterol also relates to Alzheimer's disease (AD) has been evidenced by epidemiologic studies showing a strong correlation between the prevalence of the disease and hypercholesterolemia and dietary saturated fats and that the prevalence in these circumstances increases with age.[1,2] While the association has prompted the search for cholesterol-modifying drugs to prevent AD and even stop its course, caution should be taken because cholesterol is a most essential constituent of several physiological events and therefore drugs that affect this lipid levels could lead to undesired effects, especially if they act in the central nervous system (CNS).

RSC Drug Discovery Series No. 2
Emerging Drugs and Targets for Alzheimer's Disease
Volume 1: Beta-Amyloid, Tau Protein and Glucose Metabolism
Edited by Ana Martinez

Published by the Royal Society of Chemistry, www.rsc.org

Cholesterol is a major component of the cell membrane, accounting for up to 25% of all its lipids, and also of endosomes and the Golgi apparatus.[3,4] It is essential for the fluidity of the plasma membrane and the control of proton and sodium electrochemical gradients,[5] a most critical event in membrane potential, ATP generation and synaptic transmission and propagation. Cholesterol, together with sphingolipids, also plays a crucial role in the organisation of the plasma membrane in microdomains that act as platforms for intracellular signalling involved in processes such as cell differentiation, pro and anti-apoptosis, endocytosis and proper distribution of membrane molecules.[6–8] In addition, cholesterol is the precursor of bile acids and steroid hormones.[9] In the brain, this lipid is used for the generation of a class of steroid hormones referred to as neurosteroids, which play important roles in the modulation of GABAergic responses,[10] the reaction of Purkinje cells to excitatory amino acids,[11] and the control of memory.[12] From these considerations it is easy to envision that any drug that alters cholesterol levels would need to be closely followed up in order to monitor the evolution of basic physiological activities. This is even more critical when using cholesterol-modifying drugs that cross the blood/brain barrier and induce loss of cholesterol from the plasma membrane of nerve cells. Cholesterol metabolism in the brain is obtained through *de novo* synthesis and not from the metabolism of cholesterol-rich lipoprotein particles coming from the circulation.[13] This indicates that a loss from the membrane of these cells would not be rescued by circulating cholesterol and therefore synaptic neurotransmission, signal transduction and survival will be easily perturbed.[14–17] A further complication comes from the observation that membrane cholesterol decreases with age and in AD patient brains,[18,19] thus increasing the risk of cholesterol-reducing strategies.

3.2 Alzheimer's Disease: Contribution of the Central Nervous System and Peripheral Defects

There is ample agreement in that AD is a multifactorial disorder: some people are affected because of the inheritance of a particular mutation in the amyloid precursor protein (APP) or the APP-cleaving enzymes presenilins (PSEN1 and PSEN2) and others because of previous conditions such as brain trauma, diabetes, hypercholesterolemia, hypertension or others yet to be discovered. While in the first cases the occurrence of disease is guaranteed (familial AD), the others simply constitute risk factors (late onset or sporadic AD). Irrespective of genetic or environmental, and in a most schematised manner, symptoms of disease are thought to be due to synaptic-circuit dysfunction caused by the toxic effects of the amyloid peptide (Aβ), a byproduct of APP cleavage.[20] Although consensus exists about this peptide's toxicity, it is less clear if this is due to the accumulation inside cells or after its excretion into the extracellular milieu.[21,22] Alternative to the amyloid hypothesis as the triggering event of AD symptoms, a number of pieces of evidence suggest that synaptic circuitry dysfunction is due to the formation of intracellular aggregates of the microtubule-associated protein tau, followed by perturbed axonal transport and membrane permeability.[23] For a comprehensive analysis of the data in favour of one mechanism of action or another, the interested reader should go to excellent reviews in these matters.[24–26] In any event, it is here worth mentioning that while the relevance of the amyloid peptide chain is justified by genetic data showing that the mutations associated with AD result in an excessive production of the Aβ peptide, a number of pieces of evidence suggest that, in many cases of sporadic AD, Aβ toxicity may be a consequence of vascular dysfunction. First, various findings indicate that cerebral amyloid deposits derive from soluble Aβ present in the cortical extracellular space or the cerebrospinal fluid communicating small brain vessels. In fact, cerebral amyloid deposits are always intimately associated with the smooth muscle cells of small arterioles or their surrounding pericytes.[27,28] Secondly, it has been demonstrated that blood is a major and chronic source of soluble Aβ in the brain, entering this tissue by an active transport mechanism followed by binding to selective types of neurons.[29–32] Thirdly, clearance of Aβ from the brain into the circulation is reduced in hypo-oxygenation states due to alteration in the contractility and clearance mechanisms of vascular smooth muscle cells of brain arterioles.[33,34] Thus, accumulated amyloid peptide near blood vessels would induce toxicity to the nerve terminals located in the vicinity, leading eventually to neuronal loss, which will in turn worsen the initial localised vascular phenotype. Hence, the possibility that many cases of late onset AD could be triggered by peripheral (circulatory) alterations is solidly proven. Accepting this view renders it obvious that elevated cholesterol in the bloodstream could have a most fundamental role in the development of the disease. On the other hand, being Aβ generation and degradation events occurring in cholesterol-rich membrane organelles of neurons, like the plasma membrane and endosomes, it is certainly possible that altered brain-cholesterol homeostasis could trigger excessive Aβ production and/or reduced degradation. Furthermore, CNS cholesterol may also determine disease due to alterations in Aβ clearance, which depends to a great extent on the binding of the peptide to the cholesterol-transport protein ApoE.[35] We will review here the multiple ways by which altered cholesterol homeostasis, in the circulatory or central nervous systems, could contribute to AD (see Table 3.1 for a summary).

3.3   Apolipoprotein E: The First Link with Cholesterol

Apolipoprotein E (apoE) is a small plasma protein that acts as ligand for low-density lipoprotein (LDL) receptors participating therefore in the transport of cholesterol and other lipids to different cells. A mutant form of the protein that is defective in binding to LDL receptors causes a genetic disorder characterised by elevated plasma cholesterol levels and coronary disease (familial type-III hyperlipoproteinemia).[36] In the brain, apoE is mainly synthesised and secreted by glial cells and in this form cholesterol is provided to mature neurons, in which *de novo* synthesis is reduced with age.[37] In humans, apoE has three major

Table 3.1 List of the cholesterol-related molecules currently considered as potential targets for AD treatment indicating: i) the cellular process in which they are involved; ii) the effect of their modulation in Aβ levels; iii) the compounds already used for such modulation; iv) whether clinical trials have been performed and v) if genetic links with the disease have been found.

Molecule	Cellular process involved	Effect on Aβ levels	Strategy for modulation	Clinical trial	Genetic link with AD
ApoE	Cholesterol transport/uptake	Increase with allele E4	Estrogen Indomethacin Probucol Rosiglitazione	Rogers, 1993 Poirier, 2003 Risner, 2006	Corder, 1993 Raber, 2004
LDL Receptors: LRP1 LRP1B SorLA/LR11 ApoER2	Cholesterol/APP internalisation, Interaction with APP secretases	Increase Decrease Decrease Increase	RAP SREBPs SCAP Insig		Kang, 2000 Ma, 2002
ACAT	Cholesterol intracellular distribution/ CE formation	Inhibition reduces Aβ	CP-113,818 Avasimibe		
HMG-CoA	Cholesterol synthesis	Inhibition reduces Aβ	Statins	Sepherd,2002 Yaffe, 2002 Sparks, 2005	
Seladin 1	Cholesterol synthesis/apoptosis	Inhibition increases Aβ	Estrogen Tamoxifen Raloxifene		Greeve, 2000
Cyp46	Cholesterol removal/ survival	Inhibition increases Aβ (presumably)	Histone deacetylase inhibitors		Papassotiropoulos, 2003 Borroni, 2004
ABCA1	Cholesterol removal	Enhancement reduces Aβ	Hydroxy-cholesterol Retinoic acid		
24-OHC	Cholesterol removal from brain	Inhibits Aβ	Histone deacetylase inhibitors		

protein isoforms: apoE2, apoE3, and apoE4. The association of allele four of apoE (apoE4) as a genetic risk factor for AD has been well established and accounts for 50–60% of the genetic variation in the disease.[38,39] ApoE4 AD patients have an earlier age of onset and greater amyloid deposits and reduced synaptic plasticity.[38,40–42] The reasons for this predisposition are not clear but several possibilities are envisioned. ApoE is involved in antioxidative processes,[43] and calcium mediated intracellular signalling,[44] and thus predisposition may be linked to deficiency at any of these levels. On the other hand, because of the prominent role in cholesterol transport, the E4 allele may prompt to the disease due to cholesterol-deficient metabolism. Hence, individuals with an apoE e4/e4 genotype display the lowest concentrations of plasma apoE in all humans,[45,46] and the e4 genetic variant has a poor response to physiological inducers of expression.[47] In addition, apoE levels are reduced in the brain of apoE4 mice, possibly due to faster degradation,[48] and the apoE4 protein has a low ability to release cholesterol from glial cells compared to other genotypes.[49] Hence, the presence of the apoe4 allele could predispose to AD by reducing the amount of apoE available to bind to Aβ and therefore the amount of peptide removed from the brain into the general circulation.[50] Alternatively, the apoE4 allele is a risk to suffer the disease because of defective degradation of Aβ peptides within microglia and in the extracellular space.[51] Yet another possibility would be that apoE4 carriers are more prone to develop AD because of deficient apoE-mediated synapse maintenance and cognition.[52–54] A different way to reach amyloid pathology in apoE4 bearers would be through the atherosclerosis that commonly affects these individuals. Cholesterol deposits in brain arterioles and capillaries lead to hypo-oxygenation with multiple consequences for brain function, including a less-efficient clearance of Aβ peptide towards the blood.[27] In any event, irrespective of the reasons by which the apoE4 predisposes to AD, this is a logical target for AD treatment, the current idea being to increase its levels and/or improve its function. With this aim screenings of large numbers of drugs, proteins and hormones have been performed, resulting in the identification of several enhancers of apoE synthesis and secretion in the mature rodent CNS at physiologically relevant concentrations.[47] Estrogen, the cyclooxygenase inhibitor indomethacin, the cholesterol-lowering drug probucol and the peroxisome proliferating activating receptor-g (PPAR-g) agonist rosiglitazione have been utilised in placebo-controlled, double-blind clinical trials in mild-to-moderate AD patients and the results obtained are promising. Rogers and coworkers reported years ago that indomethacin slows down the progression of the disease,[55] and a recent clinical study by Risner and colleagues revealed significant improvement of the cognitive deficit upon administration of rosiglitazione during six months.[56] In addition, standard doses of probucol used in humans to treat hypercholesterolemia produced stabilisation of cognitive symptoms.[57] These clinical benefits correlated with the increase of circulating apoE and the reduction of Aβ levels in the cerebrospinal fluid of the probucol-treated AD patients,[57] pointing in the direction of circulatory deficits as a key factor in AD predisposition.[58]

3.4 Low-Density Lipoprotein Receptor (LDLR) Family: The Uptake of Cholesterol

The LDL receptors comprise a large family of cell-surface receptors abundantly expressed in the brain. They are endocytically active and have signalling functions.[59] By binding to apoE these receptors mediate cholesterol entrance into neurons. Among the members of the family are LRP1, LRP1B, SorLA/LR11, and apoER2. Since they control the amount of neuronal cholesterol inside cells and the production of the esterified form of this lipid,[60,61] any alteration in LRP-mediated cholesterol binding or internalisation would affect membrane-occurring activities, including the amount of Aβ [62,63] (and see Chapter 5, also Section 3.5) On the other hand, members of the LDLR family interact with APP and regulate its rate of internalisation, which is a critical step in the amyloidogenic processing of this protein. Hence, while LRP1 displays a very high internalisation rate,[64] that of LRP1B and SorLA/LR11 are very slow.[65,66] Consistently, LRP1 promotes Aβ production,[67] while LRPB1 and SorLA/LR11 reduce it by stabilising APP at the cell surface.[68,69] ApoER2 interacts with APP and exhibits a low rate of endocytosis.[70] However, the effect of their coexpression results in the increase of Aβ peptide,[70] in apparent contradiction with the view that a more efficient internalisation favours amyloidogenesis. A possible explanation is that apoER2, which is abundant in cholesterol-enriched membrane microdomains, shifts the localisation of APP to these domains where its β- and γ-secretases also reside, thus favouring the amyloidogenic processing. In agreement, the expression of apoER2 significantly increases γ-secretase activity,[70] and, in the presence of apoE, induces the internalisation of APP along with its β-secretase (BACE).[71]

Still another way by which alterations in the LDLR family could lead to AD is via changes in the interaction of these proteins with components of the secretases involved in APP cleavage, influencing therefore APP access to them. Thus, LRP1 interacts with the secretases BACE and PSEN1 and it is a substrate for both, while SorLA/LR11 interacts with BACE and is proteolytically processed by the γ-secretase complex.[72,73]

Finally, LDLR alterations might lead to AD by defective Aβ clearance. This is implied from the observation that LRP1 mediates the clearance of Aβ *in vitro*,[74] and is a major efflux transporter of the peptide across the blood/brain barrier, therefore essential for its removal from the brain.[75]

While all the above indicates that a defective LDLR pathway could lead to AD by increasing brain Aβ levels, intracellular or extracellular, directly (*i.e.* altered production) or indirectly (*i.e.* reduced clearance), the question remains as to whether or not LDLR alterations could be considered a true cause of AD. The discovery that polymorphisms within the LRP1 and apoER2 genes are associated to the risk to suffer the disease,[76,77] would be consistent with this possibility.

Despite the uncertainties on whether or not alteration in the LDLR family triggers disease and how this could occur, the fact that their manipulation has consequences on Aβ levels makes these proteins interesting targets for AD therapy. That this could be a valid strategy is suggested from the observation that long-term treatment with the recombinant receptor-associated protein (RAP), which antagonises ligand binding to LDLR and assists its proper folding and trafficking along the secretory pathway,[78–80] disrupts the intracellular interaction between LRP1 and APP, increasing cell surface APP and decreasing Aβ production.[67] On the other hand, the pharmacological modulation of the sterol-responsive element-binding proteins (SREBPs), the transcription factors that regulate LDLR expression,[81] and of their activating proteins SCAP and Insig,[82] could also open therapeutic perspectives by virtue of the fact that SREBP play an essential role in LDLR-Aβ clearance from smooth muscle cells of brain vessels.[60,83]

3.5 Acyl-Coenzyme A Cholesterol Acyltransferase (ACAT): The Intracellular Distribution of Cholesterol

Cellular cholesterol is present as free cholesterol (FC) in the plasma membrane, endosomes and the Golgi apparatus or as cholesterol esters (CE) in cytoplasmic lipid droplets. The pools of FC and CE are tightly controlled by ACAT, an endoplasmic-reticulum-resident enzyme that catalyses the formation of CEs from cholesterol and long-chain fatty acids.[84] In genetically modified cell lines lacking ACAT activity there is an overproduction of FC, a paucity of CE and drastically reduced Aβ production.[62] Consistently, pharmacological inhibitors of ACAT decrease Aβ levels *in vitro*.[62] The efficacy of one such inhibitor, CP-113,818, was also tested *in vivo*. Slow-release biopolymer administration of this compound for two months drastically reduced brain CE and the accumulation of amyloid plaques and soluble and membrane-insoluble Aβ levels in mice models for AD expressing mutated human APP.[85] Importantly, spatial learning was slightly improved in these animals correlating with Aβ decrease. Hence, like for the LDLR family proteins, even if we do not know if sporadic AD could be caused by a primary defect at the level of ACAT activity (apparently a gain-of-function), the above data strengthen the notion that modifying the intracellular distribution of cholesterol and/or the FC/CE ratio through modulation of ACAT might be a good strategy for the treatment and prevention of AD. Although results on clinical trials with ACAT inhibitors for AD are not yet available, these have been tested for the treatment of hypercholesterolemia and atherosclerosis. After disappointing results of many ACAT inhibitors due to low efficacy and adrenotoxicity, today's nonadreno-toxic compound CI-1011 (Avasimibe) is considered safe for human use,[86] and opens up interesting perspectives for vascular cholesterol pathology as well as for AD therapy.

3.6 HMG-CoA: The Rate-Limiting Step of Cholesterol Synthesis

The striking reduction in the prevalence of AD found in hypercholesterolemic patients treated with the cholesterol-lowering drugs statins,[87,88] together with the early discovery of apoE4 as a risk factor led to a direct correlation between high brain cholesterol and high Aβ production being proposed.[89,90] However, this has turned out to be a simplistic assumption. First, because it implies that brain cholesterol is high in hypercholesterolemic patients. This is actually not the case as brain-cholesterol homeostasis is independent of circulating cholesterol. Secondly, because statin treatments, which reduce the amount of peptide in experimental animals,[90,91] do not change brain-cholesterol levels. Furthermore, recent evidence showed that a moderate loss of neuronal membrane cholesterol leads to increased Aβ production *in vitro* and *in vivo,*[92,93] suggesting that low brain cholesterol, and not high, favours high Aβ levels. In agreement with this possibility, Aβ production is reduced upon increase of membrane cholesterol after ACAT knockdown.[62] Given the above, it is more conceivable that reduction in blood and not brain cholesterol is at the core of the beneficial effects in cognition observed in patients treated with statins. These are in fact plasma cholesterol-reducing drugs,[94] and would therefore improve brain oxygenation and subsequently better clearance to the blood stream of Aβ peptide.[95] Statins might also improve brain function through their well-known anti-inflammatory properties.[94]

Although many consider that modulation of HMG-CoA can be beneficial, the conclusions from prospective cohort studies and clinical trials are not as clear as that aroused from retrospective studies performed in hypercholesterolemic patients.[96] In fact, subjects randomised to simvastatin, pravastatin or placebo did not demonstrate a significant improvement.[97,98] In another study better cognitive performance was demonstrated among patients receiving statins, although these effects were found to be independent of lipid levels.[99] A pilot proof-of-concept study of atorvastatin in mild to moderate AD revealed significant cognitive improvement at six months but the benefit did not persist at the end of one year.[100] Results of several ongoing randomised clinical trials with larger sample sizes will be necessary to clarify the utility of HMG-CoA inhibition in AD treatment.

3.7 Seladin 1: The Last Step of Cholesterol Synthesis

A differential display approach to identify genes with altered expression in brain regions vulnerable to AD resulted in the identification of a number of genetic variations. One of these was the marked reduction of a gene called seladin1 (for SELective AD Indicator-1) in the inferior temporal cortex of AD patients compared to the frontal cortex.[101] Seladin1 codes for the enzyme 3beta-hydroxysterol Delta24-reductase (DHCR24), which catalyses the last step in cholesterol biosynthesis.[102] These results highlighted the existence of area variations at the transcriptional level, whether intrinsic or extrinsically produced, which could explain why a phenomenon like Aβ accumulation and toxicity occurs at particular sites of the brain. These results also suggest that reduction in brain cholesterol, not its increase, could predispose to AD. Although decreased levels of Seladin1 found in postmortem tissue could be taken as a simple secondary effect of the pathology, a number of pieces of evidence suggest, however, that Seladin1 downregulation may play a causal role. Seladin1 confers protection against Aβ and oxidative-stress-induced apoptosis, by inhibiting the activation of caspase-3, a key mediator of the apoptotic process.[103] From here it seems that its downregulation would affect cell function and eventually viability leading to localised pathology. Another way by which Seladin1 deficiency could cause disease would be because of excessive production of Aβ due to the low levels of membrane cholesterol, as demonstrated to occur in Seladin-deficient neurons, both *in vivo* and *in vitro*.[93] Since the levels of plasma membrane cholesterol are critical to determine the degree of mobility of different membrane proteins, and therefore their capacity to interact with partners and signal downstream,[104,105] any change in cholesterol content at this level would affect the interaction of APP with its secretases. In fact, in hippocampal neurons in culture reduced cholesterol levels enhance the interaction of APP with BACE1.[92] Furthermore, a reduction in cholesterol content in the plasma membrane of neurons inhibits the activation of the Aβ degrading enzyme plasmin.[19,93] Moreover, the antiapoptotic role of Seladin1 mentioned above could also be explained through the control of the amount of cholesterol in the neurons' plasma membrane, which plays a key role in the level of activity of the prosurvival pathway mediated by TrkB.[106]

Altogether, these data indicate that the cholesterol reduction resulting from Seladin1 downregulation may lead to disease via a triple deleterious effect: i) reduced prosurvival response, ii) increased generation of Aβ and iii) reduced Aβ degradation. Therefore, it appears critical to set up assays to determine the levels of expression of Seladin1 during aging and to develop drugs to prevent its downregulation. In this regard, recent studies have shown that estrogens upregulate the expression of the enzyme, possibly by activation of the functional estrogen-responsive elements upstream of the coding region of the seladin1 gene.[107] In agreement, treatment of cultured cells with 17b-estradiol and the selective estrogen receptor modulators tamoxifen and raloxifene significantly increased the amount of seladin1 mRNA.[108] The fact that upon silencing of seladin1 expression by small interfering RNA (siRNA) the protective effects were lost further supported the link estrogens-seladin1- protection against amyloid.[109] Moreover, it is known that AD is more common in women and decreased estrogen levels after menopause have been considered a risk factor for the disease.[110] Despite the lack of consensus several studies have indicated that estrogen treatment may actually decrease the risk or delay the onset of AD in postmenopausal women,[111,112] and that the earlier and more prolonged administration produced the maximum benefit in terms of reduced risk.[113,114]

3.8 Cholesterol 24-Hydroxylase (Cyp46) and the ATP-Binding Cassette Transporter A1 (ABCA1): The Removal of Cholesterol

Because cholesterol is not degraded, elimination of cholesterol occurs by solubilisation after hydroxylation, a process that in brain cells is commanded by the enzyme 24(S) cholesterol hydroxylase/Cyp46.[115] Released 24S-hydroxycholesterol can freely cross the brain/blood barrier due to lipophylicity.[116] From such essential function it is logical to envision that any defect in the activity of this enzyme will pose a threat to neuron physiology. For example, cells affected by a loss of function of this enzyme, for any genetic or environmental reason, would result in cells with an excess of used up and "old" cholesterol, which may not be so efficient in performing one of cholesterol's main functions as cation permeability barrier (therefore leading to cation leakage and changes in membrane potential) or in the generation or maintenance of protein signalling platforms/rafts (therefore leading to decreased intracellular signalling accuracy/potency). In agreement with such a prediction, mice in which the enzyme is missing exhibit severe deficiencies in spatial, associative, and motor learning, and in hippocampal long-term potentiation (LTP).[117] Furthermore, it was recently shown that reduced Cyp46 expression levels block the activation of the TrkB survival pathway in stressed primary neurons *in vitro*,[106] suggesting that any deficiency in this protein would make neurons more susceptible to stress. This conclusion would be consistent with data showing that oxidative stress is a potent inductor of Cyp46 expression.[118] As with many other components of the cholesterol metabolic pathway the question applies here: can changes in Cyp46 expression or activity be responsible for the occurrence of, at least, certain cases of sporadic AD? The finding of selective expression of the enzyme around neuritic plaques in AD brains,[119] could be taken as a positive indication. On the other hand, as with most of the data from postmortem tissue, one cannot exclude that increased expression is a response to stress and not a cause of pathology. Another piece of evidence supporting a causal role is genetic: the correlation in certain families between polymorphisms in Cyp46 and higher risk to suffer late-onset AD.[120,121] In further support, it was found that this risk synergistically increases with the additional presence of 1 or 2 apoE4 alleles.[120,121] Moreover, it was shown that polymorphisms in Cyp46 gene are associated with increased Aβ load in brain tissues as well as with increased cerebrospinal fluid levels of the peptide and phosphorylated tau protein.[120] Since the polymorphisms in Cyp46 are in noncoding regions it appears quite important to establish if these result in higher or lower copy number. In any case, the critical role that Cyp46 plays in brain-cholesterol homeostasis confers to this protein a potential interest for AD therapy. At present, histone deacetylase inhibitors (*e.g.* valproate, vorinostat) are the only class of compounds known to cause increased mRNA expression of the enzyme.[122] Importantly, it was recently reported that treatment of AD mouse models with the HDACi valproate reduced Aβ production, neuritic plaque formation and behavioural deficits.[123] This suggests that human affected by loss-of-function (lower copy number) polymorphisms may benefit from this type of treatment.

Another way by which cells eliminate used cholesterol is by means of the ATP-binding cassette membrane transport proteins, which have a channel-like structure that can transport various solutes across the cell membrane.[124] ABCA1, a member of the ATP-binding cassette family of active transporters, mediates the rate-limiting step of the reverse cholesterol transport pathway, transferring cellular cholesterol onto lipid-poor apoE. The possibility that alterations in ABCA1 may lead to AD is supported by the association found between polymorphisms in its gene in a variety of affected populations.[125–128] Due to the high levels of expression of this transporter in the CNS,[129] it is likely that predisposition is due to an impact in brain-cholesterol levels. In agreement, it was shown that mice lacking ABCA1 present a dramatic reduction of apoE levels in the brain and in the cerebrospinal fluid suggesting the possibility of a central component.[130,131] In further support, a recent study showed that mice lacking ABCA1 specifically in the CNS, generated using the Cre/loxP recombination system, presented reduced plasma high-density lipoprotein (HDL) cholesterol levels associated with decreased brain cholesterol content and enhanced brain uptake of esterified cholesterol from plasma HDL. These mice displayed disturbances in motor activity and sensorimotor function and reduced synapse and synaptic vesicle numbers.[132] Hence, brain ABCA1 not only has a great impact in the regulation of CNS cholesterol levels but also in the control of plasma levels of HDL cholesterol and in the uptake by neurons of esterified plasma cholesterol. In relation to Aβ levels it has been found that induction of ABCA1 expression by activation of liver X receptors (LXRs) and retinoid X receptors (RXRs) with 22-hydroxycholesterol and retinoic acid inhibited Aβ secretion *in vitro*.[133,134] Moreover, transient transfection of ABCA1 into a CHO cell line stably expressing human APP reduced by half Aβ secretion.[135] In the brain, excess of ABCA1, induced by the LXR agonist TO-901317, significantly inhibited Aβ deposition in APP23 and Tg2576 transgenic mice,[136,137] while the genetic loss of LXR receptors,[138] or ABCA1,[130,136,139] increased Aβ load. In the same line, the overexpression of ABCA1 in the PDAPP mouse model reduced amyloid deposition.[140] All the above, together with the recent data from ABCA1 conditional knockout mice revealing the dramatic increase in HDL-cholesterol and a reduction in brain cholesterol,[132] point in the direction that ABCA1 gain and loss of function data can be interpreted by, respectively, reduction or increase in circulating HDL cholesterol. In this scenario one would have to argue that the polymorphisms of ABCA1 associated to higher risk to AD must be of the loss of function type, as these relate to increase HDL-cholesterol and high Aβ levels. If correct, enhanced brain ABCA1 expression might be a useful tool to prevent excessive Aβ production by neurons and, due to the peripheral effects, increase clearance of peptide from the brain to the circulation. More animal model studies are, however, needed to confirm the potential therapeutical value of these molecules.

3.9 24-Hydroxycholesterol (24-OHC) and 27-Hydroxycholesterol (27-OHC): The Metabolites of Cholesterol

24-OHC and 27-OHC are side-chain oxidised oxysterols with ability to cross lipophilic membranes into and out of the brain. As already mentioned, formation of 24-OHC, which is primarily found in the brain, is the quantitatively most important mechanism for elimination of cholesterol from this organ.[141] Conversely, there is a significant net uptake of 27-OHC by the brain from the circulation.[142] As it could not be otherwise, several links have been found between oxysterols and AD. It has been postulated that high blood-cholesterol levels associated with increased turnover to 27-OHC enhances the transport of this oxysterol into the brain and induce AD-like histopathological hallmarks.[143] Supporting this view the levels of 27-OHC are increased in AD patient brains.[144] On the contrary, the amount of 24-OHC is diminished.[144] In the theoretical scenario that excess circulating oxysterols appear in the brain, uptake by nerve terminals will trigger a series of effects, among them in the processing of APP. 24-OHC inhibits the production of the Aβ peptide,[119,143] and 27-OHC does the same in primary cultured neurons,[119] but not in human neuroblastoma cells, in which this oxysterol produces the opposite effect increasing APP processing.[143] Given the above, the authors of this last study proposed that 24-OHC favours the nonamyloidogenic processing of APP, while 27-OHC may upregulate APP and BACE1 enhancing Aβ production. On the other hand, it has been shown that oxysterols inhibit protein kinase C activity and APP secretion following stimulation of protein kinase C in primary neurons.[119] To what extent the changes observed correspond to the cytotoxic effects of oxysterols, due to the generation of oxidative and/or inflammatory processes,[145] rather than specific for APP remains to be determined. Still, Brown and colleagues proposed that oxysterols might modify APP processing because they are ligands for the nuclear transcription factor LXR,[119] which in turn induces ABCA1 expression and, as we described above, participates in cholesterol excretion. Compounds already described here that affect LXR or the enzymes producing these oxysterols could be envisioned as targets to modulate their levels as a strategy for AD therapy. However, because these enzymes are required for normal bile-acid synthesis this kind of treatment might require dietary bile acids. To our knowledge, no clinical trial based on the modulation of these cholesterol metabolites has been reported yet but the diagnostic value of oxysterols has been highlighted. Thus, it has been shown that 24-OHC levels in the cerebrospinal fluid have about the same diagnostic sensitivity as the standard biomarkers currently used in the diagnosis of neurodegenerative diseases and dementia, namely, levels of tau protein, phosphorylated tau and Aβ.[146] Although elevated levels may simply reflect the high stress of the brain because of AD pathology, rather than AD diagnostic marker, it is certain that they can be used for prognosis purposes to evaluate the evolution of any given treatment.

3.10 Concluding Remarks

It is nowadays indisputable that alterations in cholesterol homeostasis leading to increased levels in the circulation or to decrease or excess in the brain, will have tremendous implications for human health. In what concerns

AD this came noticeable by the discovery, more than 15 years ago, of the allele 4 of the cholesterol-transport protein apoE as a risk to suffer the disease. Since then, other genetic association studies strengthened the notion that defects in cholesterol homeostasis could play a key role in AD pathogenesis. The question arises as to how much the cholesterol-related AD occurrence is attributable to altered levels of circulating cholesterol ("peripheral" AD) and how much to altered brain-cholesterol levels ("central" AD). Without ruling out the possibility of a central component the cases of apoE or ABCA1 mutations, in which variations in peripheral cholesterol take place, speak in favour of AD predisposition by the "peripheral" mechanism. On the other hand, in the cases of ACAT or Cyp46 genetic alterations the peripheral involvement is more difficult to defend, as these are enzymes enriched in brain cells. Yet another question arising is how a defect in a most essential and universally present constituent of all cells' membranes can only affect the neurons that are responsible for AD symptoms. The most likely explanation is that cholesterol defects are minimal but severe enough to alter the function of specific neuronal populations, which are most sensitive to subtle changes in this lipid levels. Although this needs to be proven, the observation of geographically selective downregulation of Seladin 1 points in this direction.

In any event, what becomes clear is that drugs with the capacity to maintain circulating-cholesterol levels within physiological values would be beneficial to prevent the disease, minimally because of the improvement of brain oxygenation and thus better amyloid clearance. On the other hand, far more care should be taken when using blood-cholesterol-modulating drugs with the capacity to enter the CNS, especially after symptoms have started. In fact, the presence of symptoms reflects synaptic dysfunction, and reduction of cholesterol in the affected cells may worsen matters as ionic permeability would be increased, with the subsequent need to generate more ATP and hence further cellular stress. Moreover, the opposite strategy, to increase brain cholesterol, which could be envisioned from the ACAT and other studies, is not easy to support either, as it may alter the capacity of neurons to support certain antistress activities. However, these may be safer in the long run than cholesterol-lowering drugs, as neurons should be able to put to work survival pathways that do not depend on membrane-cholesterol levels. It is also possible to envision that drugs capable of reducing brain-cholesterol esters without affecting in excess the free (membrane) cholesterol pool may turn out to be beneficial. Although we still need to learn more about the relationship between cholesterol homeostasis and AD, the level of knowledge acquired so far reveals a number of target molecules at different cellular sites, which function might be enhanced or diminished (Figure 3.1). This justifies the search for more and different types of cholesterol modulating drugs, to be used in basic and animal research. At this point we need to rapidly find drugs that prevent or revert AD and cholesterol-modulating drugs could be most suitable, as cholesterol metabolism will be one of the first affected pathways in neuronal cells, no matter what caused the disease.

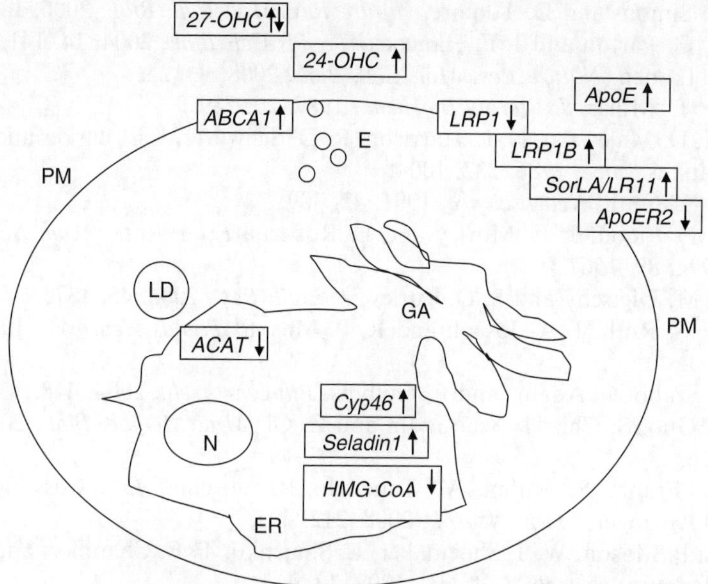

Figure 3.1 Schematized image of a cell showing the localization of the cholesterol-related molecules that are currently considered as potential targets for AD treatments (shown inside boxes). Up and down arrows indicate whether enhancement or inhibition, respectively, of the molecule's function is envisioned to diminish Aβ levels. The subcellular compartments depicted are: nucleus (N), Endoplasmic reticulum (ER), Golgi apparatus (GA), lipid droplets (LD), endosomes (E) and plasma membrane (PM).

Acknowledgements

We thank the support of the Fund for Scientific Research Flanders (FWO); Federal Office for Scientific Affairs (IUAP P6/43); SAO-FRMA Grant and Flemish Government's Methusalem Grant to C.G.D and that of Ministerio de Ciencia e Innovacion (SAF2008-01473) and Consejo Superior de Investigaciones Cientificas (PI 200820I144) to M.D.L.

References

1. M. J. Stampfer, *J. Intern. Med.*, 2006, **260**, 211.
2. C. Rosendorff, M. S. Beeri and J. M. Silverman, *Am. J. Geriatr. Cardiol.*, 2007, **16**, 143.
3. S. Mukherjee, X. Zha, I. Tabas and F. R. Maxfield, *Biophys J.*, 1998, **75**, 1915.
4. G. van Meer, D. R. Voelker and G. W. Feigenson, *Nature Rev. Mol. Cell. Biol.*, 2008, **9**, 112.
5. T. H. Haines, *Prog. Lipid Res.*, 2001, **40**, 299.

6. K. Simons and D. Toomre, *Nature Rev. Mol. Cell. Biol.*, 2000, **1**, 31.
7. R. G. Parton and J. F. Hancock, *Trends Cell Biol.*, 2004, **14**, 141.
8. E. Ikonen, *Nature Rev. Mol. Cell. Biol.*, 2008, **9**, 125.
9. W. L. Miller, *J. Steroid Biochem.*, 1987, **27**, 759.
10. M. D. Majewska, N. L. Harrison, R. D. Schwartz, J. L. Barker and S. M. Paul, *Science*, 1986, **232**, 1004.
11. S. S. Smith, *Neuroscience*, 1991, **42**, 309.
12. J. F. Flood, J. E. Morley and E. Roberts, *Proc. Natl. Acad. Sci.USA*, 1992, **89**, 1567.
13. J. M. Dietschy and S. D. Turley, *J. Lipid Res.*, 2004, **45**, 1375.
14. G. S. Roth, J. A. Joseph and R. P. Mason, *Trends Neurosci.*, 1995, **18**, 203.
15. I. Szabò, C. Adams and E. Gulbins, *Pflugers Arch.*, 2004, **448**, 304.
16. J. Guo, S. Chi, H. Xu, G. Jin and Z. Qi, *Mol. Membr. Biol.*, 2008, **25**, 216.
17. C. Frank, S. Rufini, V. Tancredi, R. Forcina, D. Grossi and G. D'Arcangelo, *Exp. Neurol.*, 2008, **212**, 407.
18. R. P. Mason, W. J. Shoemaker, L. Shajenko, T. E. Chambers and L. G. Herbette, *Neurobiol. Aging.*, 1992, **13**, 413.
19. M. D. Ledesma, J. Abad-Rodriguez, C. Galvan, E. Biondi, P. Navarro, A. Delacourte, C. Dingwall and C. G. Dotti, *EMBO Rep.*, 2003, **4**, 1190.
20. P. Wasling, J. Daborg, I. Riebe, M. Andersson, E. Portelius, K. Blennow, E. Hanse and H. Zetterberg, *J. Alzheimers Dis.*, 2009, **16**, 1.
21. Y. Ohyagi, *Curr. Alzheimer Res.*, 2008, **5**, 555.
22. B. A. Yankner and T. Lu, *J. Biol. Chem.*, 2009, **284**, 4755.
23. R. J. Wilson, *Sci. Prog.*, 2008, **91**, 65.
24. M. Goedert and M. G. Spillantini, *Science*, 2006, **314**, 777.
25. M. Blurton-Jones and F. M. Laferla, *Curr. Alzheimer Res.*, 2006, **3**, 437.
26. S. A. Small and K. Duff, *Neuron*, 2008, **60**, 534.
27. C. Iadecola, *Nature Rev. Neurosci.*, 2004, **5**, 347.
28. B. V. Zlokovic, *Neurotherapeutics*, 2008, **5**, 409.
29. B. V. Zlokovic, *Life Sci.*, 1996, **59**, 1483.
30. C. L. Martel, J. B. Mackic, E. Matsubara, S. Governale, C. Miguel, W. Miao, J. G. McComb, B. Frangione, J. Ghiso and B. V. Zlokovic, *J. Neurochem.*, 1997, **69**, 1995.
31. R. Deane, S. Du Yan, R. K. Submamaryan, B. LaRue, S. Jovanovic, E. Hogg, D. Welch, L. Manness, C. Lin, J. Yu, H. Zhu, J. Ghiso, B. Frangione, A. Stern, A. M. Schmidt, D. L. Armstrong, B. Arnold, B. Liliensiek, P. Nawroth, F. Hofman, M. Kindy, D. Stern and B. Zlokovic, *Nature Med.*, 2003, **9**, 907.
32. P. M. Clifford, S. Zarrabi, G. Siu, K. J. Kinsler, M. C. Kosciuk, V. Venkataraman, M. R. D'Andrea, S. Dinsmore and R. G. Nagele, *Brain Res.*, 2007, **1142**, 223.
33. N. Chow, R. D. Bell, R. Deane, J. W. Streb, J. Chen, A. Brooks, W. Van Nostrand, J. M. Miano and B. V. Zlokovic, *Proc. Natl. Acad. Sci. USA*, 2007, **104**, 823.

34. R. D. Bell, R. Deane, N. Chow, X. Long, A. Sagare, I. Singh, J. W. Streb, H. Guo, A. Rubio, W. Van Nostrand, J. M. Miano and B. V. Zlokovic, *Nature Cell. Biol.*, 2009, **11**, 143.
35. J. E. Donahue and C. E. Johanson, *J. Neuropathol. Exp. Neurol.*, 2008, **67**, 261.
36. W. J. Schneider, P. T. Kovanen, M. S. Brown, J. L. Goldstein, G. Utermann, W. Weber, R. J. Havel, L. Kotite, J. P. Kane, T. L. Innerarity and R. W. Mahley, *J. Clin. Invest.*, 1981, **68**, 1075.
37. F. W. Pfrieger, *Cell. Mol. Life Sci.*, 2003, **60**, 1158.
38. E. H. Corder, A. M. Saunders, W. J. Strittmatter, D. E. Schmechel, P. C. Gaskell, G. W. Small, A. D. Roses, J. L. Haines and M. A. Pericak-Vance, *Science*, 1993, **261**, 921.
39. J. Raber, Y. Huang and J. W. Ashford, *Neurobiol. Aging*, 2004, **25**, 641.
40. E. H. Corder, E. Ghebremedhin, M. G. Taylor, D. R. Thal, T. G. Ohm and H. Braak, *Ann. N Y Acad. Sci.*, 2004, **1019**, 24.
41. Y. Ji, Y. Gong, W. Gan, T. Beach, D. M. Holtzman and T. Wisniewski, *Neuroscience*, 2003, **122**, 305.
42. M. K. Lai, S. W. Tsang, M. Garcia-Alloza, S. L. Minger, J. A. Nicoll, M. M. Esiri, P. T. Wong, C. P. Chen, M. J. Ramírez and P. T. Francis, *Neurobiol. Dis.*, 2006, **22**, 555.
43. L. Lomnitski, S. Chapman, A. Hochman, R. Kohen, E. Shohami, Y. Chen, V. Trembovler and D. M. Michaelson, *Biochim. Biophys. Acta.*, 1999, **1453**, 359.
44. M. Kawahara and Y. Kuroda, *Cell. Mol. Neurobiol.*, 2001, **21**, 1.
45. G. Utermann, U. Langenbeck, U. Beisiegel and W. Weber, *Am. J. Hum. Genet.*, 1980, **32**, 339.
46. P. Bertrand, J. Poirier, T. Oda, C. E. Finch and G. M. Pasinetti, *Brain Res. Mol. Brain Res.*, 1995, **33**, 174.
47. J. Poirier, *Alzheimers Dement.*, 2008, **4**, S91.
48. D. R. Riddell, H. Zhou, K. Atchison, H. K. Warwick, P. J. Atkinson, J. Jefferson, L. Xu, S. Aschmies, Y. Kirksey, Y. Hu, E. Wagner, A. Parratt, J. Xu, Z. Li, M. M. Zaleska, J. S. Jacobsen, M. N. Pangalos and P. H. Reinhart, *J. Neurosci.*, 2008, **28**, 11445.
49. J. S. Gong, M. Kobayashi, H. Hayashi, K. Zou, N. Sawamura, S. C. Fujita, K. Yanagisawa and M. Michikawa, *J. Biol. Chem.*, 2002, **277**, 29919.
50. R. B. DeMattos, *J. Mol. Neurosci.*, 2004, **23**, 255.
51. Q. Jiang, C. Y. Lee, S. Mandrekar, B. Wilkinson, P. Cramer, N. Zelcer, K. Mann, B. Lamb, T. M. Willson, J. L. Collins, J. C. Richardson, J. D. Smith, T. A. Comery, D. Riddell, D. M. Holtzman, P. Tontonoz and G. E. Landreth, *Neuron*, 2008, **58**, 681.
52. I. Veinbergs and E. Masliah, *Neuroscience*, 1999, **91**, 401.
53. S. Chapman, T. Sabo, A. D. Roses and D. M. Michaelson, *Neuroscience*, 2000, **97**, 419.
54. J. Grootendorst, A. Bour, E. Vogel, C. Kelche, P. M. Sullivan, J. C. Dodart, K. Bales and C. Mathis, *Behav. Brain Res.*, 2005, **159**, 1.

55. J. Rogers, L. C. Kirby, S. R. Hempelman, D. L. Berry, P. L. McGeer, A. W. Kaszniak, J. Zalinski, M. Cofield, L. Mansukhani and P. Willson, *et al. Neurology*, 1993, **43**, 1609.

56. M. E. Risner, A. M. Saunders, J. F. Altman, G. C. Ormandy, S. Craft, I. M. Foley, M. E. Zvartau-Hind, D. A. Hosford and A. D. Roses, *Pharmacogenomics J.*, 2006, **6**, 246.

57. J. Poirier, *Trends Mol. Med.*, 2003, **9**, 94.

58. E. P. Helzner, J. A. Luchsinger, N. Scarmeas, S. Cosentino, A. M. Brickman, M. M. Glymour and Y. Stern, *Arch. Neurol.*, 2009, **66**, 343.

59. J. Herz and H. H. Bock, *Annu. Rev. Biochem.*, 2002, **71**, 405.

60. T. Grand-Perret, A. Bouillot, A. Perrot, S. Commans, M. Walker and M. Issandou, *Nature Med.*, 2001, **7**, 1332.

61. M. Issandou, R. Guillard, A. B. Boullay, V. Linhart and E. Lopez-Perez, *Biochem. Pharmacol.*, 2004, **67**, 2281.

62. L. Puglielli, G. Konopka, E. Pack-Chung, L. A. Ingano, O. Berezovska, B. T. Hyman, T. Y. Chang, R. E. Tanzi and D. M. Kovacs, *Nature Cell. Biol.*, 2001, **3**, 905.

63. H. Runz, J. Rietdorf, I. Tomic, M. de Bernard, K. Beyreuther, R. Pepperkok and T. Hartmann, *J. Neurosci.*, 2002, **22**, 1679.

64. Y. Li, M. P. Marzolo, P. van Kerkhof, G. J. Strous and G. Bu, *J. Biol. Chem.*, 2000, **275**, 17187.

65. C. X. Liu, Y. Li, L. M. Obermoeller-McCormick, A. L. Schwartz and G. Bu, *J. Biol. Chem.*, 2001, **276**, 28889.

66. M. P. Marzolo and G. Bu, *Semi. Cell. Dev. Biol.*, 2009, **20**, 191.

67. P. G. Ulery, J. Beers, I. Mikhailenko, R. E. Tanzi, G. W. Rebeck, B. T. Hyman and D. K. Strickland, *J. Biol. Chem.*, 2000, **275**, 7410.

68. J. A. Cam, C. V. Zerbinatti, J. M. Knisely, S. Hecimovic, Y. Li and G. Bu, *J. Biol. Chem.*, 2004, **278**, 29639.

69. K. Offe, S. E. Dodson, J. T. Shoemaker, J. J. Fritz, M. Gearing, A. I. Levey and J. J. Lah, *J. Neurosci.*, 2006, **26**, 1596.

70. R. A. Fuentealba, M. I. Barría, J. Lee, J. Cam, C. Araya, C. A. Escudero, N. C. Inestrosa, F. C. Bronfman, G. Bu and M. P. Marzolo, *Mol. Neurodegener.*, 2007, **2**, 14.

71. X. He, K. Cooley, C. H. Chung, N. Dashti and J. Tang, *J. Neurosci.*, 2007, **27**, 4052.

72. P. May, Y. K. Reddy and J. Herz, *J. Biol. Chem.*, 2002, **277**, 18736.

73. C. A. von Arnim, A. Kinoshita, I. D. Peltan, M. M. Tangredi, L. Herl, B. M. Lee, R. Spoelgen, T. T. Hshieh, S. Ranganathan, F. D. Battey, C. X. Liu, B. J. Bacskai, S. Sever, M. C. Irizarry and D. K. Strickland, *J. Biol. Chem.*, 2005, **280**, 17777.

74. R. Deane, Z. Wu, A. Sagare, J. Davis, S. Du Yan, K. Hamm, F. Xu, M. Parisi, B. LaRue, H. W. Hu, P. Spijkers, H. Guo, X. Song, P. J. Lenting, W. E. Van Nostrand and B. V. Zlokovic, *Neuron*, 2004, **43**, 333.

75. M. Shibata, S. Yamada, S. R. Kumar, M. Calero, J. Bading, B. Frangione, D. M. Holtzman, C. A. Miller, D. K. Strickland, J. Ghiso and B. V. Zlokovic, *J. Clin. Invest.*, 2000, **106**, 1489.

76. D. E. Kang, C. U. Pietrzik, L. Baum, N. Chevallier, D. E. Merriam, M. Z. Kounnas, S. L. Wagner, J. C. Troncoso, C. H. Kawas, R. Katzman and E. H. Koo, *J. Clin. Invest.*, 2000, **106**, 1159.
77. S. L. Ma, H. K. Ng, L. Baum, J. C. Pang, H. F. Chiu, J. Woo, N. L. Tang and L. C. Lam, *Neurosci. Lett.*, 2002, **332**, 216.
78. G. Bu and A. L. Schwartz, *Trends Cell. Biol.*, 1998, **8**, 272.
79. L. M. Obermoeller-McCormick, Y. Li, H. Osaka, D. J. FitzGerald, A. L. Schwartz and G. Bu, *J. Cell. Sci.*, 2001, **114**, 899.
80. J. E. Croy, W. D. Shin, M. F. Knauer, D. J. Knauer and E. A. Komives, *Biochemistry*, 2003, **42**, 13049.
81. M. S. Brown and J. L. Goldstein, *Cell*, 1997, **89**, 331.
82. C. J. Loewen and T. P. Levine, *Curr. Biol.*, 2002, **12**, R779.
83. C. G. Dotti and B. De Strooper, *Nature Cell. Biol.*, 2009, **11**, 114.
84. T. Y. Chang, C. C. Chang and D. Cheng, *Annu. Rev. Biochem.*, 1997, **66**, 613.
85. B. Hutter-Paier, H. J. Huttunen, L. Puglielli, C. B. Eckman, D. Y. Kim, A. Hofmeister, R. D. Moir, S. B. Domnitz, M. P. Frosch, M. Windisch and D. M. Kovacs, *Neuron*, 2004, **44**, 227.
86. R. K. Kharbanda, S. Wallace, B. Walton, A. Donald, J. M. Cross and J. Deanfield, *Circulation*, 2005, **111**, 804.
87. H. Jick, G. L. Zornberg, S. S. Jick, S. Seshadri and D. A. Drachman, *Lancet*, 2000, **356**, 1627.
88. B. Wolozin, W. Kellman, P. Ruosseau, G. G. Celesia and G. Siegel, *Arch. Neurol.*, 2000, **57**, 1439.
89. B. Wolozin, *Biochem. Soc. Trans.*, 2002, **30**, 525.
90. K. Fassbender, M. Simons, C. Bergmann, M. Stroick, D. Lutjohann, P. Keller, H. Runz, S. Kuhl, T. Bertsch, K. von Bergmann, M. Hennerici, K. Beyreuther and T. Hartmann, *Proc. Natl. Acad. Sci. USA*, 2001, **98**, 5856.
91. L. M. Refolo, M. A. Pappolla, J. LaFrancois, B. Malester, S. D. Schmidt, T. Thomas-Bryant, G. S. Tint, R. Wang, M. Mercken, S. S. Petanceska and K. E. Duff, *Neurobiol. Dis.*, 2001, **8**, 890.
92. J. Abad-Rodriguez, M. D. Ledesma, K. Craessaerts, S. Perga, M. Medina, A. Delacourte, C. Dingwall, B. De Strooper and C. G. Dotti, *J. Cell Biol.*, 2004, **167**, 953.
93. A. Crameri, E. Biondi, K. Kuehnle, D. Lütjohann, K. M. Thelen, S. Perga, C. G. Dotti, R. M. Nitsch, M. D. Ledesma and M. H. Mohajeri, *EMBO J.*, 2006, **25**, 432.
94. B. Cucchiara and S. E. Kasner, *J. Neurol. Sci.*, 2001, **187**, 81.
95. M. Sadowski, J. Pankiewicz, H. Scholtzova, Y. S. Li, D. Quartermain, K. Duff and T. Wisniewski, *Neurochem. Res.*, 2004, **29**, 1257.
96. N. Kandiah, H.H. Feldman, *J. Neurol. Sci.*, 2009, Mar 23.
97. J. Shepherd, G. J. Blauw, M. B. Murphy, E. L. Bollen, B. M. Buckley, S. M. Cobbe, I. Ford, A. Gaw, M. Hyland, J. W. Jukema, A. M. Kamper, P. W. Macfarlane, A. E. Meinders, J. Norrie, C. J. Packard, I. J. Perry, D. J. Stott, B. J. Sweeney, C. Twomey and R. G. Westendorp, *Lancet*, 2002, **360**, 1623.

98. Heart Protection Study Collaborative Group, *Lancet*, 2002, **360**, 7.

99. K. Yaffe, E. Barrett-Connor, F. Lin and D. Grady, *Arch. Neurol.*, 2002, **59**, 378.

100. D. L. Sparks, M. N. Sabbagh, D. J. Connor, J. Lopez, L. J. Launer, P. Browne, D. Wasser, S. Johnson-Traver, J. Lochhead and C. Ziolwolski, *Arch. Neurol.*, 2005, **62**, 753.

101. I. Greeve, I. Hermans-Borgmeyer, C. Brellinger, D. Kasper, T. Gomez-Isla, C. Behl, B. Levkau and R. M. Nitsch, *J. Neurosci.*, 2000, **20**, 7345.

102. H. R. Waterham, J. Koster, G. J. Romeijn, R. C. Hennekam, P. Vreken, H. C. Andersson, D. R. FitzPatrick, R. I. Kelley and R. J. Wanders, *Am. J. Hum. Genet.*, 2001, **69**, 685.

103. A. Peri and M. Serio, *Neuropsychiatr. Dis. Treat.*, 2008, **4**, 817.

104. G. Orr, D. Hu, S. Ozçelik, L. K. Opresko, H. S. Wiley and S. D. Colson, *Biophys. J.*, 2005, **89**, 1362.

105. D. V. Jr Nicolau, K. Burrage, R. G. Parton and J. F. Hancock, *Mol. Cell. Biol.*, 2006, **26**, 313.

106. M. G. Martin, S. Perga, L. Trovò, A. Rasola, P. Holm, T. Rantamäki, T. Harkany, E. Castrén, F. Chiara and C. G. Dotti, *Mol. Biol. Cell.*, 2008, **19**, 2101.

107. A. Peri and M. Serio, *J. Mol. Endocrinol.*, 2008, **41**, 251.

108. S. Benvenuti, P. Luciani, G. B. Vannelli, S. Gelmini, E. Franceschi, M. Serio and A. Peri, *J. Clin. Endocrinol. Metab.*, 2005, **90**, 1775.

109. P. Luciani, C. Deledda, F. Rosati, S. Benvenuti, I. Cellai, F. Dichiara, M. Morello, G. B. Vannelli, G. Danza, M. Serio and A. Peri, *Endocrinology*, 2008, **149**, 4256.

110. A. Paganini-Hill and V. W. Henderson, *Am. J. Epidemiol.*, 1994, **140**, 256.

111. H. M. Fillit, *Arch. Intern. Med.*, 2002, **162**, 1934.

112. K. Yaffe, K. Krueger, S. R. Cummings, T. Blackwell, V. W. Henderson, S. Sarkar, K. Ensrud and D. Grady, *Am. J. Psychiatry.*, 2005, **162**, 683.

113. S. M. Resnick and V. W. Henderson, *JAMA*, 2002, **288**, 2170.

114. P. P. Zandi, M. C. Carlson, B. L. Plassman, K. A. Welsh-Bohmer, L. S. Mayer, D. C. Steffens and J. C. Breitner, Cache County Memory Study Investigators, *JAMA*, 2002, **288**, 2123.

115. E. G. Lund, J. M. Guileyardo and D. W. Russell, *Proc. Natl. Acad. Sci. USA*, 1999, **96**, 7238.

116. D. Lütjohann, O. Breuer, G. Ahlborg, I. Nennesmo, A. Sidén, U. Diczfalusy and I. Björkhem, *Proc. Natl. Acad. Sci. USA.*, 1996, **93**, 9799.

117. T. J. Kotti, D. M. Ramirez, B. E. Pfeiffer, K. M. Huber and D. W. Russell, *Proc. Natl. Acad. Sci., USA*, 2006, **103**, 3869.

118. Y. Ohyama, S. Meaney, M. Heverin, L. Ekström, A. Brafman, M. Shafir, U. Andersson, M. Olin, G. Eggertsen, U. Diczfalusy, E. Feinstein and I. Björkhem, *J. Biol. Chem.*, 2006, **281**, 3810.

119. J. Brown rd, C. Theisler, S. Silberman, D. Magnuson, N. Gottardi-Littell, J. M. Lee Yager, J. Crowley, K. Sambamurti, M. M. Rahman, A. B. Reiss, C. B. Eckman and B. Wolozin, *J. Biol. Chem.*, 2004, **279**, 34674.

120. A. Papassotiropoulos, J. R. Streffer, M. Tsolaki, S. Schmid, D. Thal, F. Nicosia, V. Iakovidou, A. Maddalena, D. Lütjohann, E. Ghebremedhin, T. Hegi, T. Pasch, M. Träxler, A. Brühl, L. Benussi, G. Binetti, H. Braak, R. M. Nitsch and C. Hock, *Arch. Neurol.*, 2003, **60**, 29.

121. B. Borroni, S. Archetti, C. Agosti, N. Akkawi, C. Brambilla, L. Caimi, C. Caltagirone, M. Di Luca and A. Padovani, *Neurobiol. Aging*, 2004, **25**, 747.

122. M. Shafaati, R. O'Driscoll, I. Björkhem and S. Meaney, *Biochem. Biophys. Res. Commun.*, 2009, **378**, 689.

123. H. Qing, G. He, P. T. Ly, C. J. Fox, M. Staufenbiel, F. Cai, Z. Zhang, S. Wei, X. Sun, C. H. Chen, W. Zhou, K. Wang and W. Song, *J. Exp. Med.*, 2008, **205**, 2781.

124. L. Liu, A. E. Bortnick, M. Nickel, P. Dhanasekaran, P. V. Subbaiah, S. Lund-Katz, G. H. Rothblat and M. C. Phillips, *J. Biol. Chem.*, 2003, **278**, 42976.

125. M. A. Wollmer, J. R. Streffer, D. Lütjohann, M. Tsolaki, V. Iakovidou, T. Hegi, T. Pasch, H. H. Jung, K. Bergmann, R. M. Nitsch, C. Hock and A. Papassotiropoulos, *Neurobiol. Aging*, 2003, **24**, 421.

126. H. Katzov, K. Chalmers, J. Palmgren, N. Andreasen, B. Johansson, N. J. Cairns, M. Gatz, G. K. Wilcock, S. Love, N. L. Pedersen, A. J. Brookes, K. Blennow, P. G. Kehoe and J. A. Prince, *Hum. Mutat.*, 2004, **23**, 358.

127. L. W. Chu, Y. Li, Z. Li, A. Y. Tang, B. M. Cheung, R. Y. Leung, P. Y. Yik, D. Y. Jin and Y. Q. Song, *Am. J. Med. Genet. B. Neuropsychiatr. Genet.*, 2007, **144B**, 1007.

128. E. Rodríguez-Rodríguez, I. Mateo, J. Llorca, C. Sánchez-Quintana, J. Infante, I. García-Gorostiaga, P. Sánchez-Juan, J. Berciano and O. Combarros, *Am. J. Med. Genet. B. Neuropsychiatr. Genet.*, 2007, **144B**, 964.

129. W. S. Kim, G. J. Guillemin, E. N. Glaros, C. K. Lim and B. Garner, *Neuroreport*, 2006, **17**, 891.

130. V. Hirsch-Reinshagen, S. Zhou, B. L. Burgess, L. Bernier, S. A. McIsaac, J. Y. Chan, G. H. Tansley, J. S. Cohn, M. R. Hayden and C. L. Wellington, *J. Biol. Chem.*, 2004, **279**, 41197.

131. S. E. Wahrle, H. Jiang, M. Parsadanian, J. Legleiter, X. Han, J. D. Fryer, T. Kowalewski and D. M. Holtzman, *J. Biol. Chem.*, 2004, **279**, 40987.

132. J. M. Karasinska, F. Rinninger, D. Lütjohann, P. Ruddle, S. Franciosi, J. K. Kruit, R. R. Singaraja, V. Hirsch-Reinshagen, J. Fan, L. R. Brunham, N. Bissada, R. Ramakrishnan, C. L. Wellington, J. S. Parks and M. R. Hayden, *J. Neurosci.*, 2009, **29**, 3579.

133. R. P. Koldamova, I. M. Lefterov, M. D. Ikonomovic, J. Skoko, P. I. Lefterov, B. A. Isanski, S. T. DeKosky and J. S. Lazo, *J. Biol. Chem.*, 2003, **278**, 13244.

134. Y. Sun, J. Yao, T. W. Kim and A. R. Tall, *J. Biol. Chem.*, 2003, **278**, 27688.

135. W. S. Kim, A. S. Rahmanto, A. Kamili, K. A. Rye, G. J. Guillemin, I. C. Gelissen, W. Jessup, A. F. Hill and B. Garner, *J. Biol. Chem.*, 2007, **282**, 2851.

136. R. Koldamova, M. Staufenbiel and I. Lefterov, *J. Biol. Chem.*, 2005, **280**, 43224.
137. D. R. Riddell, H. Zhou, T. A. Comery, E. Kouranova, C. F. Lo, H. K. Warwick, R. H. Ring, Y. Kirksey, S. Aschmies, J. Xu, K. Kubek, W. D. Hirst, C. Gonzales, Y. Chen, E. Murphy, S. Leonard, D. Vasylyev, A. Oganesian, R. L. Martone, M. N. Pangalos, P. H. Reinhart and J. S. Jacobsen, *Mol. Cell. Neurosci.*, 2007, **34**, 621.
138. N. Zelcer, N. Khanlou, R. Clare, Q. Jiang, E. G. Reed-Geaghan, G. E. Landreth, H. V. Vinters and P. Tontonoz, *Proc. Natl. Acad. Sci. USA*, 2007, **104**, 10601.
139. S. E. Wahrle, H. Jiang, M. Parsadanian, R. E. Hartman, K. R. Bales, S. M. Paul and D. M. Holtzman, *J. Biol. Chem.*, 2005, **280**, 43236.
140. S. E. Wahrle, H. Jiang, M. Parsadanian, J. Kim, A. Li, A. Knoten, S. Jain, V. Hirsch-Reinshagen, C. L. Wellington, K. R. Bales, S. M. Paul and D. M. Holtzman, *J. Clin. Invest.*, 2008, **118**, 671.
141. T. H. Haines, *Prog. Lipid. Res.*, 2001, **40**, 299.
142. M. Heverin, S. Meaney, D. Lütjohann, U. Diczfalusy, J. Wahren and I. Björkhem, *J. Lipid. Res.*, 2005, **46**, 1047.
143. J. R. Prasanthi, A. Huls, S. Thomasson, A. Thompson, E. Schommer and O. Ghribi, *Mol. Neurodegener.*, 2009, **4**, 1.
144. M. Heverin, N. Bogdanovic, D. Lütjohann, T. Bayer, I. Pikuleva, L. Bretillon, U. Diczfalusy, B. Winblad and I. Björkhem, *J. Lipid. Res.*, 2004, **45**, 186.
145. A. Vejux, L. Malvitte and G. Lizard, *Braz. J. Med. Biol. Res.*, 2008, **41**, 545.
146. V. Leoni, M. Shafaati, A. Salomon, M. Kivipelto, I. Björkhem and L. O. Wahlund, *Neurosci. Lett.*, 2006, **397**, 83.

The Bimodal Features of Butyrylcholinesterase in Cholinergic Neurotransmission and Amyloid Suppression

EREZ PODOLY[1,2] AND HERMONA SOREQ[2]

[1] Wolfson Centre for Applied Structural Biology, Hebrew University of Jerusalem, Givat Ram, 91904, Jerusalem, Israel; [2] Department of Biological Chemistry, Hebrew University of Jerusalem, Givat Ram, 91904, Jerusalem, Israel

4.1 Introduction

Alzheimer's disease (AD) is a progressive neurodegenerative disorder that primarily damages cholinergic neurons in the basal nuclei, leading to irreversible dementia. Disease onset can occur as early as age 40 and by age 70 approximately 5% of the population develops AD. Currently, AD afflicts 27 million people worldwide, and this figure is expected to triple to 81 million by 2040 as trends predict a continued global increase in life expectancy.[1]

AD is associated with two neuropathological hallmarks: neurofibrillary tangles, which are composed of disorganised tubulin associated with hyper-phosphorylated tau protein in damaged nerve cells,[2] and amyloid plaques, a neuropathological hallmark. Amyloid beta (Aβ), the main constituent of these plaques, is a 39–42 amino acid amphiphilic peptide, derived from the

RSC Drug Discovery Series No. 2
Emerging Drugs and Targets for Alzheimer's Disease
Volume 1: Beta-Amyloid, Tau Protein and Glucose Metabolism
Edited by Ana Martinez

Published by the Royal Society of Chemistry, www.rsc.org

transmembrane domain and extracellular region of the Aβ precursor protein (APP).[3] At high concentrations, Aβ acquires a β-sheet structure, becomes insoluble, and accumulates as neurotoxic oligomers and fibrils[4] to compose amyloid plaques in the brain of AD patients. Recent hypotheses attribute causal roles in AD to presenilin,[5] oxidative stress,[6] metals,[7] double-hit origin[8] or mitochondrial damage.[9] The alternative theories state that Aβ represents a bystander or even a protector rather than the causative factor of disease and that Aβ amyloidogenesis is secondary to other pathogenic events.[10] Nevertheless, a wealth of evidence demonstrates a pivotal role for Aβ in the pathogenesis of AD, yielding the amyloid cascade hypothesis.[11] According to this hypothesis, the pathological accumulation of Aβ in the brain leads to oxidative stress, neuronal destruction and finally, the pathological syndrome of AD.

Postmortem studies using brains of AD patients demonstrated reduced choline uptake and acetylcholine (ACh) release, loss of cholinergic perikarya from the nucleus basalis of Meynert and a substantial presynaptic cholinergic deficit. This resulted in the dominant 'cholinergic deficit hypothesis', which was later replaced by the amyloid cascade hypothesis. According to the cholinergic hypothesis, many symptoms of dementia and especially learning difficulties were explainable by the lack of ACh. It was hence anticipated that restoring the cholinergic balance by inhibiting ACh breakdown would slow down the progression of AD and improve cognitive and general functioning, which is why cholinesterase inhibitors were developed as therapeutic agents for treating AD patients.[12]

Cholinesterases (ChEs) are polymorphic carboxylesterases that terminate neurotransmission at cholinergic synapses and neuromuscular junctions by hydrolysing acetylcholine. The two human ChEs, acetyl- and butyrylcholinesterase (AChE and BChE, respectively) share 65% of their sequence and have similar molecular forms and active centre structures.[13] AChE, the primary central nervous system ChE, is a prominent component of senile plaques.[14] Synaptic AChE-S, the primary AChE 3′- splice variant (also known as AChE-T), promotes Aβ aggregation *in vitro* and enhances amyloid toxicity in cultured neuronal cells.[15,16] The decrease in synaptic AChE-S levels and the inverse upregulation of BChE in the AD brain, raise the possibility of causal, albeit opposite involvement of these two proteins in the progression of AD.

BChE, the secondary ACh hydrolysing enzyme, is associated with the neurofibrillary tangles and amyloid plaques characteristic of AD,[17] which suggests that it functions as a potential AD modulator. BChE activity increases in the AD brain,[18–20] where it colocalises with Aβ fibrils.[14,21] It is within this context that we have studied the interactions of BChE with Aβ. Specifically, we focused on the C-terminus of BChE, which functions as a tetramerisation domain[22,23] and is responsible for its quaternary organisation. Four BChE monomers are held together by the aromatic interactions of seven highly conserved aromatic residues, termed the tryptophan amphiphilic tetramerisation domain (WAT).[22,23] The WAT domain interacts with proline-rich attachment domains (PRADs), either via proline-rich membrane anchor (PRiMA) in brain neurons[24] or, in neuromuscular junctions, with cholinesterase-associated collagen Q (ColQ).[25] In the serum, BChE tetramerisation is supported by an analogous 17-mer proline-rich peptide derived from lamellipodin.[26]

BChE purified from human serum and BSP, a synthetic peptide derived from the BChE C-terminus, were both found to associate with soluble Aβ conformers, to delay the onset and decrease the rate of Aβ fibril formation *in vitro*, at a 1:100 BChE/Aβ molar ratio and in a dose-dependent manner.[27–29] The corresponding ASP peptide, derived from the homologous C-terminus of synaptic human AChE-S, failed to significantly affect Aβ fibril formation, attributing the role of enhancing this process to a AChE domain other than the C-terminus. Molecular modelling based on circular dichroism confirmed that both peptides are α-helices. However, while ASP is amphiphilic, BSP includes a tryptophan residue in the polar side of the helical sequence. Further mutagenesis experiments, where W541R BSP showed suppressed capacity to attenuate fibril formation, supported the notion that this aromatic residue is causally involved in the attenuating effect of BChE. This indicated that in AD, BChE may have acquired an inverse role to that of AChE-S by adopting an imperfect amphiphilic C-terminus.

Analysing the quaternary organization of cholinesterases is a complicated task: to date, all biologically relevant crystal structures of cholinesterases have been truncated forms that lack the C-terminus of the protein,[30] apart from a more recent study of full-length BChE that yielded crystal packing that did not allow C-terminal interactions among subunits and lacked electron densities in the C-terminus region, indicating structural disorder.[31] Of note, the crystal structure of the homologous C-terminus of tetrameric synaptic acetylcholinesterase (AChE-S) could only be determined based on synthetic peptides derived from the sequence of the AChE-S tail and stabilised with a proline-rich attachment domain.[32]

In addition to the "usual" (BChE-U) form, BChE has nearly 40 genomic variants. The most common is the Kalow variant (BChE-K), with allelic frequencies of 0.13–0.21. BChE-K includes a single nucleotide polymorphism at position 1699 (SNP ID: rs1803274; Alleles: A/G). This leads to an alanine-to-threonine substitution at position 539, 36 residues upstream to the C-terminus end of BChE,[33] within the tetramerisation domain that we previously found to attenuate amyloid fibril formation.[27] Ample evidence supports the importance of alanine-to-threonine substitutions and their relevance to amyloidogenic processes, protein stability and quaternary organisation (Table 4.1). Thus, point mutations at the dimer interface of light chain immunoglobulins decrease their stability, so that the A34T polymorphism in this protein leads to systemic amyloidosis.[34] Also, an A25T mutant of the tetrameric human protein Transthyretin (TTR), associated with central nervous system amyloidosis, is prone to aggregation and exhibits drastically reduced tertiary and quaternary structural stabilities.[35] The reduced thermodynamic stability of the A25T TTR mutant reflects destabilisation of both monomers and tetramers of this variant. In addition, A25T TTR tetramers dissociate very rapidly (about 1200-fold faster than the dissociation of wild-type TTR), reflecting a dynamic

Table 4.1 Current BChE-K genotype studies.

#	Study	Origin	AD Cases			Normal Controls		Conc
			#Sub	DX	Age (range)	#Sub	Age (range)	
1	Alvarez-Arcaya, *Acta Neurol Scand.* 2000	Spain	249	C	$75.3 + 9.3(50–98)$	250	$79.9 + 7.9$ (63–98)	Pos
2	Beyer, *Ann N Y Acad Sci.* 2005	Spain	206	C	72.6 (48–85)	181	69.6 (45–84)	Pos
3	Blomqvist, *Hum Gent.* 2006	OL #10	–	–	–	–	–	n.a
4	Combarros, *Dement Geriatr Cog Disor.* 2005	OL #1	–	C	–	–	–	n.a
5	Combarros, *J Neurol.* 2007	OL #1	183	C	$74.2 + 8.9$ (50–97)	169	$79.7 + 7.6$ (66–98)	n.a
6	Cook, *Am J Med Gent B Neuropsy Gent.* 2005	UK	215	N	–	320	–	Neg
7	Corder, *Mech Ageing Dev.* 2006	OL #16	315	M	–	260	–	n.a.
8	Déniz-Naranjo, *Neurosci Lett.* 2007	Spain	282	C	–	–	$312 + 120(–)$	Pos
9	Emahazion, *Trends Gent.* 2001	Scotland	121	C	–	152	–	Neg
10	Grubber, *Neurosci Lett.* 1999	USA	245	C	$72.7 + 8$ (42–93)	241	$69.7 + 7.3(37–86)$	Neg
11	Helbecque, *Alz Rep.* 1998	Europe	336	C	$69.8 + 7.9$ (–)	344	$72.2 + 13.8$ (–)	Neg
12	Hiltunen, *Neurosci Lett.* 1998	Finland	78	M	–	97	$72 + 4$ (–)	Neg
13	Kehoe, *J Med Gent.* 1998	UK	181	M	–	71	$73.4 + 6.2$ (–)	Neg

14	Laws, *Alz Rep.* 1999	Australia	237	C	–	347	–	Neg
15	Lehmann, *Hum Mol Gent.* 1997	UK	149	M	76.5	104	–	Pos
16	Mateo, *Eur J Neurol.* 2008	OL #1	231	C	75.6 + 9 (50–97)	221	80.1 + 8(63–98)	n.a.
17	Mattila, *J Med Gent.* 2000	Finland	80	M	–	67	75.9 + 7.3 (–)	Pos
18	McIlroy, *J Med Gent.* 2000	Ireland	175	C	77.7 + 5.9 (67–94)	187	77.1 + 6 (67–96)	Pos
19	Panegyres, *J Neurol.* 1999	Australia	38	N	–	30	–	Pos
20	Piccardi, *Am J Med Gent B Neuropsy Gent.* 2007	Italy	158	C	76.8 + 6.9 (–)	118	65.7 + 7.8 (–)	Neg
21	Prince, *Eur J Hum Gent.* 2001	Sweden	204	M	–	186	–	Neg
22	Russ, Lancet. 1998	UK	203	C	82.4 + 6.7 (–)	122	80.4 + 4.6 (–)	Neg
23	Scacchi, *Am J Med Gent B Neuropsy Gent.* 2008	Italy	471	C	78.5 + 5.2(60–96)	254	77.5 + 8.1(60–95)	Neg
24	Singleton, *Hum Mol Gent.* 1998	UK	119	N	78.3 + 9.4 (–)	83	78.3 + 7.7 (–)	Neg
25	Tilley, *Eur J Hum Gent.* 1999	UK	177	N	81.6 + 8.6(55–101)	118	72.8 + 8.7(58–94)	Pos
26	Wiebusch, *Hum Gent.* 1999	Canada	135	N	78 + 7.8(–)	70	74 + 9.2 (–)	Pos
27	Bi, *Chin Med Sci J.* 2001	China	38	U	–	40	–	Neg
28	Ki, *Am J Med Gent.* 1999	Korea	78	C	73.2 (–)	74	73.8 (–)	Neg
29	Kim, *J Neural Transm.* 2001	Korea	164	C	71.7 + 8.7 (–)	239	69 + 5 (–)	Neg
30	Lee, *Am J Med Gent.* 2000	China	89	C	–	101	73 + 4.6 (–)	Neg

Table 4.1 (continued)

#	Study	Origin	AD Cases			Normal Controls		Conc
			#Sub	DX	Age (range)	#Sub	Age (range)	
31	Yamada, *Stroke.* 1998	Japan	48	N	84.6 + 7.3 (–)	107	85.3 + 7.9 (–)	Neg
32	Yamada, *Ann N Y Acad Sci.* 2002	Japan	82	N	86.1 + 7.9(62–104)	119	85.7 + 8(65–103)	Neg
33	Yamamoto, *J Neurol Neurosurg Psy.* 1999	Japan	203	C	68.2 + 8.5 (–)	288	68.5 + 7.4 (–)	Neg
34	Brindle, *Hum Mol Gent.* 1998	USA, Canada	188	M	73.2 (–)	165	63.4 (–)	Neg
35	Crawford, *Neurosci Lett.* 1998	USA	391	C	–	201	72.3 + 7.l(–)	Neg
36	Raygani, *Neurosci Lett.* 2004	Iran	105	C	75 + 10 (–)	129	73.5 + 11.5 (–)	Pos

Cited are currently available BChE-K genotype studies in AD patients from different ethnic origins. Note the controversial views regarding the studied involvement of BChE-K in the AD process within the different populations, which initiated our interest in approaching this question using structural and biochemical means.

OL: Overlap.

Sub: Number of subjects.

Age: Mean or median age at onset or examination, respectively.

DX: Criteria used to determine AD diagnosis – "C" (clinical AD diagnosis), "N" (neuropathological AD diagnosis), "M" (mixed, i.e. AD sample contains both clinical and neuropathological cases), "U" (unknown).

Conc: Overall conclusion reached by authors of the original publication ("positive" usually indicates significant (P<0.05) association between the K variant and AD and "negative" indicates no evidence for significant association between the K variant and AD); results obtained in duplicate or largely overlapping samples are listed as "n.a.".

(-) Either no data provided or in case of overlap, data included in original study.

Caucasian population.

Asian population.

Mixed population.

Data was derived from ALZGENE - published AD candidate genes.

destabilisation of their quaternary structure. These factors together probably contribute to the high propensity of A25T TTR to aggregate *in vitro*.

The capacity of serum BChE-K to hydrolyse butyrylthiocholine was reported to be reduced by 30% relative to BChE-U, but the reasons for that remained unclear.[33] The reduced hydrolytic activity of BChE-K predicted that aging BChE-K carriers would potentially experience sustained cholinergic transmission compared to BChE-U carriers. Supporting this notion, BChE-K expression was shown to correlate with preserved performance of attention and reduced rates of cognitive decline.[36] However, when they develop AD, BChE-K carriers are refractory to cholinesterase inhibitor therapy, the current leading AD treatment.[37] This raised the question whether BChE-K functions as an AD risk or inversely, as a protection factor. Genotype studies were controversial (Table 4.2), with some showing increased risk of AD for homozygous BChE-K carriers (*e.g.*, ref. 38), whereas others suggest a protective effect (*e.g.*, ref. 39). A recent meta-analysis concluded that in average, BChE-K is neither a risk nor a protection factor for AD.[40] Based on our previous findings of the arrest of Aβ fibril formation by BChE, and considering the accumulation of monomeric BChE in the most severe AD cases,[41] we have recently used a variety of chemical techniques to study the effect of the A539T substitution on BChE stability and tetramerisation on the one hand, and on its potency in attenuating Aβ oligomerisation and fibril formation on the other.[29] Our findings support the prevention provided by BChE of long-term amyloidogenic damages and call for further studies of alanine to thronine substitutions in protein extremities as inducers of structurally derived amyloidogenis processes.

4.2 Amyloidosis

Amyloidosis refers to a variety of conditions in which amyloid proteins are abnormally deposited in organs and/or tissues, causing disease. A protein is an amyloid if, due to an alteration in its secondary structure, it resumes an insoluble form, called the β-pleated sheet.[42]

Amyloidosis emerges as a growing group of protein misfolding diseases coupled with the formation of insoluble polypeptide depositions.[43] Approximately 25 different human proteins are known as amyloid substance structures, most of which are constituents of the plasma. Amyloids are most commonly accumulated extracellularly, as in the cases of AD and systemic amyloidoses, but they can also grow intracellularly (*e.g.* in Parkinson's and Huntington's diseases[44]). Both *in vivo* and *in vitro,* amyloids are composed of initially soluble proteins, some performing various normal functions in the body (*e.g.* insulin, glucagon, lysozyme) and others with yet incompletely resolved functions (*e.g.* APP).

When proteins are converted into amyloids, due to an alteration in their secondary structure, they become insoluble and highly protease resistant, which may inhibit their cellular and tissue clearance. The amyloid precursors have no general similarity in their sequences or secondary and tertiary structures and can be composed of predominantly α-helices, β-sheets or a mixture of the

Table 4.2 Alanine-to-thronine substitutions in amyloidogenic diseases.

Related Disease	Protein Involved	polymorphism	Amyloidogenic/Risk factor	2° structure location	Reference
Familial amyloid polyneuorpathy	TTR	A25T	Amyloidogenic	Loop	Hurshman Babbes AR, *Biochemistry* 2008
Parkinson's disease	α-Synuclein	A53T	Amyloidogenic	Helix	Heise H, *J. Mol. Biol.* 2008
Systemic amyloidosis	AL	A34T	Amyloidogenic	Loop	Karimi M, *Scand. J. Immunol.* 2003
Subarachnoid hemor-rhage (SAH)	α-1-antichymotrypsin	A15T	Risk factor	Loop	Krischek B, *Cerebrovasc. Dis.* 2007
Diabetes mellitus type II	FABP2	A54T	Risk factor	Turn	Baier LJ, *J. Clin. Invest.* 1995
Ataxia	GluR-δ2	A654T	Risk factor	Helix	Zuo J, *Nature* 1997
Ataxia	Scn8a	A1327T	Risk factor	Turn	Smith MR, *Neuroreport.* 1999
Alzheimer's disease	LDLR	A370T	Risk factor	Loop	Cheng D, *J. Med. Genet.* 2005
Alzheimer's disease	BChE	A539T	Risk factor	Helix	Wiebusch H, *Hum. Genet.* 1999

Shown are selected examples of Alanine-to-thronine substitutions which were reported to be causally correlated with the course of disease progression in the noted amyloidoses.

secondary structure elements, but amyloid fibrils share a common cross-β-sheet structure.[42]

4.2.1 Neurodegeneration and AD

Neurodegenerative diseases are a set of disorders that are characterised by a gradual and progressive loss of neural brain and spinal cord cells, leading to nervous-system dysfunction. The brain and spinal cord mediate a variety of functions, including sensory information processing and movement control. Therefore, the poor regeneration capacity of brain and spinal cord cells dictates a devastating damage. Neurodegenerative diseases are divided into two groups according to phenotypic effects: diseases causing movement disorders and diseases affecting memory and learning.[45] AD is a progressive neurodegenerative disorder of the second type, considered to be an incurable and terminal disease. The mean duration between onset of clinical symptoms and death is around 8.5 years. This brain disorder, first described by the psychiatrist Alois Alzheimer in 1906, is the most common cause of dementia among people age 65 and older.[46]

AD leads to nerve-cell death and tissue loss throughout the brain. Over time, ventricles grow larger while brain regions that are associated with higher mental functions shrink dramatically: the cortex shrivels up, damaging areas involved in thinking, planning and remembering. Shrinkage is especially severe in the hippocampus, a brain area that plays a key role in new memories formation.[47] These changes result in the development of the typical symptomology of AD, characterised by gross and progressive impairments of cognitive function and often accompanied by behavioural disturbances such as aggression, depression, and wandering.[48]

4.2.2 Amyloid Beta Production and Accumulation

Amyloid precursor protein (APP) is an integral membrane protein expressed in many tissues and concentrated in neuronal synapses. Its primary function is not completely understood, though it has been implicated as a regulator of synapse formation and neural plasticity. APP is best known and most commonly studied as the precursor molecule whose proteolysis generates Aβ, whose amyloid fibrillar form is the primary component of amyloid plaques found in the brains of AD patients.[49] In an amyloidogenic pathway, a 39–43 amino acid peptide is processed from sequential cleavage of APP by the enzymes β-secretase and γ-secretase, whereas in the nonamyloidogenic pathway a 3-kDa peptide is formed by sequential cleavage of α-secretase and γ-secretase.

4.2.3 Amyloid Beta Natural Degradation and Clearance

An imbalance between the production and clearance of Aβ is thought by many to represent the earliest event in the pathogenesis of AD. Several mechanisms have been proposed for Aβ clearance, including receptor-mediated Aβ transport across the blood/brain barrier and enzyme-mediated Aβ degradation. Additionally, pre-existing immune responses to Aβ might also be involved in Aβ clearance (see below).

Cerebrovascular dysfunction contributes to the cognitive decline and dementia in AD, and may precede cerebral amyloid angiopathy and brain accumulation of the Alzheimer's neurotoxin, Aβ. The blood/brain barrier (BBB) is critical for brain Aβ homeostasis and regulates Aβ transport via two main opposing receptors, the low-density lipoprotein receptor related protein 1 (LRP-1) and the receptor for advanced glycation end products (RAGE).[50] RAGE is the primary transporter of Aβ across the BBB from the systemic circulation into the brain, whereas LRP-1 mediates transport of Aβ out of the brain. In AD patients, RAGE is upregulated, whereas LRP-1 is downregulated, resulting in an increased Aβ concentration in the brain.[50]

Proteases, protease inhibitors and peptidases have been reported with the capability of cleaving Aβ *in vitro* or *in vivo*, contributing to endogenous mechanisms leading to plaque clearance. Examples are acyl peptide hydrolase (APH), neprilysin (NEP), endothelin-converting enzyme (ECE), insulin degrading enzyme (IDE), angiotensin-converting enzyme (ACE), cathepsin D, gelatinase A, matrix metalloendopeptidase-9 (MMP-9), coagulation factor XIa, antibody light chain c23.5 and hk14, α-1-antichymotrypsin and α2-macroglobulin complexes.[51] However, the prominent physiological functions of most of these agents suggest that increasing their doses may be detrimental.

Endogenous autoantibodies against Aβ have been found in AD patients and healthy individuals.[52] These autoantibodies tend to be depleted in AD plasma and correlated with age at onset for AD.[53] In a small-size pilot study, monthly treatment for six months with intravenous immuneoglobulins containing autoantibodies against Aβ significantly lowered the cerebro-spinal fluid (CSF) levels of total Aβ and improved the cognitive performance in AD patients.[54] Auto-antibodies isolated from immunoglobulin preparations also strongly blocked Aβ fibril formation, disrupted formation of fibrillar structures and nearly completely prevented Aβ neurotoxicity.[55] In addition, some naturally occurring proteolytic antibodies have also been found to cleave Aβ.[56] Thus, naturally occurring autoantibodies against Aβ can facilitate Aβ clearance. Although the levels of these antibodies are normally low, their persistence for many years in serum might be sufficient to protect against AD.

4.3   Treatment and Pharmacological Interventions in AD

The major barrier to treating and eventually preventing AD is insufficient understanding of the etiology and pathogenesis of neuron degeneration and loss. Disease-modifying intervention with AD would typically reduce the disease progression rate. From the patient's perspective, a disease-modifying treatment should result in long-lasting changes in the progressive disability, regardless of the drug's mechanism of action.[57] Published clinical studies do not yet show such drugs, and it may take a while before such a breakthrough in AD treatment can be expected.

4.3.1 Anticholinesterase Therapies

Organophosphate (OP) or carbamate esters are potent inhibitors of both AChE and BChE. These include drugs, pesticides, insecticides and chemical warfare agents. Phosphorylation or carabamoylation of the serine hydroxyl group in the substrate-binding domain inhibits the enzyme, causing ACh accumulation in synaptic clefts and overstimulation of cholinergic receptors. This induces various symptoms, including tension, anxiety, headaches, slurred speech, tremor, convulsions, muscle paralysis, seizure and even death by asphyxiation.[58]

ChE inhibitors (ChEIs) are the current most common drugs in use to treat AD; Tacrine, the first ChEI drug for the AD treatment, was approved in 1993, followed by other ChEIs: donepezil (Aricept®, Eisai Co., Ltd, 1996), rivastigmine (Exelon®, Novartis Pharmaceuticals, 1998) and galantamine (Razadyne®, Ortho-McNeil Neurologics, 2001). ChEIs vary widely in their pharmacological profiles and affinities for AChE and BChE. Donepezil and galantamine inhibit AChE 1000- and 50-fold, respectively, more than BChE, whereas rivastigmine inhibits both enzymes with similar efficiency.[59] These drugs do not represent a cure, as they do not stop the underlying progression of dementia, but rather lead to a temporarily slowdown of the loss of cognitive function, by decreasing the ChEs activity, resulting in higher ACh levels and improved brain function.

4.3.2 Antiamyloid Therapies

Following the development of the Aβ cascade hypothesis, research efforts commenced to develop antiamyloid therapies. Disease-modifying intervention can be achieved by decreasing Aβ production, stimulating clearance of already formed Aβ fibrils and/or intervention with Aβ aggregation.[12]

As only the combined action of β- and γ-secretase leads to Aβ formation, much research is focused on the development of specific secretase inhibitors. Unfortunately, besides APP, γ-secretase has many other substrates and cleaves several other transmembrane proteins (*e.g.* notch receptor-1).[60] Therefore, its inhibition may be counterproductive. A third way to prevent Aβ production from APP may be by stimulating the cleavage route via α-secretase. There is some evidence that α-secretase overexpression not only reduces amyloid plaque formation but also alleviates the deficits in spatial learning and synaptic plasticity observed in the control animals.[61,62]

Stimulating the clearance of Aβ load from the brain may also span different pharmacological interventions: increasing transport across the BBB by LRP-1/ RAGE manipulations,[50] facilitating Aβ-plaque clearance by immunotherapy and peripheral treatment with Aβ-bindable substances that promote Aβ efflux from the brain.[63] Immunological approaches intended to reduce Aβ load in the brain by either active or passive immunisation, have shown concomitant limitation of neuritic dystrophy and cognitive deficits in animal models.[64] Likewise, clinical trials suggested that active immunisation with the Aβ peptide is therapeutically effective, however, a significant number of patients developed autoimmune meningoencephalitis, caused primarily by the infiltration of autoreactive T lymphocytes into the brain in response to this active immunisation.[65,66]

4.3.2.1 Antiamyloid Aggregation

While therapeutic approaches aiming at decreasing Aβ production and stimulating Aβ clearance involve interference with other vital cellular cascades and native molecular processes, direct intervention with Aβ aggregation may engage fewer side effects. Most substances have only been tested *in vitro* and the majority inhibit fibril formation rather than preventing the early assembly of oligomers.

Native inhibitors of fibril formation include laminin,[67] midkine,[68] entactin,[69] Apolipoprotein J (clusterin) and Apolipoprotein E.[70] As an alternative, numerous small molecules that inhibit Aβ fibrillogenesis with IC_{50} values ranging from 0.1–3000 μM[71] have been developed. Of these, certain compounds have been described as efficient Aβ fibril-formation inhibitors, among them apomorphine and its derivatives,[72] bicyclodextrin,[73] rifamycins,[74] sulfonate dyes,[75] nicotine,[76] haemin,[77] quaternary ammonium salts,[78] carvedilol, daunomycin and rolitetracycline,[79] to name a few. Also, rationally designed peptide inhibitors of fibrillogenesis were reported.[80] Among these are short peptide fragments, based primarily on the central hydrophobic region of Aβ, with amino acid substitutions and side-chain modifications.[80,81] Other peptide inhibitors are engineered synthetic β-sheet breaker peptides[82] mimicking fibrillogenic domains that contain insertions of prolines or D-amino acids, modification of peptide termini or peptide backbone atoms (including N-methylation) and cyclic peptides.[83]

While the current medicinal chemistry approaches, which target amyloid aggregation, mainly address fibril formation, parallel approaches could potentially address all potential intermediates, as these intermediates are the neurotoxic components. This is a difficult task to perform, as the measurement of amyloid formation is complex, and only a few Aβ inhibitors prevent the formation of Aβ intermediates. One example is the synthetic glycosaminoglycan 3-amino-1-propaneosulfonic acid (3APS, tramiprosate, Alzhemed®), currently in Phase III, which is designed to prevent conformational transitions that lead to the assembly of oligomers.[84]

4.4 Cholinesterases

ChEs polymorphic carboxylesterases of broad substrate specificity, are serine hydrolases (B type) terminators of neurotransmission at cholinergic synapses and neuromuscular junctions. ChEs control the duration of ACh-mediated action on postsynaptic receptors in cholinergic synapses, and have non-hydrolytic roles in nervous-system development.

Substrate and inhibitor specificities classify ChEs into two subtypes:

- AChE, the major ChE in human blood, muscle and brain cells, which primarily hydrolyses esters with short acyl moiety, such as acetylcholine.
- BChE, the primary circulating cholinesterase; 3-fold more abundant in human blood than AChE,[85] which is also found in liver, lungs, muscles, brain and heart and hydrolyses ACh as well as esters with longer aliphatic or aromatic acyl moiety (*e.g.* butyrylcholine and benzoylcholine).

4.4.1 Physiological Roles of Cholinesterases

ChEs terminate cholinergic transmission in the central nervous system, in neuromuscular junctions (NMJs) and in the autonomic system (the para-sympathetic system, somatic motor nerves and preganglionic sympathetic nerves). A few sensory cells and the NMJ in nematodes also include ChEs.[86]

4.4.1.1 Catalytic Activities of Cholinesterases

ChEs hydrolyse choline esters and carboxylic esters. Hydrolysis involves a glutamate-histidine-serine catalytic triad,[87] similar to other serine hydrolases, through a "charge relay" process. Both enzymes can catalyse ACh hydrolysis to yield acetic acid and choline, thus maintaining proper transmission of impulses between nerve cells or from neurons to muscles and gland cells. The catalytic mechanism of AChE is extremely efficient, approaching diffusion-controlled rates, but BChE's hydrolysis rate of ACh is considerably slower. BChE deficiency, a rare genetic disorder, causes no apparent pathology, but increases carrier susceptibility to prolonged apnea under muscle-relaxant agents employed in surgery, such as succinylcholine or mivacurium, which are naturally destroyed by BChE. When administered under anesthesia to a susceptible person, these compounds are ineffectively removed, maintaining muscle relaxation. Likewise, subjects with BChE deficiency are susceptible for adverse reactions to other anticholinesterases.[85]

Both ChEs show arylacylamidase activity,[88] but BChE has a wider range of substrates: it hydrolyses several choline esters from acetyl to heptanoylcholine, as well as other aliphatic esters. High BChE activity in many first contact tissues (*e.g.* lungs, skin, blood and placenta) as well as high affinity towards a wide range of orally ingested toxic compounds and xenobiotics (*e.g.* cocaine, procainamide, acetylsalicylic acid) support the hypothesis that BChE functions as a natural bioscavenger by absorbing OP poisons (*e.g.* nerve agents) in a stoichiometric ratio before they cause neurological damage.[89]

4.4.1.2 Noncatalytic Activities of Cholinesterases

ChEs exist throughout the entire animal kingdom[90] and appear in locations where no ACh is released, suggesting that they possess additional non-hydrolytic functional properties, designated "nonclassical". These may depend on protein–protein interactions, compatible with ChE's structural homology with synaptic proteins such as neuroligins, which interact with other synaptic proteins – the neurexins.[91] Also, AChE contains two helix-loop-helix motifs, each forming a calcium-binding site (EF-hands),[92] which do not seem to play a role in its catalytic activity but may participate in protein–protein interactions.

Within the nervous system, ChEs were shown to be involved in membrane conductance and transmission of excitatory amino acids, learning and memory,[93] neurite growth,[94] neuritic translocation[95] and acute stress reactions.[96] Recent findings propose AChE's involvement in apoptosome formation.[97]

4.4.2 The Involvement of Cholinesterases in AD

Both ChEs accumulate in senile plaques and neurofibrillary tangles,[19] and colocalise with brain fibrils of Aβ.[14] BChE activity increases in the AD brain, whereas AChE activity decreases progressively.[98] The synaptic form of AChE, AChE-S, facilitates the *in vitro* and *in vivo* aggregation state of Aβ, perhaps through the noncatalytic mechanisms related to the peripheral anionic site subsite.[15,16] AChE-S may hence constitute an important cofactor in Aβ fibril-logenesis, being able to induce a conformational change of Aβ in solution and thus, having a direct role in fibril formation. Moreover, AChE-S-Aβ complexes have been shown to be more neurotoxic than Aβ peptides alone,[99,100] and may be involved in the pathogenesis of AD.[99] Thus, *in vitro* and *in vivo* data suggest that AChE-S behaves as a potent amyloid-promoting factor and potentiates the toxicity of amyloid fibrils.

The factors that contribute to the transformation of a relatively inert diffuse plaque to a neuritic or pathogenic one may involve interactions with additional plaque constituents. Such constituents include the monomeric isoforms of both ChEs.[101] AChE-S might have a crucial role in amyloid deposition at an early stage of the disease.[99] The diffuse amyloid plaques of normal ageing tend not to display BChE activity, whereas the vast majority of compact neuritic plaques do.[101]

Enhancement of cholinergic mechanisms favours nonamyloidogenic processing of APP.[98,102,103] Progressive deterioration of the widespread and dense cholinergic innervations of the human cerebral cortex therefore contributes to the salient cognitive and behavioural disturbances in AD, and is associated with decreased levels of ACh, choline acetyl transferase and AChE.[104] Both *in vitro* and *in vivo* studies demonstrated a link between cholinergic activation and APP metabolism.[105] Lesions of cholinergic nuclei cause a rapid increase in cortical and CSF levels of APP and this effect can be reversed by ChE inhibitors treatment.[106]

4.5 Butyrylcholinesterase as an AD Modulator

The C-terminal domain of BChE is encoded by an individual exon, suggesting specific biological functions. Supporting this notion, many "nonclassic" and nonhydrolytic functions of BChE were associated with its C-terminal domain. The C-terminus is exposed to the environment, and thus can readily form protein–protein interactions. Accordingly, BChE's colocalisation with the brain Aβ fibrils may require the C-terminus of BChE. Considering that BChE is under current development as a pharmacologically active natural product, due to its function as a bioscavenger of organophosphate poisons, it was tempting to assess its role in the process of Aβ aggregation.

4.5.1 Position 541 of Butyrylcholinesterase

The amphiphilic nature of the C-terminal helices forces their hydrophobic sides to turn inward, exposing the hydrophilic surface to the outside environment. In AChE-S, the hydrophilic nature of this surface would be intact, because the outward-turned arginines do not impair it. In contrast, BChE peptides would yield outward-protruding aromatic tryptophans, enabling Aβ interactions. These protruding residues can form heteroaromatic complexes with soluble, monomeric or low oligomeric, Aβ conformers. In the naturally formed tetramers, the protruding tryptophan would likely interfere with oligomerisation and/or the side-chain stabilisation of the β-sheet structure and inhibit propagation of the fibrils formation process to form toxic proto-fibrils and insoluble amyloid fibers.

Thioflavin T (ThT) fluorescence measurements showed that BChE is capable of attenuating the *in vitro* formation of amyloid fibrils from the AD amyloid peptide, unlike synaptic AChE-S, which facilitates such fibrils formation. Furthermore, a synthetic peptide having the C-terminal sequence of human BChE was found capable of performing a similarly effective attenuation of fibril formation as that of BChE. AChE-S, in comparison, promoted fibril formation and its C-terminal peptide failed to affect this process.[27]

The nonpolar substitution in position 541 of BChE is a structural evolution event that inverted the biological features of closely similar genes. Furthermore, this nonpolar substitution is unique, as it appears only in the BChE gene of primates. This suggests that primate BChE acquired this novel, promiscuous function, while not impairing its classical roles in ACh hydrolysis.[107] This distinction also explains why mouse studies overlooked this phenomenon.

4.5.2 Position 539 of Butyrylcholinesterase

The most common among BChE variants is the K variant, with allelic frequencies of 0.13–0.21, which implies that a significant fraction of the Western World's population is likely to carry this protein. BChE-K is defined by a point mutation at nucleotide 1699 (dbSNP cluster ids rs17846898; rs1803274) that leads to a change of the amino acid alanine at position 539 to threonine

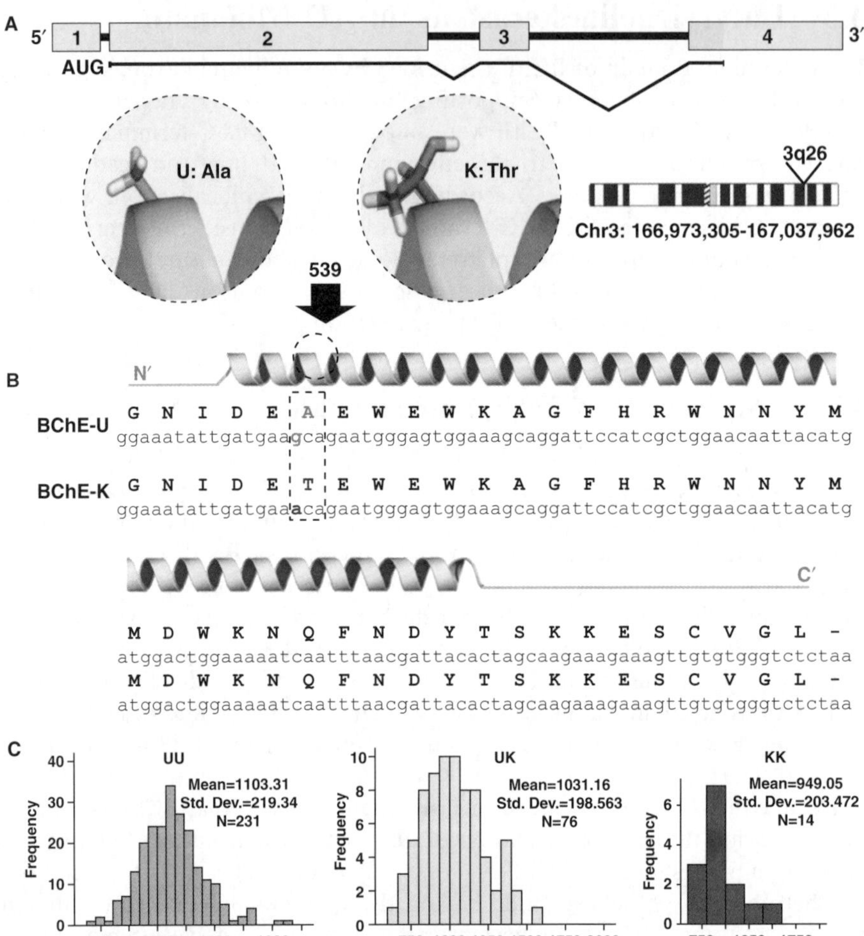

Figure 4.1 (A) BChE location on chromosome 3q26.1-q26.2 and its gene structure. (B) The C-terminal DNA and amino acid of sequences BChE-U and -K. The predicted secondary structure of the amino acid substitution (enlarged) is shown above the sequences. (C) Histograms of measured BChE activity in apparently healthy carriers of the UU, UK and KK genotypes. Units are nmole acetylthiocholine hydorlized/min/μl serum.

(GCA→ACA), 36 residues upstream to BChE's C-terminus (Figures 4.1A and B).

A potential advantage of the BCHE-K genotype for AD has been demonstrated in three independent clinical studies. First, Holmes and coworkers showed that the presence of the K allele was associated with a slower average rate of cognitive decline in subjects with severe AD.[39] O'Brien and coworkers then demonstrated that the reduced hydrolytic activity of the BChE-K variant correlates with preserved attentional performance and reduced rate of cognitive decline. Intriguingly, O'Brien *et al.* further showed that carriers of this reduced-activity enzyme do not respond to cholinesterase inhibitor therapy.[36] This was proposed to be relevant to their generally higher levels of ACh and/or to altered feedback reaction to anti-ChEs.[108] Also, a highly significant association between lower levels of BChE activity in the temporal cortex (as would be predicted for carriers of BChE-K) was reported to associate with a decreased rate of cognitive decline in patients with moderate to severe dementia with Lewy bodies.[109]

In addition to the above studies, there is an on-going debate over BChE-K's association with Apo ε4. Several reports[110–116] suggested a synergy between BChE-K and APO E4 genes in AD, while others[117–125] showed the opposite. Yet other studies argued that BCHE-K confers an increased risk of developing AD-related neuropathology that is independent of the BChE-K-ApoE association.[38] This latter observation opposes the above clinical studies and may suggest multigenic effects associated with BChE-K in various ethnic groups. Together, intriguing links emerge between BChE-K and AD, with both basic and applied implications.

While others focused on the genetics of BChE-K, we studied the effect of the A539T substitution on BChE stability and tetramerisation and on its potency in attenuating Aβ oligomerisation and fibril formation. Our findings attribute many of the features observed for BChE-K carriers in AD to structural origins.

4.5.2.1 BChE-K Shows Reduced Stability and Hydrolytic Activity

The capacity of serum BChE-K to hydrolyse butyrylthiocholine was reported to be reduced relative to BChE-U.[33] In Israeli volunteers, the mean calculated ATCh hydrolytic activities of serum BChE from UU, UK and KK carriers were 1103.31 ± 219.34, 1031.16 ± 198.56 and 949.05 ± 203.47 M/S ($p = 0.011$) respectively (Figure 4.1C), an activity decline that indicated a gene-dose dependence.

The reduced hydrolytic activity of BChE-K predicts that BChE-K carriers would potentially sustain improved cholinergic transmission compared to BChE-U carriers.[36] This and the controversial nature of genotype studies (Table 4.1) further raised the question whether BChE-K functions as an AD risk or protection factor. The location of the A539T substitution in the C-terminus that is distant from the hydrolytic site of the enzyme also indicated that the impaired stability of BChE-K caused the reduced activity. To challenge this prediction, the inherent stability of BChE-U and BChE-K was compared by incubating serum samples from genotyped subjects in increasing concentrations of urea, followed by native gel electrophoresis and activity staining (Figure 4.2A). Both BChE-U and BChE-K are enzymatically active tetramers, however, the enzymatic activity of BChE-K tetramers was greatly reduced following incubation in 1 M urea, unlike BChE-U, which was resistant to this

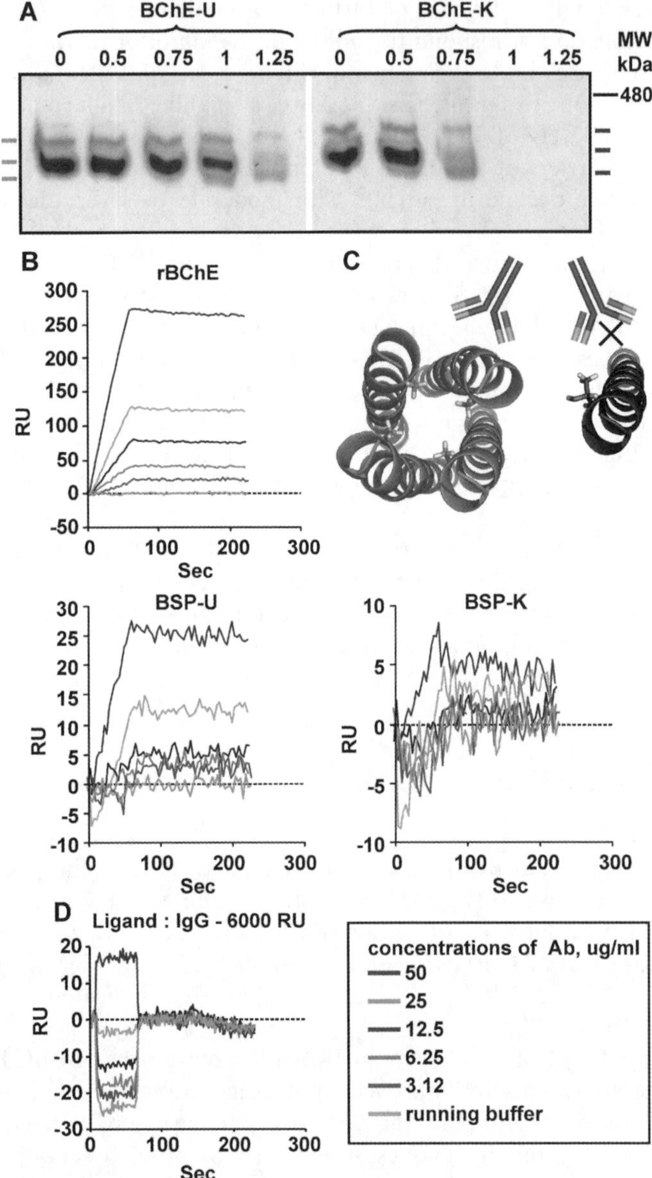

Figure 4.2   (A) Native gel subjected to activity staining of homozygous BChE-U and - K serum samples following incubation in the noted molar concentrations of urea. (B) Surface plasmon resonance analyses. Polyclonal anti-rBChE-U antibodies in the noted concentrations were injected against immobilized rBChE-U, BSP-U and BSP-K. Signals are dose-dependent. Note the 10-fold RU scale difference between rBChE-U and BSP-U signals and the lack of detectable interaction with BSP-K. (C, D) Positive surface plasmon resonance response of BSP-U but neither BSP-K nor IgG. Note the exceedingly low response scale despite adding 6000 resonance units (RU) of IgG, attesting to the specificity of the observed response with the anti-BChE antibodies (Ab).

treatment. This suggested that the hydrolytic activity was rapidly lost once BChE-K tetramers disassemble.

4.5.2.2 The Structural Origin for BChE-K's Instability

We challenged the working hypothesis regarding the structural origin for BChE-K's instability by analysing the corresponding 32-residue synthetic peptides, BSP-K and BSP-U, for their interaction with a polyclonal rabbit antibody elicited towards intact rBChE-U.[126] Surface plasmon resonance (SPR) analysis demonstrated a dose-dependent interaction of anti-rBChE-U antibodies with rBChE-U and BSP-U (albeit less profoundly), but not with BSP-K (Figures 4.2B and C). Immunoglobulin G, which does not bind to the antibodies against rBChE-U (Figure 4.2D) served as an immobilised control employed as a subtractable 'baseline' in all cases. These results supported the concept of structural differences between these two peptides and suggested that distinct tetramerisation capacities are involved. Interactions of BSP peptides were therefore studied by incubating them with the synthetic Polyproline peptide, which facilitates BChE tetramerisation.[26]

Nuclear magnetic resonance (NMR) was used to follow both interactions with Polyproline, by tracking chemical shift deviations in the amide–aliphatic interactions (fingerprint region). Polyproline fortunately has no amide protons (apart from one leucine); therefore, it does not contribute to the fingerprint region of the BSP spectra. Upon adding Polyproline, BSP-U showed mainly upfield chemical shift deviations, suggesting that the amide protons are shielded in this variant when interacting with PolyP. The amide protons of BSP-K showed the opposite, except for residues H15 and R16, which were shielded, and N2, W24, and Y31, which were deshielded in both peptides. The distinct interactions observed for these peptides with PolyP suggested inherent differences between their tetramerisation potency, compatible with the reduced stability of BChE-K compared to BChE-U (Figure 4.3).

4.5.2.3 A539T Effects on Aβ Oligomerisation

Separated fractions of BChE dimers and monomers were used to compare the effects caused by the C-terminus. Recombinant rBChE-U production in the milk of transgenic goats yielded primarily dimeric protein, with ∼10% monomeric fraction, as shown by mass spectrometry.[126] ThT fluorescence measurements showed that isolated fractions of highly purified rBChE-U monomers and dimers both attenuated fibril formation at a molar ratio of 1:100 to Aβ. However, monomeric and dimeric rBChE-U showed different durations of the lag and growth phases in the fibril-formation process (Figure 4.4A). Furthermore, calculating the rate of fibril formation revealed that recombinant BChE dimers (Figure 4.4B) and yet more so, native BChE tetramers from human serum (Figure 4.4C) attenuate this process, suggesting direct association of this rate with the number of enzyme subunits.

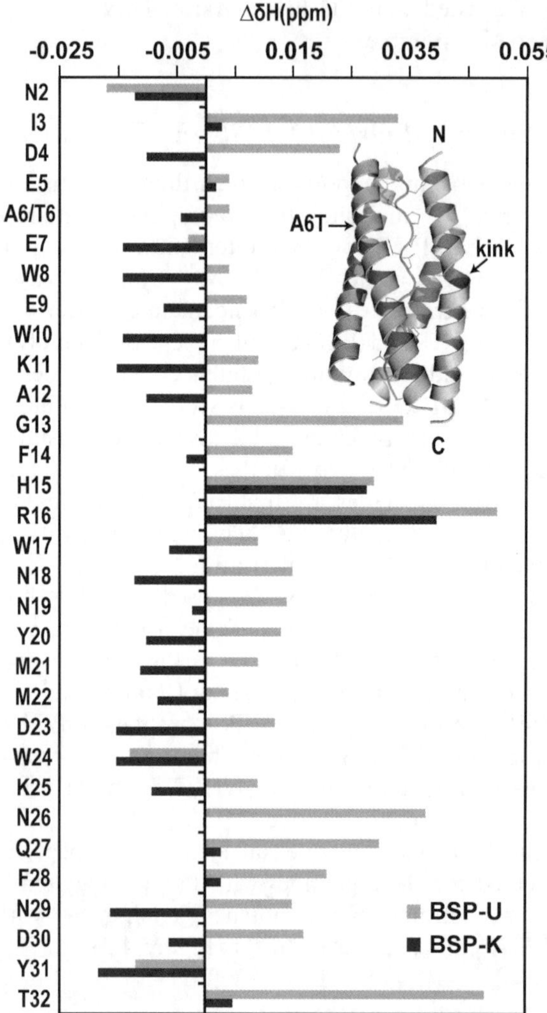

Figure 4.3   NMR-measured chemical shift differences between spectra of BSP-U and -K bound and non-bound to PolyP (shown as a central thread in the structure). Right: location of residues that showed significant chemical shifts on the BSP structure (built according to PDB structure 1VZJ) are colored in pink. The kink of the helix at Gly546 (in blue) was significantly shifted in both BSP peptides.

At the clinical level, we predicted that the distinct structural features of BSP-K could potentially impair its ability to modulate Aβ conversion to β-sheet conformation. This was assessed by incubating Aβ alone or with BSP peptides and following the CD spectrum for 24 h. Deconvoluted spectra of Aβ samples incubated alone showed a gradual increase in the content of β-sheets, from 26% to 50% within 2.5 h. In contrast, adding rBChE-U or BSP peptides

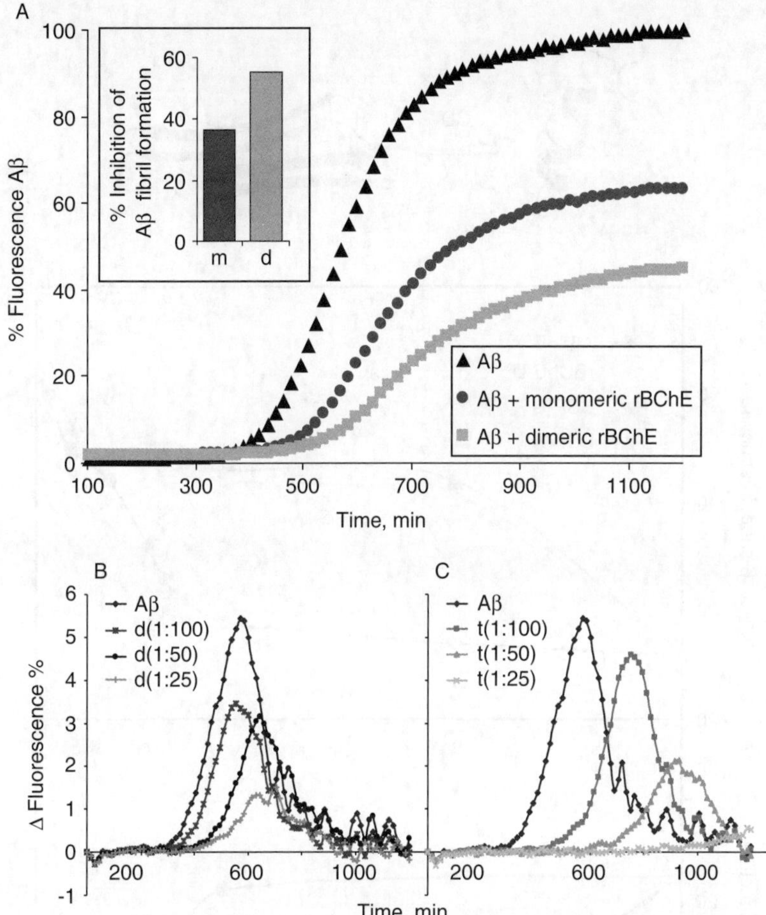

Figure 4.4   (A) ThT fluorescence demonstrating that highly purified dimers of rBChE-U suppressed Aβ-fibril formation more efficiently than monomeric rBChE-U. Inset: Percentage of attenuated Aβ-fibril formation elicited by monomers (m) and dimers (d) (B), (C) Rates of fluorescence changes demonstrating the difference in the capacity for attenuating fibril formation by human recombinant rBChE dimers (d) or serum-derived tetramers (t).

to Aβ accelerated the increase in β-sheet content, which reached an apparent plateau within seconds (50 s, to 48%) (Figure 4.5). Further studies of Aβ oligomerisation involved crosslinking of Aβ in the presence or absence of BSP-U or BSP-K, and separation of the soluble fractions by SDS-PAGE (Figure 4.6). Aβ alone precipitated rapidly, within 4 h. Both BSP peptides prolonged the time Aβ oligomers remained in solution, but BSP-U prolonged the persistence of amyloid oligomers in solution more effectively than BSP-K: after 8 h of incubation, no oligomeric forms of Aβ remained in the soluble phase in the presence of BSP-K, while in the presence of BSP-U they lasted nearly 22 h.

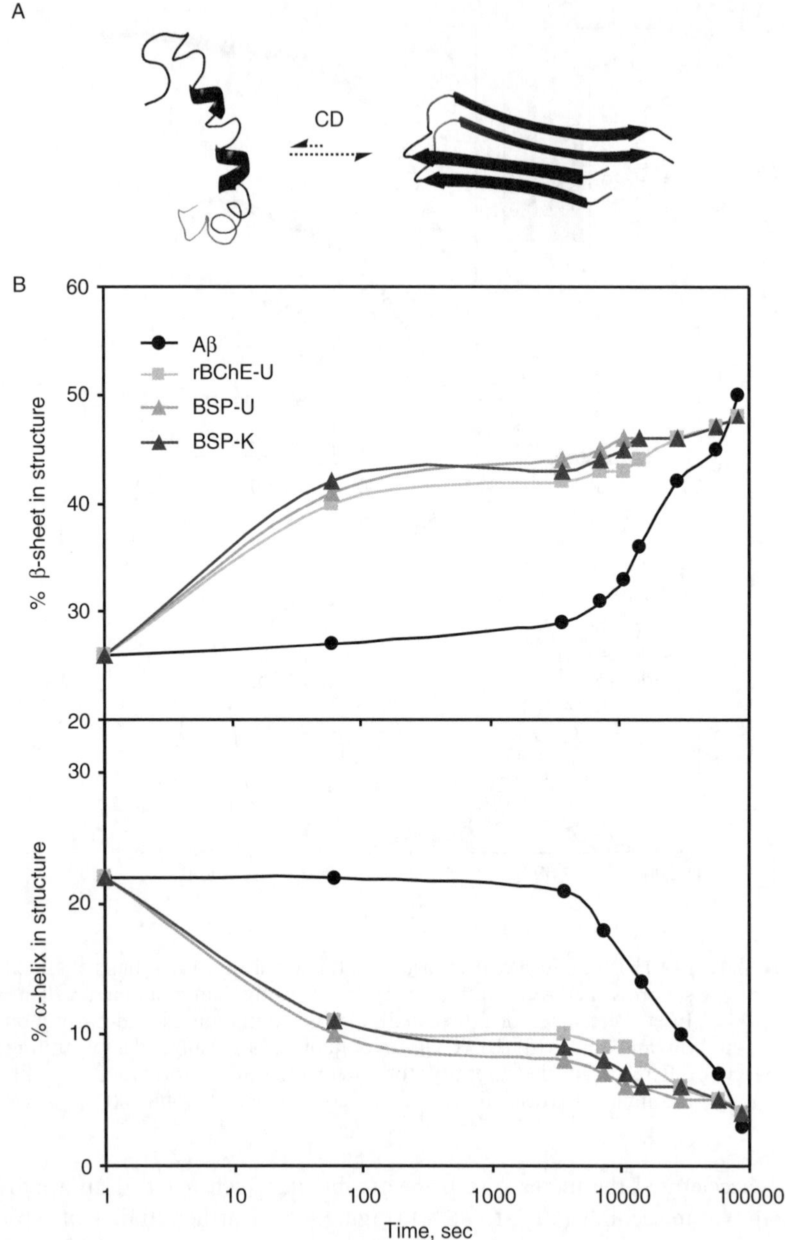

Figure 4.5 (A) Aβ transition from helical structure to β-sheet conformation. (B) CD deconvoluted spectra demonstrate a rapid increase in β-sheet content (upper plot) and rapid decrease in helical content (lower plot) with addition of rBChE-U, BSP-U and BSP-K, compared to the increase in β-sheet content seen with Aβ alone.

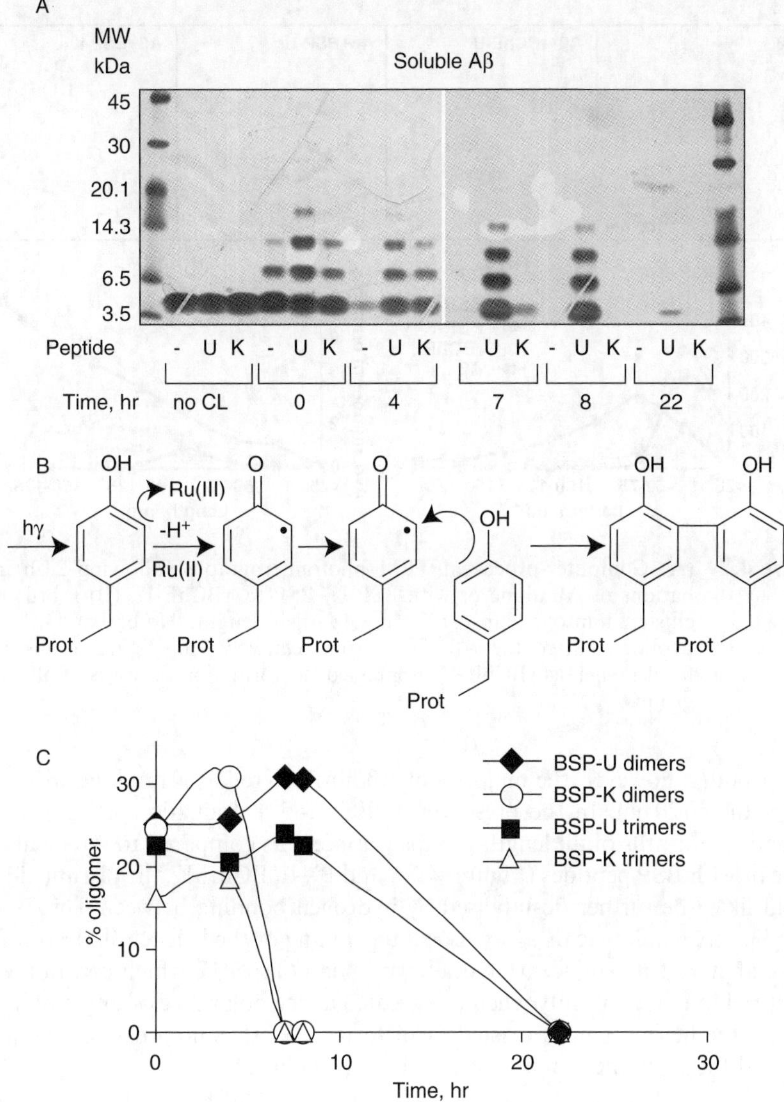

Figure 4.6 (A) 16% Tris-tricine gel BSP peptide demonstrating interference with oligomerization of Aβ. (B) The chemical reaction. (C) Quantification of Aâ dimers and trimers in solution, in the presence of BSP-K and BSP-U.

This supports the view that BSP-K is less capable of interacting both with PolyP and with Aβ than BSP-U.

Transmission electron microscopy was used to quantify longer and more developed Aβ fibrils, formed in the presence or absence of rBChE-U and BSP peptides. Following 48 h incubation, Aβ alone formed mostly 50–60 nm-long fibrils, with some fibrils reaching up to 300 nm. In the presence of either of the

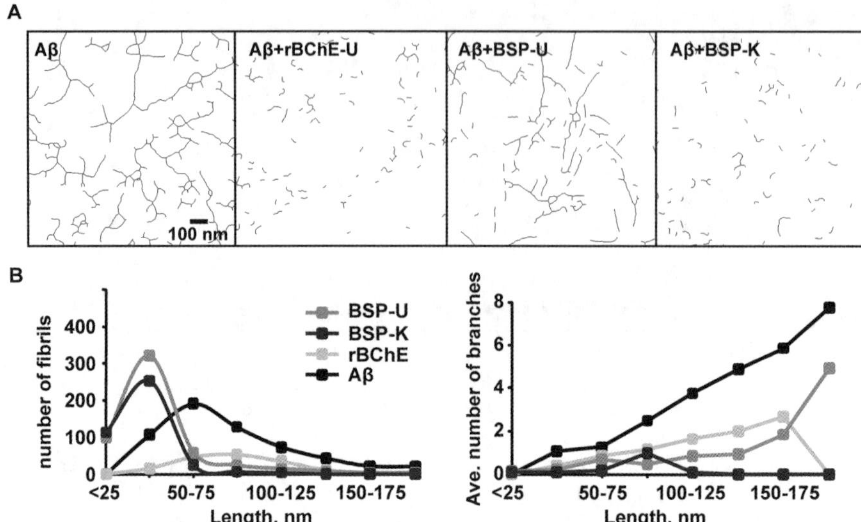

Figure 4.7   (A) Computer-processed TEM photomicrographs following 24-h incubations of Aβ alone or with BSP-U, BSP-K, rBChE-U. (B) Fibrils were clustered into groups according to their length. Note that both BSP peptides shifted the Aβ fibrils from relatively long to numerous short fibrils, whereas rBChE-U suppressed the formation of fibrils of all length groups.

BSP peptides, however, the majority of Aβ fibrils were 20–30 nm long with none longer than 120 nm. In the presence of rBChE-U a dramatic reduction in the formation of fibrils of all length groups was seen, as compared to Aβ incubated alone or with BSP peptides (Figures 4.7A and B). In BChE-K, the disrupted helix would likely be further destabilised by hydrogen bonding between Thr539 and Glu543, three amino acids away. Local unwinding of the helix could lead to loss of the ability of the adjacent pivotal tryptophan (Trp541), which contributes to structural interactions only when it is located at the polar face of an amphiphilic helix; if the helix becomes twisted or distorted, W541 is no longer at the polar face and the entire helix may cease to be amphiphilic.

4.5.2.4   *BSP Peptides are Poor Neuroprotectors Compared to rBChE-U*

Cytotoxicity of Aβ has been attributed to low molecular weight oligomers.[127] Therefore, Aβ cytotoxicity and the cytoprotection provided by BSP peptides and rBChE-U were tested on cultured human neuroblastoma SH-SY5Y cells. The cells were incubated alone, in the presence of Aβ and in the presence of Aβ with BSP peptides or rBChE-U. After 72 h, an LDH release assay for cell viability showed a 26% increase in LDH release from Aβ-treated cells compared to controls. Importantly, rBChE-U prevented $85 \pm 8\%$ of this Aβ-induced LDH increase, indicating that it protects the cells from the Aβ-dependent toxicity. In contradistinction, the BSP peptides yielded only small, variable (10–17%) changes in LDH release, suggesting that they can only function to attenuate Aβ fibril formation as an integral part of the intact BChE protein.

4.5.2.5 Alanine-to-Threonine Polymorphisms and Related Amyloidoses

Ample evidence supports the importance of alanine-to-threonine substitutions and their relevance to amyloidogenic processes, protein stability and quaternary organisation. Point mutations at the dimer interface of light-chain immunoglobulins decrease their stability, so that the A34T polymorphism in this protein leads to systemic amyloidosis.[34] An A25T mutant of the tetrameric human protein Transthyretin (TTR), associated with central nervous system amyloidosis, is prone to aggregation and exhibits drastically reduced quaternary and tertiary structural stabilities.[35] The thermodynamic instability of the A25T TTR mutant suggests that both tetramers and monomers of this variant are highly destabilised. In addition, A25T TTR tetramers dissociate very rapidly, about 1200-fold faster than the dissociation rate of wild-type TTR, reflecting a high degree of kinetic destabilisation of their quaternary structure. These factors together probably contribute to the high propensity of A25T TTR to aggregate *in vitro*. A variety of proteins with such A→T substitutions thus emerge as risk factors of neurodegenerative and amyloidogenic diseases. For example, the A54T mutant of intestinal fatty acid binding protein (FABP2) increases dietary fatty acids processing and fat oxidation, which has been shown to reduce insulin action, a major risk factor for diabetes mellitus Type II (DMII). Another case is the combination of two polymorphisms in the LDL receptor gene (LDLR), one of which involves A370T substitution, demonstrating a nine-fold increased risk among APOE4 carriers. It is noteworthy that an A370T polymorphism in APOE4 was identified in patients with familial hypercholesterolaemia and was also suspected as a risk for coronary heart disease. Several disease-associated A→T polymorphisms are listed in Table 4.2.

The theoretical Chou–Fasman parameters[128] demonstrate strong tendencies among individual amino acids to prefer one type of secondary structure over others. Alanine was identified as one of the strongest helix facilitators, whereas threonine was recognised as one of the strongest β-sheet supporters. The preference of threonine to form β-sheet is crucial in the context of amyloidogenic processes, as all amyloid fibrils share a common cross-β-sheet structure. The above discussed A→T substitutions may hence facilitate β-sheet conversion thanks to these structural preferences of the mutated domains.

4.6 Discussion and Perspectives

Various chemical and biological approaches were combined to explore the role of the BChE variants in Aβ accumulation, which is characteristic of the AD pathophysiology. Our findings attribute many of the features observed for BChE-K carriers in AD to structural origins. We found BChE-K to be inherently unstable compared to BChE-U, which may induce elevated ACh levels in its carriers, thereby delaying AD onset. Yet, BChE-K was considerably less effective in attenuating the accumulation of Aβ fibrils than BChE-U. While bearing in mind the alternative hypotheses for the role played by Aβ in AD, our findings suggest that BChE-K may pose either a risk or a protective factor in AD, in a context-dependent manner affected by other variables. It is conceivable that distinct population studies reflect these variables, so that the null differences observed in the meta-analysis are less meaningful than the different population studies. Our findings attribute the association between BChE-K and AD progression to a structural origin, with dual activity: impaired stability and quaternary organisation of BChE-K compared with BChE-U and the consequently suppressed hydrolytic activity, counterbalanced by aberrant interactions with Aβ. This statement is supported by several considerations, as briefly listed below.

Our findings suggest that the BChE-K C-terminal substitution impairs both its stability and its intersubunit interactions. This is supported by findings where tetrameric serum BChE-K is thermounstable, compared to BChE-U demonstrating that monomeric BChE-K loses hydrolytic activity within 10 min when incubated at 48 °C.[129] The genetic implications of BChE-K in AD progression were studied comprehensively by others (Table 4.2), but the role of the A539T polymorphism in the tetramerisation and stability of BChE[22,23] has not been considered.

NMR detected chemical shift differences between BSP peptides upon adding Polyproline. Nine residues in three regions of BSP-U showed chemical shift deviations ($\Delta\delta \geq 0.02$), compared to only two BSP-K residues. Also, the changes observed in the BSP-U spectra upon binding Polyproline were larger than those in the BSP-K spectra, indicating that Polyproline induces greater organisation in BSP-U. Changes in the NMR chemical shift largely indicate those residues that undergo structural changes upon binding (*e.g.*, see ref. 130). The distribution of BSP regions whose adjacent chemical environment changed, including the middle region and both termini, may indicate a change in the relative tilt of the helices or a motion in the hinge of the helix (at Gly546/G13) due to rearrangement of the kink. One possible explanation of our findings is that threonine may slightly disrupt the helix or the amphiphilic distribution.

Based on our findings, AChE-S and BChE have seemingly opposite effects on fibril formation: Human and recombinant BChE and its derived C-terminal peptides associate with the soluble Aβ conformers, delay their accumulation, and decrease the rate of Aβ fibril formation *in vitro*. The intact protein provides protection to neural cells in culture. Therefore, BChE can be considered as one of those few Aâ inhibitors, which prevent the formation of Aβ intermediates and protect cells from their neurotoxicity.

We found BChE-K to be inherently unstable compared to BChE-U, which may maintain higher ACh levels in its aging carriers compared to BChE-U carriers, thereby delaying AD onset. Yet, BChE-K was considerably less effective in attenuating the accumulation of Aβ fibrils than BChE-U, suggesting that once the disease commences it may accelerate the progression of AD pathology. BChE-K can therefore pose either a risk or a protective factor in a context-dependent manner affected by other variables. It is conceivable that distinct population studies reflect these variables, so that the null differences observed in the meta-analysis are less meaningful than the different population studies. BChE-K was recently found to associate with reduced tau phosphorylation. This suggests that BChE-K may exert a protective effect against tau-related development of dystrophic neurite processes characteristic of AD. As our work was focused on the amyloidogenic pathway alone, it is yet to be explored whether tau phosphorylation is one of those variables.

Our findings attribute the association between BChE-K and AD progression to a structural origin, with dual activity: impaired stability and quaternary organisation of BChE-K compared with BChE-U and the consequently suppressed hydrolytic activity, counterbalanced by aberrant interactions with Aβ. In BChE-K, the disrupted helix would likely be further destabilised by hydrogen bonding between Thr539 and Glu543, three amino acids away. Local unwinding of the helix could lead to loss of the ability of the adjacent pivotal tryptophan (W541), which we previously demonstrated disrupts this amphiphilic helix, to attenuate Aβ fibril formation.[27] W541 contributes to structural interactions only when it is located at the polar face of an amphiphilic helix; if the helix becomes twisted or distorted, W541 is no longer at the polar face and the entire helix may cease to be amphiphilic. Another example of a disrupted helix originating from an alanine-to-threonine substitution is the solid-state NMR study that showed that the familial Parkinson's disease-associated mutation A53T, which occurs in the N-terminus of the amphiphilic helical α-synuclein, is more prone to forming β-structures than wild-type α-synuclein.[131]

The "peripheral sink" hypothesis suggests that Aβ-bindable substances sequester plasma Aβ, leading to clearance of Aβ by promoting a net efflux of a rapidly mobilised soluble pool of Aβ. Peripheral treatment with an agent that has high affinity for Aβ reduces the level of Aβ in the brain, probably because of a peripheral action.[132] Based on this hypothesis, sequestering of plasma Aβ by antibodies and other peripherally administered substances that can bind Aβ (*e.g.* curcumin and enoxaparin) can establish a gradient favouring the efflux of Aβ from the brain.[133] Plasmatic BChE can act to minimise Aβ damages and thus, BChE administered intravenously could be a potential therapeutic approach, as in the case of other antiamyloidogenic substances (*e.g.* see ref. 134).

Various Aβ aggregation pathways may exist simultaneously and distinct inhibitors may affect only some of these. For instance, if oligomers are essential intermediates on the pathway to fibril formation, inhibitors that block oligomer formation would ultimately block fibril formation as well. Alternatively, if fibrils and oligomers represent distinct aggregation pathways, some inhibitors might block oligomerisation but not fibril formation and *vice versa*. We followed Aβ oligomerisation by crosslinking, and fibril formation by TEM. In both approaches, rBChE-U and BSP peptides attenuated both oligomer and

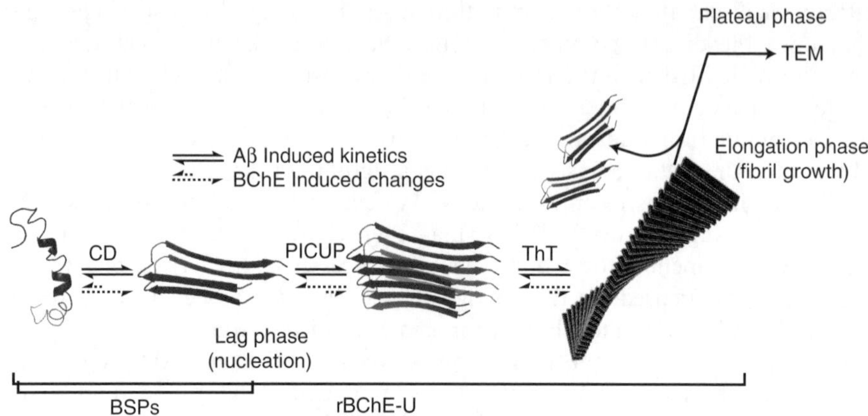

Figure 4.8 The Aβ aggregation pathway involves a set of mutually dependent reactions in complex equilibria (Monomer ⇆ Dimer ⇆ Oligomer ⇆ Fibril). The transition from helical structure to β-sheet conformation was studied by CD, Aβ oligomerization was followed by crosslinking and fibril formation was tracked by ThT fluorescence and TEM. These reactions along the Aβ aggregation pathway are differentially affected by rBChE-U, BSP-U and BSP-K.

fibril formation, albeit with different efficiencies. This tentatively implies that under the experimental conditions employed, oligomers are fibril intermediates. One possible explanation of the observed attenuation effects is that both rBChE-U and the BSP peptides can induce a shift of Aβ accumulation towards a stabilised Aβ oligomer, as large as 100-mer, if judged by the 1:100 molar ratio of effective inhibition (Figure 4.8).

Penetration of Aβ-bindable substances into the brain enables inhibition of soluble Aβ aggregation and/or resolubilisation of Aβ fibrils that may shift brain equilibrium towards soluble Aβ fibrils and facilitate Aβ clearance. BChE upregulation in the brain can hence delay the onset of Aβ accumulation and possibly enable transport of soluble Aβ to the blood. BChE would potentially be administered either intravenously or intraventricularly depending on the dynamics of these processes. It still has to be studied whether the BBB disruption that is apparently induced by cholinergic imbalances in AD[135,136] can act like statins to optimise Aβ transport over the BBB[137] or alternatively to bypass the LRP1/RAGE-mediated transport.

The amyloid hypothesis postulates that increased Aβ production and deposition plays a key role in triggering neuronal dysfunction and death in AD.[11,138] Many argue that the toxicity of Aβ lies in its soluble oligomeric intermediates[127] rather than the insoluble fibrils that accumulate in AD plaques. Recent data indicate that small soluble oligomeric forms, as short as dimers, are the main neurotoxic species involved in the pathogenesis of AD.[23,24] Our current findings indicate that neuroprotection may be achieved by shifting the equilibria of Aβ accumulation away from the potentially more hazardous oligomeric species, such as by BChE. Also, BSP-K displayed less effective attenuation than BSP-U, possibly because it interacted with the Aβ monomers and/or intermediates less tightly. Nevertheless, in crosslinking and electron microscopy experiments, BSP peptides affected Aβ assembly less efficiently than rBChE-U. The differences in interactions are likely to be involved in the rBChE-U protection from Aβ neurotoxicity in cultured cells, where the peptides failed to provide protection.

The specific association between Aβ and the cell membrane has been shown to be the initial step in the chain of events that leads to toxicity.[139] Our findings demonstrate that rBChE-U interacts with the neurotoxic Aβ oligomers, whereas Aβ-BSP interactions are much less effective. This is compatible with the hypothesis that BChE-U blocks the association of Aβ oligomers with the cell membrane. Therefore, BSP peptides are essential, but unable to promote such protection on their own. Furthermore, we demonstrated that the rBChE-U disruption extended to active fibril disassembly, compatible with a mechanism suggested by others,[140] whereby inhibitor binding to the edge of the fibril leads to disassembly of the fibril by gradual strand removal. However, this mechanism requires high ratios of inhibitor to Aβ and therefore does not fit the 1:100 ratio we observed, in which a single rBChE-U molecule can initiate breakdown of large Aβ oligomers. Rather, our results suggest that the symmetry between the layers is broken by steric hindrance, which induces structural instabilities due to the A to T mutation and leads to collapse of the entire structure of the fibril. The same structural features that destablises BChE-K are therefore those that make it less effective as a neuroprotector.

Acknowledgments

The authors are grateful to Drs. Deborah Shalev, Sophie Diamant, Estelle R. Bennett (Hebrew University) and Tsafrir Bravman (BioRad, Haifa, Israel) for collaborative efforts. This study was supported by the Hurwitz Fund (to H.S.). E.P. was an incumbent of the national PhD Eshkol Fellowship.

References

1. C. P. Ferri, M. Prince, C. Brayne, H. Brodaty, L. Fratiglioni, M. Ganguli, K. Hall, K. Hasegawa, H. Hendrie, Y. Huang, A. Jorm, C. Mathers, P. R. Menezes, E. Rimmer and M. Scazufca, *Lancet*, 2005, **366**, 2112–2117.
2. K. Iqbal and I. Grundke-Iqbal, *J. Alzheimers Dis.*, 2006, **9**, 219–242.
3. P. T. Lansbury Jr., *Biochemistry*, 1992, **31**, 6865–6870.
4. A. Lomakin, D. S. Chung, G. B. Benedek, D. A. Kirschner and D. B. Teplow, *Proc. Natl. Acad. Sci. USA*, 1996, **93**, 1125–1129.
5. J. Shen and R. J. Kelleher 3rd, *Proc. Natl. Acad. Sci. USA*, 2007, **104**, 403–409.
6. D. Pratico, *Trends Pharmacol. Sci.*, 2008, **29**, 609–615.
7. A. I. Bush, *J. Alzheimers Dis.*, 2008, **15**, 223–240.

8. X. Zhu, A. K. Raina, G. Perry and M. A. Smith, *Lancet Neurol.*, 2004, **3**, 219–226.

9. C. Mancuso, G. Scapagini, D. Curro, A. M. Giuffrida Stella, C. De Marco, D. A. Butterfield and V. Calabrese, *Front Biosci.*, 2007, **12**, 1107–1123.

10. H. G. Lee, G. Casadesus, X. Zhu, A. Takeda, G. Perry and M. A. Smith, *Ann N Y Acad. Sci.*, 2004, **1019**, 1–4.

11. J. Hardy and D. J. Selkoe, *Science*, 2002, **297**, 353–356.

12. R. J. van Marum, *Fundam. Clin. Pharmacol.*, 2008, **22**, 265–274.

13. A. Chatonnet and O. Lockridge, *Biochem. J.*, 1989, **260**, 625–634.

14. C. I. Wright, C. Geula and M. M. Mesulam, *Ann. Neurol.*, 1993, **34**, 373–384.

15. M. Bartolini, C. Bertucci, V. Cavrini and V. Andrisano, *Biochem. Pharmacol.*, 2003, **65**, 407–416.

16. G. V. De Ferrari, M. A. Canales, I. Shin, L. M. Weiner, I. Silman and N. C. Inestrosa, *Biochemistry*, 2001, **40**, 10447–10457.

17. P. Gomez-Ramos and M. A. Moran, *Mol. Chem. Neuropathol.*, 1997, **30**, 161–173.

18. K. A. Carson, C. Geula and M. M. Mesulam, *Brain Res.*, 1991, **540**, 204–208.

19. A. L. Guillozet, J. F. Smiley, D. C. Mash and M. M. Mesulam, *Ann. Neurol.*, 1997, **42**, 909–918.

20. E. K. Perry, R. H. Perry, G. Blessed and B. E. Tomlinson, *Neuropathol. Appl. Neurobiol.*, 1978, **4**, 273–277.

21. M. A. Moran, E. J. Mufson and P. Gomez-Ramos, *Acta Neuropathol.*, 1993, **85**, 362–369.

22. C. V. Altamirano and O. Lockridge, *Chem. Biol. Interact.*, 1999, **119–120**, 53–60.

23. R. M. Blong, E. Bedows and O. Lockridge, *Biochem. J.*, 1997, **327**(Pt 3), 747–757.

24. A. L. Perrier, J. Massoulie and E. Krejci, *Neuron.*, 2002, **33**, 275–285.

25. S. Bon, A. Ayon, J. Leroy and J. Massoulie, *Neurochem. Res.*, 2003, **28**, 523–535.

26. H. Li, L. M. Schopfer, P. Masson and O. Lockridge, *Biochem. J.*, 2008, **411**, 425–432.

27. S. Diamant, E. Podoly, A. Friedler, H. Ligumsky, O. Livnah and H. Soreq, *Proc. Natl. Acad. Sci. USA*, 2006, **103**, 8628–8633.

28. E. Podoly, T. Bruck, S. Diamant, N. Melamed-Book, A. Weiss, Y. Huang, O. Livnah, S. Langermann, H. Wilgus and H. Soreq, *Neurodegener. Dis.*, 2008, **5**, 232–236.

29. E. Podoly, D. E. Shalev, S. Shenhar-Tsarfaty, E. R. Bennett, E. Ben Assayag, H. Wilgus, O. Livnah and H. Soreq, *J. Biol. Chem.*, 2009.

30. Y. Nicolet, O. Lockridge, P. Masson, J. C. Fontecilla-Camps and F. Nachon, *J. Biol. Chem.*, 2003, **278**, 41141–41147.

31. M. N. Ngamelue, K. Homma, O. Lockridge and O. A. Asojo, *Acta Crystallogr. Sect. F. Struct. Biol. Cryst. Commun.*, 2007, **63**, 723–727.

32. H. Dvir, M. Harel, S. Bon, W. Q. Liu, M. Vidal, C. Garbay, J. L. Sussman, J. Massoulie and I. Silman, *Embo. J.*, 2004, **23**, 4394–4405.
33. C. F. Bartels, F. S. Jensen, O. Lockridge, A. F. van der Spek, H. M. Rubinstein, T. Lubrano and B. N. La Du, *Am. J. Hum. Genet.*, 1992, **50**, 1086–1103.
34. E. M. Baden, B. A. Owen, F. C. Peterson, B. F. Volkman, M. Ramirez-Alvarado and J. R. Thompson, *J. Biol. Chem.*, 2008, **283**, 15853–15860.
35. A. R. Hurshman Babbes, E. T. Powers and J. W. Kelly, *Biochemistry*, 2008, **47**, 6969–6984.
36. K. K. O'Brien, B. K. Saxby, C. G. Ballard, J. Grace, F. Harrington, G. A. Ford, J. T. O'Brien, A. G. Swan, A. F. Fairbairn, K. Wesnes, T. del Ser, J. A. Edwardson, C. M. Morris and I. G. McKeith, *Pharmacogenetics*, 2003, **13**, 231–239.
37. O. L. Lopez, J. T. Becker, S. Wisniewski, J. Saxton, D. I. Kaufer and S. T. DeKosky, *J. Neurol. Neurosurg. Psychiatry*, 2002, **72**, 310–314.
38. E. Ghebremedhin, D. R. Thal, C. Schultz and H. Braak, *Neurosci. Lett.*, 2002, **320**, 25–28.
39. C. Holmes, C. Ballard, D. Lehmann, A. David Smith, H. Beaumont, I. N. Day, M. Nadeem Khan, S. Lovestone, M. McCulley, C. M. Morris, D. G. Munoz, K. O'Brien, C. Russ, T. Del Ser and D. Warden, *J. Neurol. Neurosurg. Psychiatry*, 2005, **76**, 640–643.
40. L. Bertram, M. B. McQueen, K. Mullin, D. Blacker and R. E. Tanzi, *Nature Genet.*, 2007, **39**, 17–23.
41. T. Arendt, M. K. Bruckner, M. Lange and V. Bigl, *Neurochem. Int.*, 1992, **21**, 381–396.
42. M. Fandrich, *Cell. Mol. Life Sci.*, 2007, **64**, 2066–2078.
43. F. Chiti and C. M. Dobson, *Annu. Rev. Biochem.*, 2006, **75**, 333–366.
44. M. G. Spillantini and M. Goedert, *Ann. N Y Acad. Sci.*, 2000, **920**, 16–27.
45. C. Soto and L. D. Estrada, *Arch. Neurol.*, 2008, **65**, 184–189.
46. D. H. Small and R. Cappai, *J. Neurochem.*, 2006, **99**, 708–710.
47. D. M. Mann, P. O. Yates and B. Marcyniuk, *J. Neurol. Sci.*, 1985, **69**, 139–159.
48. S. I. Finkel, *J. Clin. Psychiatry*, 2001, **62**(Suppl 21), 3–6.
49. T. E. Golde, S. Estus, L. H. Younkin, D. J. Selkoe and S. G. Younkin, *Science*, 1992, **255**, 728–730.
50. R. Deane and B. V. Zlokovic, *Curr. Alzheimer Res.*, 2007, **4**, 191–197.
51. N. N. Nalivaeva, L. R. Fisk, N. D. Belyaev and A. J. Turner, *Curr. Alzheimer Res.*, 2008, **5**, 212–224.
52. V. Geylis, V. Kourilov, Z. Meiner, I. Nennesmo, N. Bogdanovic and M. Steinitz, *Neurobiol. Aging*, 2005, **26**, 597–606.
53. R. D. Moir, K. A. Tseitlin, S. Soscia, B. T. Hyman, M. C. Irizarry and R. E. Tanzi, *J. Biol. Chem.*, 2005, **280**, 17458–17463.
54. R. C. Dodel, Y. Du, C. Depboylu, H. Hampel, L. Frolich, A. Haag, U. Hemmeter, S. Paulsen, S. J. Teipel, S. Brettschneider, A. Spottke, C. Nolker, H. J. Moller, X. Wei, M. Farlow, N. Sommer and W. H. Oertel, *J. Neurol. Neurosurg. Psychiatry*, 2004, **75**, 1472–1474.

55. Y. Du, X. Wei, R. Dodel, N. Sommer, H. Hampel, F. Gao, Z. Ma, L. Zhao, W. H. Oertel and M. Farlow, *Brain*, 2003, **126**, 1935–1939.
56. R. Liu, C. McAllister, Y. Lyubchenko and M. R. Sierks, *Biochemistry*, 2004, **43**, 9999–10007.
57. B. Vellas, S. Andrieu, C. Sampaio and G. Wilcock, *Lancet Neurol.*, 2007, **6**, 56–62.
58. M. O'Malley, *Lancet*, 1997, **349**, 1161–1166.
59. T. Thomsen, B. Kaden, J. P. Fischer, U. Bickel, H. Barz, G. Gusztony, J. Cervos-Navarro and H. Kewitz, *Eur. J. Clin. Chem. Clin. Biochem.*, 1991, **29**, 487–492.
60. M. S. Wolfe, *Curr. Top Med. Chem.*, 2008, **8**, 2–8.
61. K. F. Bell, L. Zheng, F. Fahrenholz and A. C. Cuello, *Neurobiol. Aging*, 2008, **29**, 554–565.
62. U. Schmitt, C. Hiemke, F. Fahrenholz and A. Schroeder, *Behav. Brain Res.*, 2006, **175**, 278–284.
63. A. Woodhouse, T. C. Dickson and J. C. Vickers, *Drugs Aging*, 2007, **24**, 107–119.
64. J. C. Dodart, K. R. Bales, K. S. Gannon, S. J. Greene, R. B. DeMattos, C. Mathis, C. A. DeLong, S. Wu, X. Wu, D. M. Holtzman and S. M. Paul, *Nature Neurosci.*, 2002, **5**, 452–457.
65. I. Ferrer, M. Boada Rovira, M. L. Sanchez Guerra, M. J. Rey and F. Costa-Jussa, *Brain Pathol.*, 2004, **14**, 11–20.
66. J. A. Nicoll, D. Wilkinson, C. Holmes, P. Steart, H. Markham and R. O. Weller, *Nature Med.*, 2003, **9**, 448–452.
67. F. C. Bronfman, J. Garrido, A. Alvarez, C. Morgan and N. C. Inestrosa, *Neurosci. Lett.*, 1996, **218**, 201–203.
68. G. S. Yu, J. Hu and H. Nakagawa, *Neurosci. Lett.*, 1998, **254**, 125–128.
69. Y. Kiuchi, Y. Isobe and K. Fukushima, *Neurosci. Lett.*, 2001, **305**, 119–122.
70. R. D. Bell, A. P. Sagare, A. E. Friedman, G. S. Bedi, D. M. Holtzman, R. Deane and B. V. Zlokovic, *J. Cereb. Blood Flow Metab.*, 2007, **27**, 909–918.
71. M. Necula, R. Kayed, S. Milton and C. G. Glabe, *J. Biol. Chem.*, 2007, **282**, 10311–10324.
72. H. A. Lashuel, D. M. Hartley, D. Balakhaneh, A. Aggarwal, S. Teichberg and D. J. Callaway, *J. Biol. Chem.*, 2002, **277**, 42881–42890.
73. P. Camilleri, N. J. Haskins and D. R. Howlett, *FEBS Lett.*, 1994, **341**, 256–258.
74. T. Tomiyama, S. Asano, Y. Suwa, T. Morita, K. Kataoka, H. Mori and N. Endo, *Biochem. Biophys. Res. Commun.*, 1994, **204**, 76–83.
75. S. J. Pollack, Sadler II, S. R. Hawtin, V. J. Tailor and M. S. Shearman, *Neurosci. Lett.*, 1995, **197**, 211–214.
76. A. R. Salomon, K. J. Marcinowski, R. P. Friedland and M. G. Zagorski, *Biochemistry*, 1996, **35**, 13568–13578.
77. D. Howlett, P. Cutler, S. Heales and P. Camilleri, *FEBS Lett.*, 1997, **417**, 249–251.

78. S. J. Wood, L. MacKenzie, B. Maleeff, M. R. Hurle and R. Wetzel, *J. Biol. Chem.*, 1996, **271**, 4086–4092.
79. D. R. Howlett, A. R. George, D. E. Owen, R. V. Ward and R. E. Markwell, *Biochem. J.*, 1999, **343**(Pt 2), 419–423.
80. K. L. Sciarretta, D. J. Gordon and S. C. Meredith, *Methods Enzymol.*, 2006, **413**, 273–312.
81. H. Okuno, K. Mori, T. Okada, Y. Yokoyama and H. Suzuki, *Chem. Biol. Drug Des.*, 2007, **69**, 356–361.
82. M. A. Chacon, M. I. Barria, C. Soto and N. C. Inestrosa, *Mol. Psychiatry*, 2004, **9**, 953–961.
83. K. L. Sciarretta, D. J. Gordon, A. T. Petkova, R. Tycko and S. C. Meredith, *Biochemistry*, 2005, **44**, 6003–6014.
84. H. Geerts, *Curr. Opin. Investig. Drugs*, 2004, **5**, 95–100.
85. Y. Loewenstein-Lichtenstein, M. Schwarz, D. Glick, B. Norgaard Pedersen, H. Zakut and H. Soreq, *Nature Med.*, 1995, **1**, 1082–1085.
86. J. B. Rand, *WormBook*, 2007, 1–21.
87. J. L. Sussman, M. Harel, F. Frolow, C. Oefner, A. Goldman, L. Toker and I. Silman, *Science*, 1991, **253**, 872–879.
88. S. T. George and A. S. Balasubramanian, *Eur. J. Biochem.*, 1980, **111**, 511–524.
89. J. Grunwald, D. Marcus, Y. Papier, L. Raveh, Z. Pittel and Y. Ashani, *J. Biochem. Biophys. Methods*, 1997, **34**, 123–135.
90. A. Silver, *Frontiers of Biology: The Biology of Cholinesterases,* North-Holland Publishing Company, Amsterdam, 1974.
91. K. Ichtchenko, Y. Hata, T. Nguyen, B. Ullrich, M. Missler, C. Moomaw and T. C. Sudhof, *Cell*, 1995, **81**, 435–443.
92. I. Tsigelny, I. N. Shindyalov, P. E. Bourne, T. C. Sudhof and P. Taylor, *Protein Sci.*, 2000, **9**, 180–185.
93. R. Beeri, C. Andres, E. Lev-Lehman, R. Timberg, T. Huberman, M. Shani and H. Soreq, *Curr. Biol.*, 1995, **5**, 1063–1071.
94. M. Sternfeld, J. D. Patrick and H. Soreq, *J. Physiol. Paris*, 1998, **92**, 249–255.
95. E. Meshorer, C. Erb, R. Gazit, L. Pavlovsky, D. Kaufer, A. Friedman, D. Glick, N. Ben-Arie and H. Soreq, *Science*, 2002, **295**, 508–512.
96. D. Kaufer, A. Friedman, S. Seidman and H. Soreq, *Nature*, 1998, **393**, 373–377.
97. S. E. Park, N. D. Kim and Y. H. Yoo, *Cancer Res.*, 2004, **64**, 2652–2655.
98. N. H. Greig, T. Utsuki, Q. Yu, X. Zhu, H. W. Holloway, T. Perry, B. Lee, D. K. Ingram and D. K. Lahiri, *Curr. Med. Res. Opin.*, 2001, **17**, 159–165.
99. A. Alvarez, R. Alarcon, C. Opazo, E. O. Campos, F. J. Munoz, F. H. Calderon, F. Dajas, M. K. Gentry, B. P. Doctor, F. G. De Mello and N. C. Inestrosa, *J. Neurosci.*, 1998, **18**, 3213–3223.
100. C. Opazo and N. C. Inestrosa, *Mol. Chem. Neuropathol.*, 1998, **33**, 39–49.
101. C. Geula and M. M. Mesulam, *Alzheimer Dis. Assoc. Disord.*, 1995, **9**(Suppl 2), 23–28.

102. T. G. Beach, D. G. Walker, P. E. Potter, L. I. Sue and A. Fisher, *Brain Res.*, 2001, **905**, 220–223.
103. A. Fisher, R. Brandeis, R. Haring, N. Bar-Ner, M. Kliger-Spatz, N. Natan, H. Sonego, I. Marcovitch and Z. Pittel, *J. Neural. Transm. Suppl.*, 2002, 189–202.
104. P. Etienne, Y. Robitaille, P. Wood, S. Gauthier, N. P. Nair and R. Quirion, *Neuroscience*, 1986, **19**, 1279–1291.
105. E. Giacobini, *Neurochem. Res.*, 2003, **28**, 515–522.
106. V. Haroutunian, N. Greig, X. F. Pei, T. Utsuki, R. Gluck, L. D. Acevedo, K. L. Davis and W. C. Wallace, *Brain Res. Mol. Brain Res.*, 1997, **46**, 161–168.
107. A. Aharoni, L. Gaidukov, O. Khersonsky, Q. G. S. Mc, C. Roodveldt and D. S. Tawfik, *Nature Genet.*, 2005, **37**, 73–76.
108. T. Darreh-Shori, O. Almkvist, Z. Z. Guan, A. Garlind, B. Strandberg, A. L. Svensson, H. Soreq, E. Hellstrom-Lindahl and A. Nordberg, *Neurology*, 2002, **59**, 563–572.
109. E. Perry, I. McKeith and C. Ballard, *Neurology*, 2003, **60**, 1852–1853.
110. D. J. Lehmann, C. Johnston and A. D. Smith, *Hum. Mol. Genet.*, 1997, **6**, 1933–1936.
111. D. J. Lehmann, J. Williams, J. McBroom and A. D. Smith, *Neuroscience*, 2001, **108**, 541–554.
112. K. M. Mattila, J. O. Rinne, M. Roytta, P. Laippala, T. Pietila, H. Kalimo, T. Koivula, H. Frey and T. Lehtimaki, *J. Med. Genet.*, 2000, **37**, 766–770.
113. S. P. McIlroy, V. L. Crawford, K. B. Dynan, B. M. McGleenon, M. D. Vahidassr, J. T. Lawson and A. P. Passmore, *J. Med. Genet.*, 2000, **37**, 182–185.
114. A. V. Raygani, M. Zahrai, A. Soltanzadeh, M. Doosti, E. Javadi and T. Pourmotabbed, *Neurosci. Lett.*, 2004, **371**, 142–146.
115. L. Tilley, K. Morgan, J. Grainger, P. Marsters, L. Morgan, J. Lowe, J. Xuereb, C. Wischik, C. Harrington and N. Kalsheker, *Eur. J. Hum. Genet.*, 1999, **7**, 659–663.
116. H. Wiebusch, J. Poirier, P. Sevigny and K. Schappert, *Hum. Genet.*, 1999, **104**, 158–163.
117. F. Crawford, D. Fallin, Z. Suo, L. Abdullah, M. Gold, A. Gauntlett, R. Duara and M. Mullan, *Neurosci. Lett.*, 1998, **249**, 115–118.
118. J. M. Grubber, A. M. Saunders, A. R. Crane-Gatherum, W. K. Scott, E. R. Martin, C. S. Haynes, P. M. Conneally, G. W. Small, A. D. Roses, J. L. Haines and M. A. Pericak-Vance, *Neurosci. Lett.*, 1999, **269**, 115–119.
119. M. Hiltunen, A. Mannermaa, S. Helisalmi, A. Koivisto, M. Lehtovirta, M. Ryynanen, P. Riekkinen Sr and H. Soininen, *Neurosci. Lett.*, 1998, **250**, 69–71.
120. P. G. Kehoe, H. Williams, P. Holmans, G. Wilcock, N. J. Cairns, J. Neal and M. J. Owen, *J. Med. Genet.*, 1998, **35**, 1034–1035.
121. C. S. Ki, D. L. Na, J. W. Kim, H. J. Kim, D. K. Kim and B. K. Yoon, *Am. J. Med. Genet.*, 1999, **88**, 113–115.

122. D. W. Lee, H. C. Liu, T. Y. Liu, C. W. Chi and C. J. Hong, *Am. J. Med. Genet.*, 2000, **96**, 167–169.

123. A. B. Singleton, G. Smith, A. M. Gibson, R. Woodward, R. H. Perry, P. G. Ince, J. A. Edwardson and C. M. Morris, *Hum. Mol. Genet.*, 1998, **7**, 937–939.

124. N. Sodeyama, M. Yamada, Y. Itoh, E. Otomo, N. Suematsu and M. Matsushita, *J. Neurol. Neurosurg. Psychiatry*, 1999, **67**, 693–694.

125. Y. Yamamoto, M. Yasuda, E. Mori and K. Maeda, *J. Neurol. Neurosurg. Psychiatry*, 1999, **67**, 94–96.

126. Y. J. Huang, Y. Huang, H. Baldassarre, B. Wang, A. Lazaris, M. Leduc, A. S. Bilodeau, A. Bellemare, M. Cote, P. Herskovits, M. Touati, C. Turcotte, L. Valeanu, N. Lemee, H. Wilgus, I. Begin, B. Bhatia, K. Rao, N. Neveu, E. Brochu, J. Pierson, D. K. Hockley, D. M. Cerasoli, D. E. Lenz, C. N. Karatzas and S. Langermann, *Proc. Natl. Acad. Sci. USA*, 2007, **104**, 13603–13608.

127. W. L. Klein, W. B. Stine Jr. and D. B. Teplow, *Neurobiol. Aging*, 2004, **25**, 569–580.

128. P. Y. Chou and G. D. Fasman, *Biochemistry*, 1974, **13**, 211–222.

129. J. A. Edwards and S. Brimijoin, *Biochim. Biophys. Acta*, 1983, **742**, 509–516.

130. H. Tidow, R. Melero, E. Mylonas, S. M. Freund, J. G. Grossmann, J. M. Carazo, D. I. Svergun, M. Valle and A. R. Fersht, *Proc. Natl. Acad. Sci. USA*, 2007, **104**, 12324–12329.

131. H. Heise, M. S. Celej, S. Becker, D. Riedel, A. Pelah, A. Kumar, T. M. Jovin and M. Baldus, *J. Mol. Biol.*, 2008, **380**, 444–450.

132. Y. Matsuoka, M. Saito, J. LaFrancois, M. Saito, K. Gaynor, V. Olm, L. Wang, E. Casey, Y. Lu, C. Shiratori, C. Lemere and K. Duff, *J. Neurosci.*, 2003, **23**, 29–33.

133. B. Solomon, *Curr. Opin. Investig. Drugs*, 2007, **8**, 519–524.

134. W. V. Nikolic, H. Hou, T. Town, Y. Zhu, B. Giunta, C. D. Sanberg, J. Zeng, D. Luo, J. Ehrhart, T. Mori, P. R. Sanberg and J. Tan, *Stem Cells Dev.*, 2008, **17**, 423–439.

135. D. Glick, L. Ben Moyal and H. Soreq, *Genetic variation in butyrylcholinesterase and the physiological consequences for acetylcholinesterase function,* Martin Dunitz, London, 2003.

136. E. Meshorer, I. E. Biton, Y. Ben-Shaul, S. Ben-Ari, Y. Assaf, H. Soreq and Y. Cohen, *Faseb. J.*, 2005, **19**, 910–922.

137. R. Deane, Z. Wu and B. V. Zlokovic, *Stroke*, 2004, **35**, 2628–2631.

138. C. Glabe, *J. Mol. Neurosci.*, 2001, **17**, 137–145.

139. A. Demuro, E. Mina, R. Kayed, S. C. Milton, I. Parker and C. G. Glabe, *J. Biol. Chem.*, 2005, **280**, 17294–17300.

140. P. Soto, M. A. Griffin and J. E. Shea, *Biophys. J.*, 2007.

CHAPTER 5

scyllo-Inositol: A Potential Therapeutic for Alzheimer's Disease

DANIELA FENILI,[1,2] KERAN MA[1,3] AND JOANNE MCLAURIN[1,2]

[1] Department of Laboratory Medicine and Pathobiology; [2] Centre for Research in Neurodegenerative Diseases; [3] Department of Physiology, University of Toronto, University of Toronto, 6 Queen's Park Crescent West, Toronto, Ontario, Canada, M5S 3H2

5.1   Introduction

Although neurodegenerative diseases such as, Alzheimer's, Parkinson's, polyglutamine diseases and amyotrophic lateral sclerosis involve neuronal vulnerability in specific brain regions, they have in common aggregation of peptides/proteins leading to amyloid and inclusion body formation.[1–3] Furthermore, these diseases share features both at the protein structure level and mechanistically at the cellular level.[1–3] For Alzheimer's disease, amyloid is formed by the Aβ peptide, while paired helical filaments (PHFs) are formed by hyper-phosphorylated tau. Both affect predominantly the hippocampus and entorhinal cortex. Aβ and tau are vastly different in molecular weight and sequence yet both undergo a structural change under pathological conditions, in which the final protein structure is predominantly β-sheet. This structural change is key to the initiation of aggregation (Figure 5.1).[3] The self-assembly of proteins

RSC Drug Discovery Series No. 2
Emerging Drugs and Targets for Alzheimer's Disease
Volume 1: Beta-Amyloid, Tau Protein and Glucose Metabolism
Edited by Ana Martinez

Published by the Royal Society of Chemistry, www.rsc.org

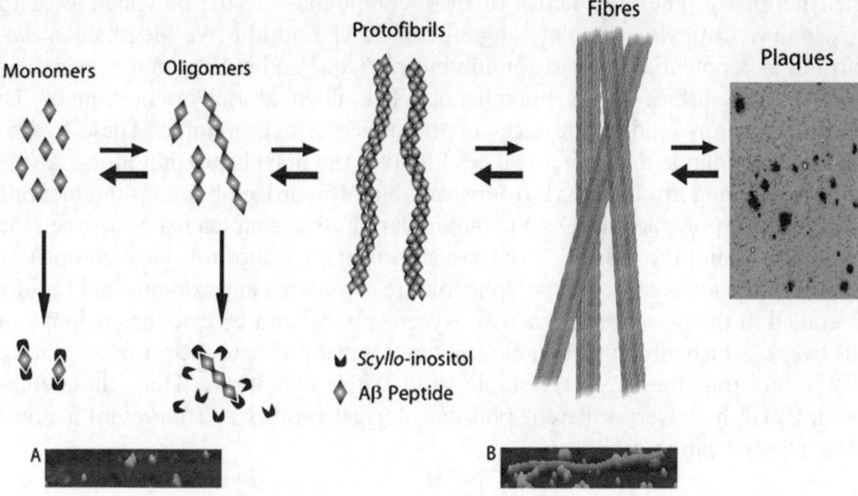

Figure 5.1 Aβ aggregation pathway. Aβ aggregation begins as Aβ monomers bind to each other to form dimers, trimers and larger oligomers. These aggregates then elongate into longer and wider amyloid protofibril. Further lateral aggregation creates Aβ fibres, which deposit into plaques. *scyllo*-Inositol binds Aβ monomers and oligomers to inactivate Aβ-induced effects. Forces of equilibrium then drive larger aggregates to dissociate into smaller Aβ species. Atomic force microscopy images in A and B show *in vitro* aggregation of Aβ in the presence and absence of *scyllo*-inositol, respectively.

or peptides into oligomers precedes the formation of amyloid fibres and PHFs. There are several kinds of aggregates, including disordered amorphous aggregates, but all contain β-structure at least in the core of the protein and combine to form fibres and filaments.[3] The tinctorial properties of this diverse array of aggregates are common, *i.e.* binding Congo red and thioflavin. Congo red intercalates into β-sheets and thus labels all amyloid structures.[4] Furthermore, data from *in vitro* studies point to common initiation factors that might 'seed' misfolding and aggregation. Although aggregation can be self-driven by increasing concentrations, covalent modification such as oxidation as seen in Aβ, phosphorylation seen for tau or interaction with exogenous 'seeds', such as fatty acids and acidic lipids, are a common thread in amyloid and PHF formation. Recent data has demonstrated that it is the oligomers that are neurotoxic. In light of the data, the inhibition of oligomer formation or stabilisation of monomeric or oligomeric peptides/proteins would be one common strategy to treat pathologies associated with AD (Figure 5.1).

Literature reports of screening libraries of over 200 000 small molecules have revealed the common physical characteristics necessary to inhibit tau and Aβ aggregation.[5,6] Interestingly, similar compounds have been identified as tau and Aβ aggregation inhibitors in separate library screens.[5,7] The active compounds for Aβ and tau are planar aromatic compounds, which display polar heteroatoms at their periphery. The most active of these compounds are the polyphenols, with IC_{50} values in the low μmolar range for both Aβ and tau. We identified *scyllo*-inositol as a potential aggregation inhibitor of Aβ,[8] and it also exhibits a similar amphipathic surface to the polyphenols. The all-equatorial orientation of the hydroxyl-groups renders the faces of the molecule hydrophobic. These hydrophobic faces mimic the aromatic faces found in the polyphenol inhibitors. At the edge of the ring are hydroxyl groups capable of hydrogen-bonding interactions similar to the polyphenols. Despite the similar physical characteristics between the polyphenols and the inositols, we have shown that the inositols have completely different stability *in vivo*. The polyphenols are considered antioxidants, are rapidly degraded in the presence of reactive oxygen species[9] and by cytochrome p450 in the liver,[10] which inhibit further clinical development. Due to the saturated nature of the inositols, they are very stable to oxidative conditions. Thus, these compounds will have very different pharmacological profiles and represent a novel class of aggregation inhibitors.

5.2   The Inositol Stereoisomers

Inositol was first isolated in the mid-19th century from muscle extracts and was accordingly named inosit, from the Greek root word *inos* for muscle.[11] Eight years later, a second, related compound was isolated from the shark *Scyllium canicula*, the skates *Raja batis* and *Raja clavata*, and was given the name Scyllit.[12] These compounds would later be renamed *myo*-inositol and *scyllo*-inositol and identified as two of the nine possible stereoisomers of inositol.[13]

The inositol isomers belong to the class of compounds known as cyclitols, 5–7 carbon ring compounds, containing at least 3 carbons with an attached hydroxyl group. Inositols are cyclohexanehexols and members of the polyol family. The inositol isomers are 6-carbon-ring molecules, with a hydroxyl group attached to each carbon of the ring. The structure of *scyllo*-inositol was first determined by Posternakin in 1941,[14,15] followed by the structure for *myo*-inositol a year later.[16] The 9 stereoisomers differ from each other based on the spatial orientation of their hydroxyl groups along axial or equatorial planes (Figure 5.2). Of the nine, seven are optically inactive, whereas D- and L-*chiro*-inositol are enantiomers. While nine stereoisomers are possible, only six are found in nature (*myo*-, *scyllo*-, D-*chiro*, *epi*-, *muco*-, and *neo*-).

5.3   *scyllo*-Inositol in Nature

The biosynthesis of inositol is an evolutionarily conserved pathway in nature, found across phylogenic kingdoms, including plants, animals, parasites, bacteria and archaea.[17] The majority of inositol present in humans is in the form of *myo*-inositol, with the next most abundant isomer, *scyllo*-inositol, present at 5–12% the concentration of *myo*-inositol.[18] *scyllo*-Inositol has been documented in a number of other organisms, including potential dietary sources.[19–21] In the skate (*Raja erinacea*), *scyllo*-inositol was recorded at higher concentrations

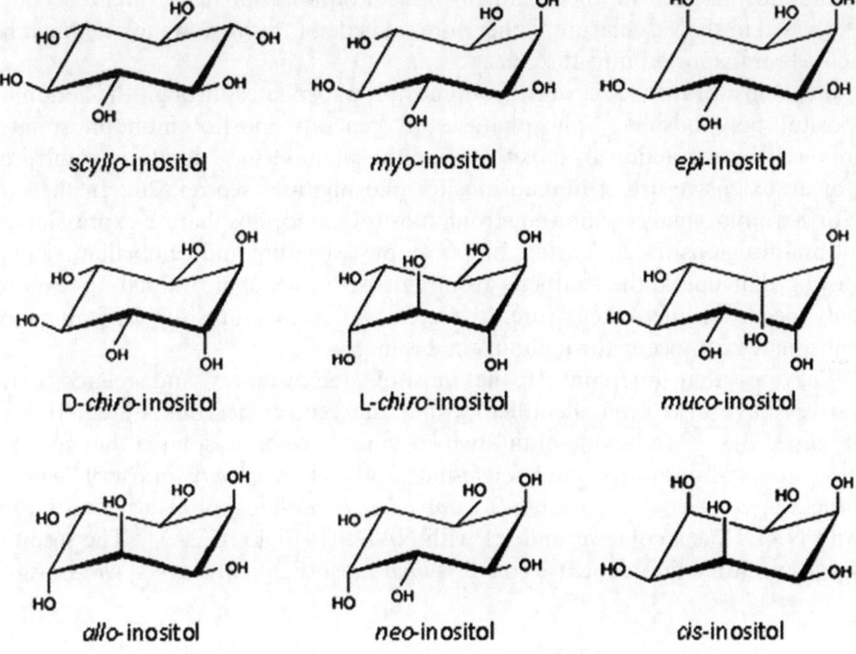

Figure 5.2 The inositol stereoisomers. There are 9 possible stereoisomers for inositol, which differ from each other based on the spatial orientation of their hydroxyl groups along axial or equatorial planes.

than *myo*-inositol, with the highest levels reported in the kidney.[19] In fruit, *scyllo*-inositol has been found in trace amounts in grapes and citrus fruits, with the highest levels reported in grapefruit.[20] *scyllo*-Inositol has also been reported in the *Apiaceae* family of vegetables, which includes carrots, fennel, parsley and coriander.[21] The parts per million levels of *scyllo*-inositol in dietary sources preclude the use of dietary intervention alone to alter *scyllo*-inositol levels within the central nervous system. In humans, it is unclear what function *scyllo*-inositol serves, however, the presence of enzymes for the interconversion between *scyllo*-inositol and *myo*-inositol would suggest that *scyllo*-inositol does have a function within the cell.

5.4 Inositol Synthesis and Degradation Pathways

Inositol levels within the human body are maintained through a combination of diet, synthesis from D-glucose-6-phosphate, recycling from inositol phosphates and through the interconversion between inositol derivatives.[22] The average human ingests 1 g/day of inositol in the diet, predominantly as *myo*-inositol and phytic acid[23] and can synthesise up to 4 g/day in the kidneys.[24] *myo*-Inositol is synthesised from D-glucose-6-phosphate, through the activity of two enzymes, *myo*-inositol-3-phosphate synthase and inositol monophosphatase. In the brain, *myo*-inositol-3-phosphate synthase is only expressed in the vasculature,[25] therefore, outside of this region inositol must be actively transported into the cells.

myo-Inositol can be recycled from inositol phosphates through the action of inositol polyphosphate phosphatases to generate inositol monophosphate, followed by the action of inositol monophosphatase to generate *myo*-inositol (for an extensive list of human inositol phosphatases see ref. 26). In the rat, Northern blot analysis showed strong inositol monophosphatase expression in all brain regions tested: cortex, hippocampus, striatum and cerebellum,[27] suggesting that unlike the synthesis from glucose-6-phosphate, which appears to only occur in the vasculature, the creation of *myo*-inositol from inositol phosphates can occur throughout the brain.

Enzymes that interconvert the inositol stereoisomers and related derivatives have also been identified within the central nervous system (CNS) (Figure 5.3).[22,28] In bovine brain, two enzymes have been isolated that convert *myo*-inosose-2 into *myo*- and *scyllo*-inositol.[22] The first, *myo*-inositol oxidoreductase, converts *myo*-inosose-2 to *myo*- or *scyllo*-inositol at a ratio of 10:1 with NADH as a cofactor and 1:1 with NADPH (Figure 5.3A).[22] The second, *scyllo*-inositol oxidoreductase converts *myo*-inosose-2 to *myo*- or *scyllo*-inositol

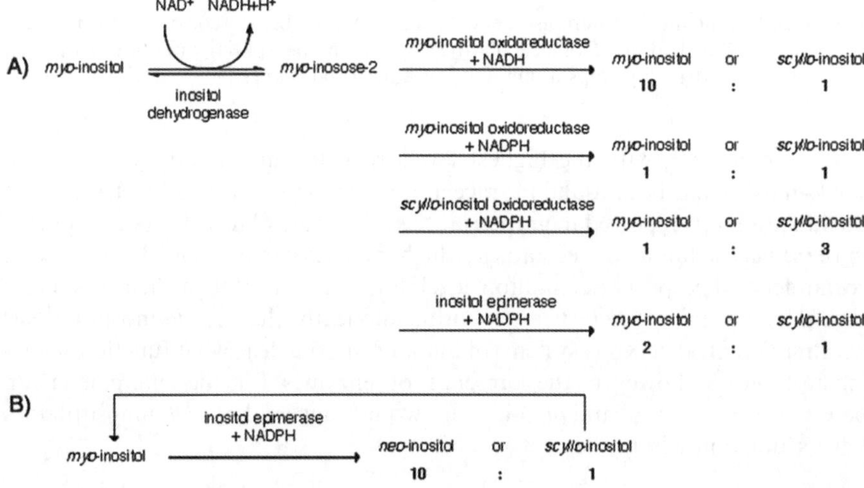

Figure 5.3 Endogenous sources of *scyllo*-Inositol. (A) A number of enzymes have been isolated from bovine brain that convert *myo*-inositol to *scyllo*-inositol, via *myo*-inosose-2.[22,28] The first, *myo*-inositol oxidoreductase, converts *myo*-inosose-2 to *myo*- or *scyllo*-inositol, using either NADH or NADPH as a cofactor. In contrast, *scyllo*-inositol oxidoreductase and inositol *epi*merase only function in the presence of NADPH to convert *myo*-inosose-2 to *scyllo*-inositol. (B) In addition, another NADPH-dependent epimerase has been isolated that converts *myo*-inositol to *neo*-inositol and *scyllo*-inositol and can convert *scyllo*-inositol back into *myo*-inositol.

at a ratio of 1:3 with NADPH and showed no activity with NADH.[22] In addition, a NADP$^+$-dependent inositol epimerase, which was isolated from bovine brain, was also found to convert *myo*-inosose-2 to *myo*- or *scyllo*-inositol, at a ratio of 2:1, in the presence of NADP$^+$.[28] *scyllo*-Inositol and *myo*-inositol are also interconverted by the NADP$^+$ dependent inositol epimerase.[28] This epimerase converts *myo*-inositol to either *scyllo*- or *neo*-inositol, at a ratio of 1:10, and converts *scyllo*-inositol back to *myo*-inositol (Figure 5.3B).

Removal of inositol from the body occurs by direct excretion in the kidney or by degradation predominantly in the kidney and the liver. *myo*-Inositol oxygenase, which is the enzyme required for the first step in *myo*-inositol catabolism, is mainly expressed in the proximal tubular epithelial cells of the kidney cortex.[29] Western blot analysis of mouse tissue, showed no detectable levels of *myo*-inositol oxygenase in the brain, heart, lung, liver, spleen, intestines or muscles.[29] Trace amounts of the mRNA were detected in the sciatic nerve, the liver and the heart.[29] This enzyme cleaves the 6-carbon ring of inositol to form D-glucuronate, which can be metabolised in the liver. These combined studies suggest that inositol is actively transported throughout the body for both function and degradation. Three inositol transporters have been identified and characterised in mammals, one proton-*myo*-inositol transporter (HMIT) and two sodium-*myo*-inositol transporters (SMIT1, SMIT2).

5.5 The Inositol Transporters

5.5.1 H(+)-*myo*-Inositol Transporter

HMIT, alternatively known as solute carrier family 2 (facilitated glucose transporter), member 13, is a member of the major facilitator superfamily, a group of secondary transporters that includes uniporters, symporters and antiporters. These transporters are characterised by 12 transmembrane domains and N- and C-terminal tails located on the cytoplasmic side of the cellular membrane[30] (reviewed in ref. 31). HMIT is predominantly expressed in the brain, with limited expression also found in adipose tissue and in kidney.[32] In the brain, mRNA expression was observed in both neurons and glia with high levels in all brain regions examined: cerebral cortex, hippocampus, hypothalamus, cerebellum and brainstem.[32] Rat HMIT transports *myo*-inositol preferentially followed by *scyllo*-inositol > *chiro*-inositol > *muco*-inositol at a 1:1 ratio with H$^+$.[32] *allo*-Inositol was not transported by HMIT and no other inositols were examined. Despite being labelled a facilitated glucose transporter, rat HMIT did not transport D- or L-glucose or other related hexoses (galactose, fructose, mannose, 2-deoxy-glucose, glucosamine or maltose). Homology between rat and human HMIT is 90%, which suggests that similar transport activities are present.

HMIT is typically internalised in the cell. Cell surface translocation and regulation of expression are stimulated by membrane depolarisation, changes in protein kinase C, internal calcium concentrations and acidification of the extracellular environment,[32,33] all of which occur with synaptic activity.[34]

In light of this, the primary role of HMIT may be to adjust the intracellular inositol pools involved in cellular signalling pathways or phosphatidylinositol synthesis, which are required in areas with high rates of signalling and endo/exocytosis.[33]

5.5.2 Na(+)-*myo*-Inositol Transporter 1

SMIT1, also known as solute carrier family 5 (sodium/glucose cotransporter), member 3, is a member of the sodium/solute symporter family,[35] which are characterised by 13–14 transmembrane domains, with an N-terminal domain located in the extracellular space and a C-terminal domain, either located in the cytoplasm in members with 13 transmembrane domains, or encased in the membrane as the 14th transmembrane domain, as is the case for SMIT1.[35] The human, mouse, rat, canine and bovine SMIT1 amino acid sequences are more than 92% homologous.[36–39] In humans, SMIT1 mRNA expression was found in the kidney, brain, placenta, pancreas, heart, skeletal muscle and the lung, but not in the liver,[37] although its expression has been reported in the HepG2 human liver cell line.[40] In brain, SMIT1 mRNA expression was highest in the choroid plexus.[41] High SMIT1 mRNA expression was also observed in the hippocampus, the locus coeruleus, the suprachiasmatic nucleus, the olfactory bulb and the Purkinje and granule cell layers of the cerebellum.[41] Across the brain, SMIT1 was expressed in almost all neurons and small glia-like cells.[41] In the hippocampus, SMIT1 mRNA was localised to pyramidal cells in areas CA1 to CA3 and to granule cells in the dentate gyrus.[41] In the choroid plexus, SMIT1 has been specifically localised to the basolateral side of cells, indicating that it is responsible for transporting inositol from the blood into cells.[42]

SMIT1 appears to be the main transporter responsible for inositol transport into cells; SMIT1$^{-/-}$ mice, show a 92% reduction in inositol levels in the brain and an 84% reduction in inositol levels in the periphery.[43] Heterozygous mice show a 15% reduction in inositol levels in the frontal cortex and a 25% decrease in the hippocampus,[44] which would suggest a dependence on SMIT1 transport in these areas or that the demand for inositol is higher than elsewhere. In support of the latter, when rat astrocytes were cultured from the cortex, hippocampus, cerebellum, diencephalon or tegmentum were compared, the K_m and V_{max} for uptake in the cortex and hippocampus were higher than those for the other three regions.[38]

SMIT1 is unique among the inositol transporters because it shows an equal affinity for *myo*- and *scyllo*-inositol, as demonstrated in Xenopus oocytes transfected with the canine SMIT1 gene.[45] Overall, the sugar selectivity of the transporter was *myo*-inositol = *scyllo*-inositol > L-fucose > L-xylose > L-glucose = D-glucose = α-methyl-D-glucopyranoside > D-galactose = D-fucose = 3-O-methyl-D-glucose = 2-deoxy-D-glucose > D-xylose.[45] The ability of *scyllo*-inositol to strongly compete out *myo*-inositol transport was confirmed in murine neuroblastoma, murine cerebral microvessel endothelial and bovine aortic endothelial cell-lines.[46]

Inositol transport through SMIT1 is dependent on the cotransport of sodium with a stochiometry ratio of two Na^+ ions per molecule of *myo*-inositol.[45] Due to this dependence on Na^+, SMIT1 transport was inhibited by the sodium-dependent transport inhibitor, phlorizin.[45] The two Na^+ ions bind to the transporter first followed by *myo*-inositol, this sequential binding may explain the sodium leak currents observed in the absence of substrate.[45]

Cells react to changes in osmolality by adjusting the levels of compatible organic osmolytes in the cell. In cultured rat cortical astrocytes, *myo*-inositol accounts for 56–100% of the osmolyte(s) utilised for adaptation to hypertonicity.[47] It has been shown that the accumulation of *myo*-inositol in the cell, as a response to hypertonicity, requires the presence of *myo*-inositol in the extracellular space, indicating that this accumulation is not due to synthesis from glucose-6-phosphate or conversion from other inositol derivatives or phosphates, but from the transportation of inositol into the cell.[48] SMIT1 activity is upregulated by hypertonicity[38,46,48–60] and downregulated by hypotonicity.[46]

SMIT1 mRNA levels and activity can also be regulated by TNF-α,[58,59] possibly via NFκB, protein kinase C and ceramide activation.[59] Treatment of bovine aorta, pulmonary endothelial or cerebral microvessel endothelial cells with TNF-α caused a significant decrease in SMIT1 mRNA levels and activity.[59] The effects of TNF-α on SMIT1 expression and activity required RNA synthesis and were inhibited by actinomycin D.[58] TNF-α decreased SMIT1 V_{max} without altering the K_m, suggesting a reduction in SMIT1 transporters at the cell surface. In contrast, IGF-1, platelet-derived growth factor, TGF-β, IL-1α, IL-1β, IL-2 or IL-6 did not affect SMIT1 activity.[58]

As shown for HMIT, SMIT1 activity is also regulated by changes in pH.[60,61] However, unlike HMIT, SMIT1 is more active at physiological pH and inhibited when pH is reduced. This is the opposite pattern to that observed for HMIT[32] and is thought to be due to a reduction in the transporter's affinity for sodium.[61] SMIT1 activity was also reduced by depolarising concentrations of potassium,[62] again in direct opposition to HMIT regulation.[33] Therefore, it would appear that the body has adapted multiple methods for regulating inositol concentrations at the cellular level.

5.5.3   Na(+)-*myo*-Inositol Transporter 2

The last known member of the sodium/solute symporter family that transports inositol is SMIT2, also known as solute carrier family 5 (sodium/glucose cotransporter), member 11. When comparing the human amino acid sequences, SMIT2 is 43% homologous to SMIT1[63] and like SMIT1, the SMIT2 amino acid sequence encodes for 14 transmembrane domains.[64] In humans, SMIT2 expression is highest in the kidney, liver, heart, skeletal muscles and placenta, with much lower expression observed in the brain and minimal levels detected in the spleen, small intestine, lungs and lymphocytes.[64] Interestingly, in polarised cells, such as Madin–Darby canine kidney cells, SMIT1 is preferentially expressed at the basolateral membrane and SMIT2 is preferentially expressed at the apical membrane.[65,66] These studies suggest that SMIT1/2 might work in concert to regulate inositol within tissues.

In transport studies, SMIT2 displays an equal affinity for *myo*-inositol and its stereoisomer, D-*chiro*-inositol.[63,67,68] SMIT2 in HepG2 human liver cells showed a transport preference for D-*chiro*-inositol and D-glucose over their L-stereoisomers[69] but did not transport α-methylglucose and L-fucose.[63,70] This is in contrast to SMIT1, which shows an equal preference for both D- and L-glucose[63] and transports α–methylglucose and L-fucose.[63,70] The remaining inositol stereoisomers have not been examined for transport by SMIT2.

As observed for SMIT1, inositol transport by SMIT2 is dependent on the cotransport of two sodium ions per substrate molecule.[63,71] As was observed with the other members of the sodium/substrate symporter family, SMIT2 displays a sodium leak current in the absence of substrate,[63] and SMIT2 activity is upregulated by hyperosmotic conditions.[65,66] This upregulation is the result of an increase in SMIT2 transcription and translation, since upregulation was inhibited by actinomycin B and cycloheximide treatment.[66] SMIT2 activity can also be regulated by the mitogen-activated protein kinases p38 and c-Jun amino-terminal kinase, but not the extracellular signal-regulated kinase (ERK).[66] Inhibiting p38 and c-Jun amino-terminal kinase resulted in a 40% and 32% reduction in SMIT2 activity, respectively.[66] In addition, SMIT2 activity can be upregulated by insulin.[68] Insulin treatment for 24 h caused an 18-fold increase in D-*chiro*-inositol-^{3}H transport in rat L6 skeletal muscle cells transiently transfected with human SMIT2[68] and insulin treatment in diabetes has been linked to increased D-*chiro*-inositol levels (reviewed in ref. 72). In contrast to HMIT and SMIT1, SMIT2 does not appear to be sensitive to changes in pH.[61]

The expression and activity of the inositol transporters both in the periphery and central nervous system place them in juxtiposition for the uptake of inositol as a therapeutic agent. With regard to Alzheimer's disease, the high expression and activity of the transporters in the cortex and hippocampus further suggest that compounds utilising these transporters will be targeted to the appropriate areas of the brain.

5.6 Inositol Efflux

Inositol efflux from cells is not a well-understood process although fast and slow efflux currents have been detected. Efflux of inositol may involve both volume-sensitive and volume-insensitive components,[73–75] which are both active following hypertonicity-induced cell swelling, though perhaps not under basal conditions. The fast component appears to be mediated by volume-sensitive organic osmolyte-anion channels.[74–78] Both fast and slow *myo*-inositol efflux currents were observed in bovine lenses exposed to hypertonic solution.[73] The fast efflux current was the result of the activation of a common anionic (chloride) channel. These two efflux currents were observed in primary astrocyte cultures but only the fast component was inhibitable by anion membrane transport inhibitors.[75] In human teratocarcinoma-derived Ntera2/D1 neuron-like cells, inositol efflux was blocked by a Cl^- channel blocker, supporting the hypothesis that the volume-sensitive organic osmolyte-anion channel is a common chloride channel.[77]

5.7 Inositol Pools

Experimental data suggest that inositol in the cell exists as two or three separate pools.[79,80] This was first suggested by Diringer and Rott,[79] who examined the effects of inositol metabolism in chick embryo cells and discovered differentially regulated inositol pools, one smaller pool and one or more larger pools. Only the smaller pool appeared to regulate the synthesis of phosphatidylinositol.[79] Behavioral evidence has also suggested the presence of inositol compartmentalisation.[81] Lithium-pilocarpine administration to rats leads to seizures, through the combined actions of lithium, which prevents inositol resynthesis and pilocarpine, which increases the use of inositol for second messenger systems by stimulating cholinergic activity.[82] These effects could be prevented by intracerebroventricular administration of *myo*-inositol[82] but not by osmotically induced increases in inositol levels,[81] suggesting that the inositol used to sustain phosphatidylinositol levels must be different from that used to maintain osmotic function, strengthening the argument of different inositol pools.[81]

Separate inositol pools have also been suggested by inositol uptake and efflux kinetics.[83,84] Uptake of *myo*-inositol-^{3}H into the soluble fraction of cells showed no evidence of saturation, however, incorporation into the lipid fraction had a K_m of 0.28 mmol/L and was inhibited by phlorizin, a SMIT1/2 inhibitor. It was proposed that intracellular inositol exists as one larger, metabolically inert, pool and one smaller pool that equilibrates with external inositol levels and is utilised for the synthesis of phosphoinositides.[83] In astrocytes exposed to *myo*-inositol-^{3}H, three inositol pools were identified; the largest showed slow efflux kinetics, while the two smaller pools had faster efflux kinetics.[84] Unlike the findings in hepatocytes, the largest pool was membrane-associated and influenced the phosphatidylinositide second messenger system, while the two smaller pools were located in the cytosol and were thought to be involved in osmotic regulation.[84] These results suggest that it may be possible to alter certain pools of inositol without altering signal-transduction pathways.

5.8 Inositol Incorporation into Phosphatidylinositol

One potential concern with using *scyllo*-inositol as a therapeutic is that elevating levels might alter normal function or alter the levels of phosphatidylinositols and other inositol-containing compounds within the cell. In lower organisms such as mycobacteria, tetrahymena cells and barley seeds, *scyllo*-inositol containing phosphatidylinositols, phosphatidylinositol-linked glycans and polyphosphoinositols have been observed.[85–88] However, *scyllo*-inositol incorporation into phosphatidylinositols has not been observed in higher organisms.[89,90] This may be due to the reduced binding affinity of phosphatidylinositol synthesising enzymes for *scyllo*-inositol, even when levels are elevated.[86,89] The final step in the *de novo* synthesis of phosphatidylinositol is catalysed by CDP-diglyceride:inositol transferase, which has been found to bind *myo*-inositol with a K_m of 2.5 mM, a binding for which *scyllo*-inositol does not appear to compete.[89] When phosphatidylinositol:inositol phosphatidyl transferase activity was examined in rat liver microsomes, *scyllo*-inositol inhibition of *myo*-inositol-^{3}H incorporation into phosphatidylinositol was two orders of magnitude lower than that of *myo*-inositol and had no inhibitory effect if given at micromolar concentrations, as would be found in most tissues.[91] A lack of incorporation of *scyllo*-inositol into phosphatidylinositols was also observed in mice, both under basal conditions and in animals with elevated brain levels of *scyllo*-inositol.[90] These combined results demonstrate that *scyllo*-inositol will not directly affect PI pathways via replacement of *myo*-inositol. *scyllo*-Inositol-containing sialyloligosaccharides have, however, been isolated from human urine, indicating that *scyllo*-inositol can incorporate into some glycosides, at least in the periphery.[92] The ratio of *scyllo*-inositol-containing glycosides to *myo*-inositol glycosides was 2:1–4:1, which would suggest a preference of β-galactosidase or galactosyltransferase for *scyllo*-inositol.[92] Overall, *scyllo*-inositol displays limited incorporation into inositol-containing compounds in humans.

5.9   Inositol in Health and Disease

While *scyllo*-inositol's role in the body is unknown, its levels within the CNS have been linked to *myo*-inositol concentrations and thus have been speculated to be either an internal pool for *myo*-inositol synthesis or a product of *myo*-inositol metabolism. Proton NMR has identified *scyllo*-inositol within the CNS of human brain as a function of aging and disease state.[93–95] Evidence for the tight link between endogenous *scyllo*- and *myo*-inositol is supported by the simultaneous fluctuation of *scyllo*- with *myo*-inositol in patients with brain pathologies.[94] An increase in inositol levels with maintenance of the correlation between *scyllo*- and *myo*-inositol has been reported as a function of age within the white matter of healthy human subjects.[94] In contrast, no changes were detected in either *myo*- or *scyllo*-inositol within the posterior cingulate gyrus of healthy subjects as a function of age.[95] This report further suggests that *scyllo*-inositol/creatine ratios are increased in patients with AD and that these elevations are positively correlated with *myo*-inositol/creatine ratios.[95] These studies support the hypothesis that *scyllo*-inositol is directly linked to *myo*-inositol within the CNS and suggest that *myo*-inositol:*scyllo*-inositol ratios are maintained at a 12:1 ratio. However, the identification of high *scyllo*-inositol concentrations, 5:1 ratio to *myo*-inositol, in the brain of a healthy human subject with normal neurological function, suggest that it may be possible to increase *scyllo*-inositol concentrations while maintaining normal *myo*-inositol levels.[18] In contrast to previous studies, this data further suggests the independent metabolism of *scyllo*-inositol.

Inositol drug therapy has been extensive over the last 20 years. *myo*-Inositol has been used to control many psychiatric disorders without deleterious effects to hematology, kidney, liver or heart function.[96] These disorders include depression, panic disorder, obsessive-compulsive disorder and eating disorders, which suggests that inositol has therapeutic benefits for the spectrum of illnesses responsive to serotonin-selective re-uptake inhibitors.[82,96–99] *myo*-Inositol supplementation is also effective for treating respiratory distress syndrome in premature infants, the infants show increased survival, a decreased incidence of retinopathy and a lack of bronchopulmonary dysplasia.[100,101] *epi*-Inositol has been shown to be even more effective than *myo*-inositol at treating anxiety in rats.[102,103] D-*chiro*-Inositol is more effective than *myo*-inositol in preventing folate-resistant mouse neural-tube defects.[104] Decreased *chiro*-inositol has been speculated to play a role in insulin resistance in Type-II diabetes and polycystic ovary syndrome[40,105,106] and its supplementation has been shown to improve ovarian function and metabolic factors in women with polycystic ovary syndrome.[107–109] *epi*-Inositol on the other hand has been used to treat depression and is effective at reversing lithium effects on cytidine monophosphorylphosphatidate.[110] To date, our studies are the first to examine the use of *scyllo*-inositol as a therapeutic agent.[8,111,112]

A double blind, crossover placebo-controlled clinical trial of *myo*-inositol treatment for Alzheimer's disease (AD) has previously been reported.[113] In this study 11 AD patients ranging from mild to severe AD, as determined by the cognitive section of the CAMCOG test, were administered 6 g of *myo*-inositol or placebo (dextrose) per day for 4 weeks and switched for an additional 4 weeks. These results demonstrated that *myo*-inositol does not improve AD as measured by CAMDEX. The lack of effect may be the result of many factors, however, our preclinical studies demonstrated that high-dose *myo*-inositol treatment for 4 weeks in the TgCRND8 mouse model of AD did not alter cerebrospinal fluid or brain *myo*-inositol levels.[90] These results suggest that due to the extremely important role of *myo*-inositol within the CNS, its concentration is tightly regulated.

5.10 *scyllo*-Inositol as a Therapeutic for Alzheimer's Disease

Multiple lines of evidence suggest that one of the central events in the pathogenesis of AD is the accumulation of neurotoxic oligomeric/protofibrillar aggregates of Aβ (Figure 5.1).[114,115] The discovery of *scyllo*-inositol as a possible treatment for AD started with the investigation of Aβ-lipid interactions, as a mechanism for promotion of fibril formation and as a possible mechanism for Aβ-mediated toxicity.[111,116–121] Acidic phospholipids were shown to induce a β-structural transition in Aβ40 and Aβ42, with a concomitant disruption of the bilayer.[116,117] The most potent phospholipid was phosphatidylinositol (PI), therefore components, *i.e.* the headgroup, phosphorylation state or fatty acyl chains of PI were examined to determine the crucial element(s) for β-structure induction.[111] *myo*-Inositol, the headgroup of PI, mediated a transition to β-structure of Aβ42, while the addition of a single phosphate group abolished this transition.[111] This suggested that *myo*-inositol was responsible for inducing Aβ42 β-structural transition. The effect of *myo*-inositol on Aβ42 conformational change was an immediate, not time-dependent, effect. Interestingly, even though *myo*-inositol induced a β-structure formation, it did not induce fibrillisation but maintained Aβ42 in a soluble form.[111] Negative-stain electron microscopy showed Aβ42 in buffer alone formed short thin fibrils, whereas no fibrils were observed in the presence of *myo*-inositol.[111] Aβ42–*myo*-inositol interactions suggested that even though β-structure is required for fibril formation, Aβ is also able to form stable nonfibrillar β-structures.[111]

Since *myo*-inositol induced stable micelles of Aβ42, the effects of other inositol stereoisomers on Aβ42 aggregation were similarly investigated.[8] *epi*-Inositol, *scyllo*-inositol and *chiro*-inositol were examined as potential inhibitors of fibrillogenesis. These stereoisomers differ in the position of their hydroxyl groups; *myo*-inositol has five equatorial hydroxyl groups and one axial hydroxyl group. Compared to *myo*-inositol, *epi*-inositol and D-*chiro*-inositol have one or two extra hydroxyls in the axial position, respectively. *scyllo*-Inositol, on the other hand, has all its hydroxyl groups in the equatorial position. *epi*-Inositol and *scyllo*-inositol but not *chiro*-inositol induced a random to β-structure transition in Aβ42, without formation of fibrils.[8] Aβ42 incubated with *chiro*-inositol formed fibrils indistinguishable from Aβ42 in buffer alone, displaying an inactive isomer.[8]

Stabilisation of Aβ42-inositol conformers may have increased Aβ-mediated neuronal toxicity, yet it protected PC-12 and primary neuronal cultures from death.[8] Normally, Aβ42 accumulates on the cell surface of PC-12 cells. Aβ42 in the presence of *myo*-, *scyllo*-, and *epi*-inositol resulted in decreased cell surface Aβ accumulation, while *chiro*-inositol showed no change in the amount of Aβ accumulation.[8] The ability of *myo*-, *scyllo*-, and *epi*-inositols to decrease Aβ accumulation on neuronal membranes offers a possible mechanism to the attenuation of Aβ-induced neurotoxicity.

In vitro studies showed that inositol stereoisomers stabilise Aβ conformers, inhibit Aβ fibril assembly, accelerate disassembly of preformed fibrils, and protect primary cultured neurons from Aβ-induced toxicity.[8,111] The Aβ conformers stabilised by inositol are small β-structured spherical micelles that are nontoxic *in vitro*. These compounds exhibit stereoisomer-specific differences in their ability to inhibit Aβ aggregation and cytotoxicity. Aβ aggregation and toxicity are more efficiently inhibited by *scyllo*-inositol than by *myo*-inositol.[8]

To determine the importance of this structure–function relationship, a series of *scyllo*-inositol derivatives were synthesised in which one or two hydroxyl groups were replaced with fluoro, chloro, methoxy or hydrogen substituents. This approach has been previously demonstrated to be an effective method to garner information about hydrogen-bonding requirements of a given hydroxyl group in the carbohydrate binding sites of lectins or antibodies.[122–124] The hydroxyl groups at positions C-1 and C-4 were modified in light of evidence that vicinal diols at positions 1, 3, 4 and 6 are recognised by the epimerases that interconvert inositol stereoisomers.[28,125] Therefore, replacement of epimerase-targeted hydroxyl groups at positions 1 and 4 with stable substituents was hypothesised to increase *in vivo* compound stability. We identified and characterised these sites on the inositol backbone and showed maintenance of Aβ activity with some substitutions and enhanced *in vivo* stability with respect to epimerisation. The data on the effects of these compounds on Aβ-aggregation suggest that only the most conservative single hydroxyl substitutions are tolerated, thus 1-deoxy-1-fluoro-*scyllo*-inositol behaves similarly to, but not as well as, the parent compound.[126]

The potency of various inositol stereoisomers *in vivo* was investigated in a transgenic model of AD, the TgCRND8 mouse.[112] TgCRND8 mice express a human amyloid precursor protein transgene (APP_{695}) bearing two missense mutations that cause AD in humans (KM670/671NL and V717F). At about three months of age, the TgCRND8 mice display progressive spatial learning deficits that are accompanied both by rising cerebral Aβ levels and by increasing numbers of cerebral amyloid plaques.[127] By six months of age, the levels of Aβ and the morphology, density and distribution of the amyloid plaques in the brain of TgCRND8 mice are similar to those seen in the brains of humans with well-established AD.[128–130] As in patients with AD, the biochemical, behavioural and neuropathological features of this mouse are accompanied by accelerated mortality.[128–130] *myo*-Inositol had no beneficial effects, while *epi*-inositol had early effects that were not sustained with disease progression.[112] *scyllo*-Inositol treatment of TgCRND8 mice increased the survival of treated mice from 42% to 72% at 6 months of age ($p = 0.02$). This increase in survival was accompanied by total ablation of cognitive deficits using the Morris water maze test of spatial memory.[112] *scyllo*-Inositol treatment decreased total Aβ40 ($p < 0.001$) and total Aβ42 ($p < 0.05$) and maintained a 25% reduction in Aβ42 concentrations over 6 months ($p < 0.05$). Plaque load was uniformly decreased by 35% across the entire brain, indicating that inositol action is not region specific ($p < 0.05$). A similar reduction was seen in the percent brain area covered by vascular amyloid and in the size of cerebrovascular Aβ deposits. This decrease in deposited Aβ was due to *scyllo*-inositol-induced alterations of Aβ species in treated transgenic mice. Prophylactic treatment reduced soluble Aβ oligomers of mass greater than 40 kDa by 40% at 4 and 6 months of age. *scyllo*-Inositol treated 4-month-old TgCRND8 showed a significant decrease in high-molecular weight Aβ oligomers and an increase in trimeric and monomeric species.

This cognitive benefit is reflected in *scyllo*-inositol's reduction of synaptic toxicity in transgenic animals, as evidenced by a 146% increase in synatophysin immunoreactive boutons and cell bodies by 6 months of age. Improvements in neuroinflammatory status were marked by a reduction in astrogliosis and microgliosis. Both synaptic and inflammatory changes were likely due to *scyllo*-inositol blockage of Aβ oligomer-induced toxicity.

Aβ oligomer-induced inhibition of long-term potentiation (LTP) was studied using mouse hippocampal slices and rescue of this phenotype was shown with preincubation of *scyllo*-inositol with naturally occurring oligomers prior to perfusing brain slices.[131] However, application of *scyllo*-inositol after Aβ perfusion did not confer LTP protection. This effect is due to the direct binding of *scyllo*-inositol to Aβ trimers and neutralisation of toxicity. Taken together, binding and neutralising Aβ trimers could explain the increased amount of trimeric Aβ species observed in *scyllo*-inositol-treated TgCRND8 animals; stabilising Aβ trimers would also cause the decrease in high-molecular weight Aβ oligomers observed in the same animals. Coapplication of *scyllo*-inositol and Aβ oligomers also prevented oligomer-induced decreases in dendritic spine density.[132] The protective effect of *scyllo*-inositol on synaptic function may be partially the result of preventing oligomer-induced reductions in phosphatidylinositol-4,5-bisphosphate levels.[133]

These results demonstrate that selected inositols can significantly inhibit the development of AD-like phenotype in TgCRND8 mice, when given prior to the onset of disease (Table 5.1).[112] However, most AD patients will seek treatment only after their disease state is significantly advanced, *i.e.* at a time when Aβ oligomerisation, deposition, toxicity and plaque formation are already well advanced. To assess whether *scyllo*-inositol could abrogate an established disease state, we delayed the start of treatment in TgCRND8 mice until five months of age.[112] At this age, TgCRND8 mice have significant behavioural deficits, accompanied by significant Aβ peptide and plaque burdens.[127]

Spatial learning in these mice was compared at six month's of age between mice treated or untreated with *scyllo*-inositol for 28 days. *scyllo*-Inositol treatment resulted in significantly better behavioural performance in treated compared to untreated TgCRND8 mice ($p = 0.01$). The cognitive performance of these *scyllo*-inositol-treated TgCRND8 mice was not significantly different from non-Tg littermates ($p = 0.11$). This beneficial effect of inositol treatment was not due to nonspecific effects because *scyllo*-inositol had no effect on the cognitive performance of non-Tg mice ($p = 0.39$). A 28-day course of *scyllo*-inositol at 5 months of age also: 1) reduced brain levels of Aβ40 and Aβ42 (*e.g.* insoluble Aβ40 $p < 0.05$; insoluble Aβ42 $p < 0.05$) and 2) significantly reduced plaque burden ($p < 0.05$). The decrease in plaque burden was accompanied by a decrease in soluble, high molecular weight, oligomers. These results were comparable in effect size to those of the prophylactic studies.

In vivo studies in rats further confirmed the efficacy of *scyllo*-inositol in rescuing cognitive deficits caused by Aβ. Preincubation of *scyllo*-inositol with Aβ oligomers, prior to intracerebroventricular (ICV) injection into rats improved delayed alternation and complex reference memory, measured by the lever cyclic ratio assay as switching and perseveration errors.[131] Oral administration of *scyllo*-inositol via drinking water at least 3 days prior to ICV Aβ injection in rats also restored switching and perseveration errors to baseline levels.[131] These combined results suggest that *scyllo*-inositol is effective in multiple model systems of AD (Table 5.1).

Table 5.1 A list of common pathological changes found in mouse models of Alzheimer's disease, and those rescued by *scyllo*-inositol treatment.

Pathological Changes in Alzheimer's Disease Mice			*scyllo-Inositol Efficacy*
Aβ accumulation	Plaque size, number	increased	decreased[90,112]
	Aβ42 and Aβ40 levels	increased	decreased[90,112]
	Soluble oligomers	increased	decreased[90,112]
	Vascular amyloid load	increased	decreased[90,112]
	Secretase activity		no change
Neuronal Activity	Long-term potentiation (LTP)	impaired	restored[131]
	Synaptic transmission	decreased	increased[112,132]
	Excitotoxicity	increased	not examined to date
	Function of cholinergic neurons	impaired	ongoing
	Neuronal survival	reduced	not examined to date
	Cholinergic innervation	reduced	ongoing
Neuroinflammation	Number of activated microglia	increased	Decreased[112]
	Number of activated astrocytes	increased	decreased[112]
	State(s) of microglial activation	altered	not examined to date
Tau abnormalities	Tau hyperphosphorylation	increased	not examined to date
	Tau aggregation	increased	not examined to date
Behavioral changes	Spatial memory	impaired	restored[112]
	Object memory	impaired	not examined to date
	Olfactory memory	impaired	not examined to date
	Alternation memory	Impaired	restored[131]
	Reference memory	impaired	restored[131]
Neurogenesis	Growth factors BDNF, NGF	altered	not examined to date
	Rate of subventricular zone neurogenesis	increased	not examined to date
	Rate of dentate gyrus neurogenesis	increased	not examined to date
Survival		decreased	increased[112]

In order to demonstrate that the effect of *scyllo*-inositol was exerted within the CNS, the concentration of *scyllo*- and *myo*-inositol in TgCRND8 mice, after treatment was quantified. Gas chromatography-mass spectrometry reliably measures inositol in CSF and brain tissue, and effectively separates the inositol stereoisomers.[90] This assay relies on the gas chromatography of acetylated compounds, which increases the volatility and thermal stability of the compounds, while improving chromatographic separation and efficient analyte detection using single-ion monitoring in a quadrupole mass spectrometer. To demonstrate the bioavailability of inositol after oral delivery, mice were administered *scyllo*-inositol as in the preclinical efficacy studies.

scyllo-Inositol treatment did not significantly alter CSF *myo*-inositol concentration in comparison to untreated controls ($p = 0.35$ and $p = 0.7$, respectively). However, *scyllo*-inositol treatment caused a 16-fold increase in *scyllo*-inositol within the CSF ($p < 0.001$) and showed a 7.6-fold increase in brain *scyllo*-inositol over control ($p < 0.001$). These results suggest that sustained increases in *scyllo*-inositol intake are reflected in a subsequent increase in the brain.

5.11 Human Clinical Trials

During the various stages of clinical development, *scyllo*-inositol has been referred to as *scyllo*-cyclohexanehexol,[112] AZD-103 (Transition Therapeutics Inc.) and ELND005 (Elan Pharmaceuticals Inc.). After successful preclinical and toxicology studies on *scyllo*-inositol, human clinical trials were approved in Canada in the first quarter of 2006; specifically a Phase-I trial designed as a single-blind, randomised, placebo-controlled study.[134] In the fall of 2006, thirteen healthy volunteers were dosed to evaluate the pharmacokinetics, safety and tolerability of a single ascending dose of *scyllo*-inositol. The study showed that *scyllo*-inositol had a favourable pharmacokinetic profile, was well tolerated with no safety concerns or significant adverse events. Subsequently, clearance was received from the United States Food and Drug Administration (FDA) to initiate a Phase-I clinical trial in the United States. This Phase-I trial, in which healthy volunteers received placebo or an escalating acute dose of *scyllo*-inositol, was designed to evaluate the safety, tolerability and pharmacokinetics of *scyllo*-inositol in a larger population.[134] The Phase-I data confirmed that *scyllo*-inositol was safe and well tolerated at all doses examined but also that it achieved levels in the CSF and brain that have been shown to be effective in preclinical studies. In total, approximately 110 subjects were exposed to *scyllo*-inositol in multiple Phase-I studies, including single and multiple ascending dosing; pharmacokinetic evaluation of levels in the brain, cerebrospinal fluid and plasma. *scyllo*-Inositol was also shown to be orally bioavailable, to cross the blood/brain barrier and to achieve levels in the human brain and CSF that were effective in animal models for Alzheimer's disease.[90,112,134]

Before the completion of the Phase-I clinical trials in North America, Transition Therapeutics Inc., in collaboration with Elan Corporation, announced that the United States FDA had granted Fast Track designation to investigational drug candidate *scyllo*-inositol/AZD-103/ELND005.[135] Under the FDA Modernization Act of 1997, Fast Track designation is intended to facilitate the development and expedite the review of a drug or biologic if it is intended for the treatment of a serious or life-threatening condition, and it demonstrates the potential to address unmet medical needs for such a condition.

After submission of data supporting Phase-II studies to the FDA and Health Canada, a Phase-II clinical trial was announced in the last quarter of 2007, with the first patient dosed in December 2007.[134] The study is a randomised, double-blind, placebo-controlled, dose-ranging, safety and efficacy study in patients with mild to moderate Alzheimer's disease.[136] Oral administration of *scyllo*-inositol in male and female participants aged 50–85 years with mild to moderate AD is underway. Approximately 340 patients were enrolled at 65 study sites with a treatment duration of 18 months. The inclusion criteria for the Phase-II trial included: Mini-Mental Status Exam score of 16–26, brain magnetic resonance imaging (MRI) scan consistent with the diagnosis of AD, fluency in English, French, or Spanish for the cognitive tests, stable doses of medications (cholinesterase inhibitors and memantine allowed) and the caregiver is able to attend all study visits.[136] However, exclusion criteria included: significant neurological disease other than AD, major psychiatric disorders, significant medical illness, history of stroke, seizure or heart attack within the last 2 years and the presence of pacemakers or foreign metal objects in the eyes, skin, or body that would prevent an MRI scan. The trial is underway but slated to end in the second quarter of 2010. The potential of *scyllo*-inositol as a disease-modifying drug will need to await the outcome of these and future clinical trials.

References

1. J. P. Taylor, J. Hardy and K. H. Fischbeck, *Science*, 2002, **296**, 1991–5.
2. E. Bossy-Wetzel, R. Schwarzenbacher and S. A. Lipton, *Nature Med.*, 2004, **10**, S2–9.
3. C. A. Ross and M. A. Poirier, *Nature Med.*, 2004, **10**, S10–7.
4. J. D. Sipe and A. S. Cohen, *J. Struct. Biol.*, 2000, **130**, 88–98.
5. S. Taniguchi, N. Suzuki, M. Masuda, S. Hisanaga, T. Iwatsubo, M. Goedert and M. Hasegawa, *J. Biol. Chem.*, 2005, **280**, 7614–23.
6. M. Pickhardt, M. von Bergen, Z. Gazova, A. Hascher, J. Biernat, E. M. Mandelkow and E. Mandelkow, *Curr. Alzheimer Res.*, 2005, **2**, 219–26.
7. D. R. Howlett, A. R. George, D. E. Owen, R. V. Ward and R. E. Markwell, *Biochem. J.*, 1999, **343**, 419–23.
8. J. McLaurin, R. Golomb, A. Jurewicz, J. P. Antel and P. E. Fraser, *J. Biol. Chem.*, 2000, **275**, 18495–502.
9. H. A. Lashuel, D. M. Hartley, D. Balakhaneh, A. Aggarwal, S. Teichberg and D. J. Callaway, *J. Biol. Chem.*, 2002, **277**, 42881–90.
10. F. P. Guengerich, *Chem. Res. Toxicol.*, 2001, **14**, 611–50.
11. J. Scherer, *Liebigs Ann. Chem.*, 1850, **73**, 322–8.
12. G. Staedeler and F. T. Frerichs, *J. prakt. Chem.*, 1858, **73**, 48–55.
13. L. Bouveault, in *Bull. Soc. Chim. Paris*, 1894, **11**, 44–147.
14. T. Posternak, *Helv. Chim. Acta.*, 1941, **24**, 1045–58.
15. T. Posternak, *Helv. Chim. Acta.*, 1942, **25**, 746–52.
16. G. Dangschat, *Naturwissenschaften*, 1942, **30**, 146.
17. A. L. Majumder, A. Chatterjee, K. Ghosh Dastidar and M. Majee, *FEBS Lett.*, 2003, **553**, 3–10.

18. E. R. Seaquist and R. Gruetter, *Magn. Reson. Med.*, 1998, **39**, 313–6.
19. W. R. Sherman, P. C. Simpson and S. L. Goodwin, *Comp. Biochem. Physiol. B.*, 1978, **59**, 201–2.
20. M. L. Sanz, M. Villamiel and I. Martinez-Castro, *Food Chem.*, 2004, **87**, 325–8.
21. A. C. Soria, M. L. Sanz and M. Villamiel, *Food Chem.*, 2009, **114**, 758–62.
22. P. P. Hipps, M. R. Eveland, M. H. Laird and W. R. Sherman, *Biochem. Biophys. Res. Commun.*, 1976, **68**, 1133–8.
23. B. J. Holub, *Annu. Rev. Nutr.*, 1986, **6**, 563–97.
24. R. S. Clements Jr and A. G. Diethelm, *J. Lab. Clin. Med.*, 1979, **93**, 210–9.
25. Y. H. Wong, S. J. Kalmbach, B. K. Hartman and W. R. Sherman, *J. Neurochem.*, 1987, **48**, 1434–42.
26. K. K. Caldwell, M. Sosa and C. T. Buckley, *Cell. Commun. Signal.*, 2006, **4**, 2.
27. G. McAllister, P. Whiting, E. A. Hammond, M. R. Knowles, J. R. Atack, F. J. Bailey, R. Maigetter and C. I. Ragan, *Biochem. J.*, 1992, **284**, 749–54.
28. P. P. Hipps, W. H. Holland and W. R. Sherman, *Biochem. Biophys. Res. Commun.*, 1977, **77**, 340–6.
29. R. J. Arner, K. S. Prabhu, V. Krishnan, M. C. Johnson and C. C. Reddy, *Biochem. Biophys. Res. Commun.*, 2006, **339**, 816–20.
30. M. Mueckler, C. Caruso, S. A. Baldwin, M. Panico, I. Blench, H. R. Morris, W. J. Allard, G. E. Lienhard and H. F. Lodish, *Science*, 1985, **229**, 941–5.
31. F. Q. Zhao and A. F. Keating, *Curr. Genomics*, 2007, **8**, 113–28.
32. M. Uldry, M. Ibberson, J. D. Horisberger, J. Y. Chatton, B. M. Riederer and B. Thorens, *EMBO. J.*, 2001, **20**, 4467–77.
33. M. Uldry, P. Steiner, M. G. Zurich, P. Béguin, H. Hirling, W. Dolci and B. Thorens, *EMBO. J.*, 2004, **23**, 531–40.
34. M. Chesler and K. Kaila, *Trends Neurosci.*, 1992, **15**, 396–402.
35. E. M. Wright and E. Turk, *Pflugers Arch.*, 2004, **447**, 510–8.
36. H. M. Kwon, A. Yamauchi, S. Uchida, A. S. Preston, A. Garcia-Perez, M. B. Burg and J. S. Handler, *J. Biol. Chem.*, 1992, **267**, 6297–301.
37. G. T. Berry, J. J. Mallee, H. M. Kwon, J. S. Rim, W. R. Mulla, M. Muenke and N. B. Spinner, *Genomics*, 1995, **25**, 507–13.
38. B. Lubrich, O. Spleiss, P. J. Gebicke-Haerter and D. van Calker, *Neuropharmacology*, 2000, **39**, 680–90.
39. K. E. McVeigh, J. J. Mallee, A. Lucente, B. L. Barnoski, S. Wu and G. T. Berry, *Cytogenet. Cell. Genet.*, 2000, **88**, 153–8.
40. R. E. Ostlund Jr, J. B. McGill, I. Herskowitz, D. M. Kipnis, J. V. Santiago and W. R. Sherman, *Proc. Natl. Acad. Sci. USA*, 1993, **90**, 9988–92.
41. K. Inoue, S. Shimada, Y. Minami, H. Morimura, A. Miyai, A. Yamauchi and M. Tohyama, *Neuroreport*, 1996, **7**, 1195–8.
42. A. Hakvoort, M. Haselbach and H. J. Galla, *Brain Res.*, 1998, **795**, 247–56.

43. G. T. Berry, S. Wu, R. Buccafusca, J. Ren, L. W. Gonzales, P. L. Ballard, J. A. Golden, M. J. Stevens and J. J. Greer, *J. Biol. Chem.*, 2003, **278**, 18297–302.

44. A. Shaldubina, R. Buccafusca, R. A. Johanson, G. Agam, R. H. Belmaker, G. T. Berry and Y. Bersudsky, *Genes Brain Behav.*, 2007, **6**, 253–9.

45. K. Hager, A. Hazama, H. M. Kwon, D. D. Loo, J. S. Handler and E. M. Wright, *J. Membr. Biol.*, 1995, **143**, 103–13.

46. T. J. Wiese, J. A. Dunlap, C. E. Conner, J. A. Grzybowski, W. L. Lowe Jr and M. A. Yorek, *Am. J. Physiol.*, 1996, **270**, C990–7.

47. K. Strange, F. Emma, A. Paredes and R. Morrison, *Glia*, 1994, **12**, 35–43.

48. T. Nakanishi, R. J. Turner and M. B. Burg, *Proc. Natl. Acad. Sci. USA*, 1989, **86**, 6002–6.

49. H. M. Kwon, A. Yamauchi, S. Uchida, R. B. Robey, A. Garcia-Perez, M. B. Burg and J. S. Handler, *Am. J. Physiol.*, 1991, **260**, F258–63.

50. A. Paredes, M. McManus, H. M. Kwon and K. Strange, *Am. J. Physiol.*, 1992, **263**, C1282–8.

51. P. R. Cammarata and H. Q. Chen, *Invest. Ophthalmol. Vis. Sci.*, 1994, **35**, 1223–35.

52. P. R. Cammarata, H. Q. Chen, C. Zhou and R. Reeves, *Exp. Eye Res.*, 1994, **59**, 83–9.

53. A. Miyai, A. Yamauchi, T. Nakanishi, M. Sugita, Y. Takamitsu, K. Yokoyama, T. Itoh, A. Andou, T. Kamada, N. Ueda and Y. Fujiwara, *Kidney Int.*, 1995, **47**, 473–80.

54. J. J. Mallee, M. G. Atta, V. Lorica, J. S. Rim, H. M. Kwon, A. D. Lucente, Y. Wang and G. T. Berry, *Genomics*, 1997, **46**, 459–65.

55. C. Denkert, U. Warskulat, F. Hensel and D. Häussinger, *Arch. Biochem. Biophys.*, 1998, **354**, 172–80.

56. Y. Matsuoka, A. Yamauchi, T. Nakanishi, T. Sugiura, H. Kitamura, M. Horio, Y. Takamitsu, A. Ando, E. Imai and M. Hori, *Nephrol. Dial. Transplant.*, 1999, **14**, 1217–23.

57. M. A. Yorek, J. A. Dunlap and W. L. Lowe Jr, *Kidney Int.*, 1999, **55**, 215–24.

58. M. A. Yorek, J. A. Dunlap and W. L. Lowe Jr, *Biochem. J.*, 1998, **336**, 317–25.

59. M. A. Yorek, J. A. Dunlap, M. J. Thomas, P. R. Cammarata, C. Zhou and W. L. Lowe Jr, *Am. J. Physiol.*, 1998, **274**, C58–71.

60. J. Matskevitch, C. A. Wagner, T. Risler, H. M. Kwon, J. S. Handler, S. Waldegger, A. E. Busch and F. Lang, *Pflugers Arch.*, 1998, **436**, 854–7.

61. D. Eladari, R. Chambrey, F. Pezy, R. A. Podevin, M. Paillard and F. Leviel, *Kidney Int.*, 2002, **62**, 2144–51.

62. H. Wiesinger, *J. Neurochem.*, 1991, **56**, 1698–704.

63. M. J. Coady, B. Wallendorff, D. G. Gagnon and J. Y. Lapointe, *J. Biol. Chem.*, 2002, **277**, 35219–24.

64. P. Roll, A. Massacrier, S. Pereira, A. Robaglia-Schlupp, P. Cau and P. Szepetowski, *Gene*, 2002, **285**, 141–8.

65. P. Bissonnette, M. J. Coady and J. Y. Lapointe, *J. Physiol.*, 2004, **558**, 759–68.

66. P. Bissonnette, K. Lahjouji, M. J. Coady and J. Y. Lapointe, *Am. J. Physiol. Cell Physiol.*, 2008, **295**, C791–9.

67. R. Aouameur, S. Da Cal, P. Bissonnette, M. J. Coady and J. Y. Lapointe, *Am. J. Physiol. Gastrointest. Liver Physiol.*, 2007, **293**, G1300–7.

68. X. Lin, L. Ma, R. L. Fitzgerald and R. E. Ostlund Jr, *Arch. Biochem. Biophys.*, 2009, **481**, 197–201.

69. R. E. Ostlund Jr, R. Seemayer, S. Gupta, R. Kimmel, E. L. Ostlund and W. R. Sherman, *J. Biol. Chem.*, 1996, **271**, 10073–8.

70. K. Lahjouji, R. Aouameur, P. Bissonnette, M. J. Coady, D. G. Bichet and J. Y. Lapointe, *Biochim. Biophys. Acta*, 2007, **1768**, 1154–9.

71. F. Bourgeois, M. J. Coady and J. Y. Lapointe, *J. Physiol.*, 2005, **563**, 333–43.

72. J. Larner, *Int. J. Exp. Diabetes Res.*, 2002, **3**, 47–60.

73. P. R. Cammarata, G. T. Xu, L. Huang, C. Zhou and M. Martin, *Exp. Eye Res.*, 1997, **64**, 745–57.

74. A. Karihaloo, K. Kato, D. A. Greene and T. P. Thomas, *Am. J. Physiol.*, 1997, **273**, C671–8.

75. R. E. Isaacks, A. S. Bender, C. Y. Kim, Y. F. Shi and M. D. Norenberg, *J. Neurosci. Res.*, 1999, **57**, 866–71.

76. P. S. Jackson and K. Strange, *Am. J. Physiol.*, 1993, **265**, C1489–500.

77. J. E. Novak, B. W. Agranoff and S. K. Fisher, *Neurochem. Res.*, 2000, **25**, 561–6.

78. D. Loveday, A. M. Heacock and S. K. Fisher, *J. Neurochem.*, 2003, **87**, 476–86.

79. H. Diringer and R. Rott, *Eur. J. Biochem.*, 1977, **79**, 451–7.

80. M. A. Yorek, J. A. Dunlap and M. R. Stefani, *Diabetes*, 1991, **40**, 240–8.

81. Y. Bersudsky, Z. Kaplan, Y. Shapiro, G. Agam, O. Kofman and R. H. Belmaker, *Eur. Neuropsychopharmacol.*, 1994, **4**, 463–7.

82. O. Kofman, Y. Bersudsky, I. Vinnitsky, C. Alpert and R. H. Belmaker, *Isr. J. Med. Sci.*, 1993, **29**, 580–6.

83. S. H. Sigal, J. R. Yandrasitz and G. T. Berry, *Metabolism*, 1993, **42**, 395–401.

84. M. Wolfson, Y. Bersudsky, E. Hertz, V. Berkin, E. Zinger and L. Hertz, *Neurochem. Res.*, 2000, **25**, 977–82.

85. R. L. Kinnard, B. Narasimhan, G. Pliska-Matyshak and P. P. Murthy, *Biochem. Biophys. Res. Commun.*, 1995, **210**, 549–55.

86. M. Salman, J. T. Lonsdale, G. S. Besra and P. J. Brennan, *Biochim. Biophys. Acta.*, 1999, **1436**, 437–50.

87. P. E. Ryals and M. C. Kersting, *Arch. Biochem. Biophys.*, 1999, **366**, 261–6.

88. B. M. Riggs, T. A. Lansley and P. E. Ryals, *J. Eukaryot. Microbiol.*, 2007, **54**, 119–24.

89. T. Takenawa and K. Egawa, *J. Biol. Chem.*, 1977, **252**, 5419–23.

90. D. Fenili, M. Brown, R. Rappaport and J. McLaurin, *J. Mol. Med.*, 2007, **85**, 603–11.

91. R. F. Irvine, *Biochim. Biophys. Acta.*, 1998, **1393**, 292–8.

92. J. Parkkinen, *FEBS. Lett.*, 1983, **163**, 10–3.

93. J. Frahm, H. Bruhn, W. Hänicke, K. D. Merboldt, K. Mursch and E. Markakis, *J. Comput. Assist. Tomogr.*, 1991, **15**, 915–22.

94. L. G. Kaiser, N. Schuff, N. Cashdollar and M. W. Weiner, *NMR. Biomed.*, 2005, **18**, 51–5.

95. H. R. Griffith, J. A. den Hollander, C. C. Stewart, W. T. Evanochko, S. D. Buchthal, L. E. Harrell, E. Y. Zamrini, J. C. Brockington and D. C. Marson, *NMR. Biomed.*, 2007, **20**, 709–16.

96. J. Levine, *Eur. Neuropsychopharmacol.*, 1997, **7**, 147–55.

97. G. Agam, Y. Shapiro, Y. Bersudsky, O. Kofman and R. H. Belmaker, *Pharmacol. Biochem. Behav.*, 1994, **49**, 341–3.

98. D. Gelber, J. Levine and R. H. Belmaker, *Int. J. Eat. Disord.*, 2001, **29**, 345–8.

99. J. Benjamin, G. Agam, J. Levine, Y. Bersudsky, O. Kofman and R. H. Belmaker, *Psychopharmacol. Bull.*, 1995, **31**, 167–75.

100. M. Hallman, A. L. Järvenpää and M. Pohjavuori, *Arch. Dis. Child.*, 1986, **61**, 1076–83.

101. M. Hallman, K. Bry, K. Hoppu, M. Lappi and M. Pohjavuori, *N. Engl. J. Med.*, 1992, **326**, 1233–9.

102. H. Einat, Z. Elkabaz-Shwortz, H. Cohen, O. Kofman and R. H. Belmaker, *Int. J. Neuropsychopharmacol.*, 1998, **1**, 31–34.

103. Y. Bersudsky, H. Einat, Z. Stahl and R. H. Belmaker, *Curr. Psychiatry Rep.*, 1999, **1**, 141–7.

104. P. Cogram, S. Tesh, J. Tesh, A. Wade, G. Allan, N. D. Greene and A. J. Copp, *Hum. Reprod.*, 2002, **17**, 2451–8.

105. J. P. Baillargeon, E. Diamanti-Kandarakis, R. E. Ostlund Jr, T. Apridonidze, M. J. Iuorno and J. E. Nestler, *Diabetes Care*, 2006, **29**, 300–5.

106. J. P. Baillargeon, J. E. Nestler, R. E. Ostlund, T. Apridonidze and E. Diamanti-Kandarakis, *Hum. Reprod.*, 2008, **23**, 1439–46.

107. J. E. Nestler, D. J. Jakubowicz, P. Reamer, R. D. Gunn and G. Allan, *N. Engl. J. Med.*, 1999, **340**, 1314–20.

108. M. J. Iuorno, D. J. Jakubowicz, J. P. Baillargeon, P. Dillon, R. D. Gunn, G. Allan and J. E. Nestler, *Endocr. Pract.*, 2002, **8**, 417–23.

109. S. Gerli, M. Mignosa and G. C. Di Renzo, *Eur. Rev. Med. Pharmacol. Sci.*, 2003, **7**, 151–9.

110. M. H. Richards and R. H. Belmaker, *J. Neural Transm.*, 1996, **103**, 1281–5.

111. J. McLaurin, T. Franklin, A. Chakrabartty and P. E. Fraser, *J. Mol. Biol.*, 1998, **278**, 183–94.

112. J. McLaurin, M. E. Kierstead, M. E. Brown, C. A. Hawkes, M. H. Lambermon, A. L. Phinney, A. A. Darabie, J. E. Cousins, J. E. French, M. F. Lan, F. Chen, S. S. Wong, H. T. Mount, P. E. Fraser, D. Westaway and P. St George-Hyslop, *Nature Med.*, 2006, **12**, 801–8.

113. Y. Barak, J. Levine, A. Glasman, A. Elizur and R. H. Belmaker, *Prog. Neuropsychopharmacol. Biol. Psychiatry*, 1996, **20**, 729–35.

114. C. A. McLean, R. A. Cherny, F. W. Fraser, S. J. Fuller, M. J. Smith, K. Beyreuther, A. I. Bush and C. L. Masters, *Ann. Neurol.*, 1999, **46**, 860–6.

115. S. Sinha, *Med. Clin. North Am.*, 2002, **86**, 629–39.

116. J. McLaurin and A. Chakrabartty, *J. Biol. Chem.*, 1996, **271**, 26482–9.

117. J. McLaurin and A. Chakrabartty, *Eur. J. Biochem.*, 1997, **245**, 355–63.

118. C. M. Yip, E. A. Elton, A. A. Darabie, M. R. Morrison and J. McLaurin, *J. Mol. Biol.*, 2001, **311**, 723–34.
119. L. P. Choo-Smith and W. K. Surewicz, *FEBS. Lett.*, 1997, **402**, 95–8.
120. J. J. Kremer, M. M. Pallitto, D. J. Sklansky and R. M. Murphy, *Biochemistry*, 2000, **39**, 10309–18.
121. C. C. Curtain, F. E. Ali, D. G. Smith, A. I. Bush, C. L. Masters and K. J. Barnham, *J. Biol. Chem.*, 2003, **278**, 2977–82.
122. C. P. J. Glaudemans, *Chem. Rev.*, 1991, **91**, 25–33.
123. F. I. Auzanneau, H. R. Hanna and D. R. Bundle, *Carbohydr. Res.*, 1993, **240**, 161–81.
124. C. P. Swaminathan, D. Gupta, V. Sharma and A. Surolia, *Biochemistry*, 1997, **36**, 13428–34.
125. Y. Pak, L. C. Huang, K. J. Lilley and J. Larner, *J. Biol. Chem.*, 1992, **267**, 16904–10.
126. Y. Sun, G. Zhang, C. A. Hawkes, J. E. Shaw, J. McLaurin and M. Nitz, *Bioorg. Med. Chem.*, 2008, **16**, 7177–84.
127. M. A. Chishti, D. S. Yang, C. Janus, A. L. Phinney, P. Horne, J. Pearson, R. Strome, N. Zuker, J. Loukides, J. French, S. Turner, G. Lozza, M. Grilli, S. Kunicki, C. Morissette, J. Paquette, F. Gervais, C. Bergeron, P. E. Fraser, G. A. Carlson, P. S. George-Hyslop and D. Westaway, *J. Biol. Chem.*, 2001, **276**, 21562–70.
128. J. Wang, D. W. Dickson, J. Q. Trojanowski and V. M. Lee, *Exp. Neurol.*, 1999, **158**, 328–37.
129. J. Näslund, V. Haroutunian, R. Mohs, K. L. Davis, P. Davies, P. Greengard and J. D. Buxbaum, *JAMA*, 2000, **283**, 1571–7.
130. R. Li, K. Lindholm, L. B. Yang, X. Yue, M. Citron, R. Yan, T. Beach, L. Sue, M. Sabbagh, H. Cai, P. Wong, D. Price and Y. Shen, *Proc. Natl. Acad. Sci. USA*, 2004, **101**, 3632–7.
131. M. Townsend, J. P. Cleary, T. Mehta, J. Hofmeister, S. Lesne, E. O'Hare, D. M. Walsh and D. J. Selkoe, *Ann. Neurol.*, 2006, **60**, 668–76.
132. G. M. Shankar, B. L. Bloodgood, M. Townsend, D. M. Walsh, D. J. Selkoe and B. L. Sabatini, *J. Neurosci.*, 2007, **27**, 2866–75.
133. D. E. Berman, C. Dall'Armi, S. V. Voronov, L. B. McIntire, H. Zhang, A. Z. Moore, A. Staniszewski, O. Arancio, T. W. Kim and G. Di Paolo, *Nature Neurosci.*, 2008, **11**, 547–54.
134. www.transitiontheraputics.com.
135. www.elan.com.
136. www.clinicaltrials.gov.

CHAPTER 6

Immunotherapeutic Strategies Towards Treatment of Alzheimer's Disease

BEKA SOLOMON

Professor of Biotechnology, Chair for Biotechnology of Neurodegenerative Diseases, Department of Molecular Microbiology & Biotechnology, George S. Wise Faculty of Life Sciences, Tel-Aviv University, Israel

This chapter is dedicated to Professor Ephraim Katchalsky-Katzir, former President of the State of Israel, and outstanding scientist.

6.1 Introduction

Alzheimer's disease (AD) is characterised by progressive loss of memory and cognitive function and is the most common form of dementia in the elderly. According to the World Health Organization, an estimated 37 million people word-wide currently have dementia; Alzheimer's disease affects about 18 million of them. Increasing age is the greatest risk factor for AD. Its prevalence approximately doubles every 5 years after the age of 60 – one in ten individuals over 65 years and nearly half of those over 85 are affected by the disease.[1] The disease is characterised by massive cell loss, especially of cholinergic neurons, deposition of fibrillar amyloid-β (Aβ) peptides as senile plaques, and

RSC Drug Discovery Series No. 2
Emerging Drugs and Targets for Alzheimer's Disease
Volume 1: Beta-Amyloid, Tau Protein and Glucose Metabolism
Edited by Ana Martinez

Published by the Royal Society of Chemistry, www.rsc.org

intracellular accumulation of hyperphosphorylated tau protein as neurofibrillary tangles.[2]

Although AD pathogenesis is complex and remains unclear, aggregation of
Aβ peptides is considered to be the earliest event in the disease development,
leading to neurodegeneration.[3]

The most studied hypothesis, amyloid cascade,[3–5] states that overproduction
of Aβ peptide, or failure to clear this peptide, leads to AD primarily through
amyloid deposition that is presumed to be involved in neurofibrillary tangles
formation; these lesions are then associated with cell death, which is reflected in
memory impairment, the hallmarks of this dementia.[6]

During the past ten years, the amyloid cascade hypothesis has gained strength
and remains the most attractive explanation as the underlying cause of AD based
on the general occurrence of amyloid pathology in the brain of all AD patients.[3–6]
In strong support are the mutations in familial cases of early onset AD in the
genes coding for amyloid precursor protein (APP) and presenilins (PS).[7–9]

The hypothesis of a "cascade" has exerted considerable attraction since that
process would allow for intervention at multiple and different points to slow or
halt the disease process. The most effective treatment for AD is accepted to be
directed towards (i) decreasing beta-amyloid production, (ii) increasing Aβ
removal, or (iii) preventing Aβ fibril formation. Modulating events or
mechanisms that are downstream of Aβ formation and deposition, *e.g.* free
radicals, inflammation, calcium homeostasis, excitotoxicity, apoptosis, *etc.*
would be less fundamental or even symptomatic.[10]

6.2   Chaperone-Like Activity of Site-Directed Antibodies in Modulation of Protein Conformation

Classic molecular chaperones are involved in the conformational change between
the soluble and insoluble forms, perhaps through stabilisation of a certain
folding intermediate. The molecular chaperones bind reversibly to aggregation
intermediate until folding or assembly are completed, enabling folding and stabilisation of unrelated proteins.[11–13] However, despite the existence of chaperones, aggregation of protein still occurs *in vivo*. The central problem remaining
in vivo folding is how to efficiently prevent aggregation without blocking the
forward pathway of correct folding and biological activity of the native state.

The availability of monoclonal antibodies (mAbs) has facilitated understanding how highly specific antigen-antibody interactions may affect antigen
stability and folding. The complementary conformation between the interacting regions of the antibody and its antigen may confer high specificity and
stability to the immunocomplex formed.[14] Antibodies known as reporting
probes for the detection of antigens are able to play an active role in inducing
changes in the antigen molecule. Such antibody–antigen interactions involve
conformational changes that can range from insignificant to considerable.
Binding of high-affinity monoclonal antibodies to regions of high flexibility and
antigenicity may alter the molecular dynamics of the whole antigen and may

induce structural rearrangements in the molecular edifice.[15–17] Like the ubiquitous chaperones,[11,13] mAbs raised against specific native antigens may assist in refolding by recognising incompletely folded epitopes and inducing their native conformation. Appropriate mAbs interact at strategic sites where protein aggregation is initiated, stabilising the protein, preventing further aggregation and resolubilising already formed protein aggregates.[17–19]

6.3 *In vitro* Modulation of Beta-Amyloid Conformation-Related Neurotoxicity by Site-Directed Antibodies

Beta-amyloid formation is a complex kinetic and thermodynamic process and the reversibility of amyloid plaque growth *in vitro* suggests a steady-state equilibrium between Aβ peptide in plaques and in solution.[20] The dependence of Aβ peptide polymerisation on peptide–peptide interactions to form a β-pleated sheet fibril and the stimulatory influence of other proteins on the reaction suggest that amyloid formation may be subject to modulation.

We investigated a large panel of mAbs against various regions of Aβ peptide regarding their anti-aggregating properties. We showed that whole and/or fragments of antibodies towards the N-terminal region of the Aβ peptide bind to preformed Aβ fibrils, leading to their disaggregation and inhibition of their neurotoxic effect.[21–23] Antibodies directed to other regions of Aβ peptide only prevent aggregation but do not dissolve aggregates. Binding of such antibodies to aggregated AβP interfered with noncovalent interactions between the amyloid fibrils and led to deterioration of amyloid fibrillar assembly. Disaggregation, as well as prevention of amyloid formation, was found to be dependent on the location of the epitopes on the Aβ and on the binding characteristics of the respective antibodies.[21–24]

Using the phage-peptide library composed of filamentous phage displaying random combinatorial peptides we defined the EFRH residues located at positions 3–6 of the N-terminal Aβ peptide as the epitope of the anti-aggregating antibodies studied.[25] Locking of the EFRH epitope by the respective antibodies was found to modulate the dynamics of aggregation as well as to resolubilise already formed aggregates, confirming its key role in modulation of conformational changes of the whole Aβ peptide molecule.

Identification of the 'aggregating epitopes' as strategic positions related to sites where protein aggregation is initiated, and preparing antibodies against these regions, became the basis of the immunological concept, which seems to have therapeutic importance.

6.4 Clearance of Alzheimer-Like Beta-Amyloid Plaques by Mobilising the Immune Response

The abundant evidence that Aβ peptide aggregation is an essential early event in AD pathogenesis has prompted an intensive search for therapeutics that target Aβ. This search has been aided immeasurably by transgenic mouse models of AD. Several labs have bred AD diseased models of transgenic mice that produce Aβ and develop plaques and neuron damage in their brains (as reviewed in refs. 26,27). Although they do not develop the widespread neuron death and severe dementia seen in the human disease they are used as models for the study of AD.

The immunological concept in the treatment of AD has gained more attention and immunisation approaches are being pursued in order to stimulate clearance of brain Aβ plaques.[28–31] They include both active and passive immunisation techniques. Active immunisation approaches employ various routes of administration, types of adjuvants, the use of modified Aβ peptide epitopes and/or immunogenic Aβ peptide conjugates. Passive immunisation approaches include monoclonal antibodies or specific antibody fractions (Fabs) directed against specific Aβ epitopes.

6.4.1   Active Immunisation

Production of anti-Aβ antibodies by immunisation with the fibrillar Aβ of the PDAPP mouse model of AD, led to inhibition of the formation of amyloid plaques and the associated dystrophic neurites in the mouse brain.[32] Immunisation of older animals with established pathology also showed clearance of beta-amyloid plaque in the brain.

In an alternate approach, nasal administration of Aβ 1-42 was effective in stimulating peripheral antibody production and decreasing Aβ 1-42 in the brain as well plaque burden in the hippocampus of PDAPP mice.[33]

In spite of the encouraging results reported, an important note for concern with the use of beta-amyloid as an immunisation agent has been the possibility that these peptide species may form toxic fibrils and/or seed the formation of amyloid filaments in plaque deposits. Even without this extra risk, the notion of immunisation with a toxic peptide does not sound very attractive.

To overcome these potential difficulties, some investigators are exploring alternative strategies with nonamyloidogenic Aβ immunogens. Sigurdsson *et al.*[34] reported a novel immunisation based on a construct consisting of the first thirty amino acids of Aβ with a further six lysine residues at the N-terminus to induce immunogenicity. Immunisation with Aβ 1-30 in Tg2576 mice[35] developed by K. Hsiao's laboratory led to a robust antibody response against the immunogen and, after seven months of inoculations, to a dramatic reduction of the Aβ burden in the hippocampus and neocortex.

We developed an immunisation procedure for raising anti-Aβ peptide antibodies using as antigen the only EFRH peptide displayed on the filamentous phage. The EFRH epitope[25] is available for antibody binding when Aβ peptide is either in solution or is an aggregate, and the locking of this epitope by antibodies affects the dynamics of all the molecules, preventing self-aggregation as well as enabling resolubilisation of already formed aggregates.[36]

We have shown that immunisation of mice, as well as of guinea pigs that share homology with human Aβ peptide sequences, with a filamentous bacteriophage, carrying about 300 copies of the EFRH-epitope, elicited high titers of IgG antibodies with high affinity against the beta-amyloid peptide in a relatively short period of time.[37–41] The anti-EFRH-phage antisera were demonstrated to label the amyloid plaques effectively, both in brain sections of human AD patients and in brain sections of the transgenic mice. These antibodies were operationally similar or identical in their *in vitro* and *in vivo* antiaggregating properties to antibodies raised by injection of whole Aβ peptide.

Reduction of Aβ plaques *in vivo* was demonstrated using APP[717I] transgenic mice[39] aged 16 months, immunised with the EFRH-phage and analysed at age 21 months. At this age the amyloid plaque pathology is maximally and stably established. Age-matched APP[V717I] transgenic mice were injected with vehicle only and served as untreated controls. Five out of eight (65%) APP [V717I] transgenic mice that were immunised with the EFRH-phage developed and maintained serum titers of antibodies against beta-amyloid that varied between 1:100–1:1000. Three (35%) of the APP[V717I] transgenic mice developed only a low titer (∼ 1:10) over the 5-month immunisation period. The amyloid burden in the brain was significantly reduced in the immunised APP [V717I] transgenic mice that developed anti-Aβ titers of at least 1:100, indicating a dose–response relation between antibody-titer and reduced amyloid load.

6.4.2 Passive Immunisation

Another set of experiments showed that peripheral administration of antibodies against Aβ peptide was sufficient to reduce amyloid burden in the AD-like affected mice brains.[42] Circulating antibodies were able to cross the blood/brain barrier and bind to brain Aβ deposits. Despite their relatively modest serum levels, the passively administered antibodies were able to enter the central nervous system, either to decorate plaques and induce clearance of pre-existing amyloid in old mice or to prevent plaque formation in young mice.[42–47]

6.4.2.1 Antibodies Against N-terminal of Aβ Peptide

Peripheral administration of antibodies directed against the N-terminal region of Aβ peptide was reported to clear amyloid burden in the brain of PDAPP Tg mice.[42] Some of these antibodies enter the brain, bind to plaques and trigger microglial-dependent clearance amyloid plaques. Of the antibodies tested, only mAbs 10D5, 3D6 and polyclonal anti-AβP 1-42, directed to the N-terminal regions of Aβ peptide, demonstrated antiaggregating properties *in vivo*. Peripheral administration of antibodies against Aβ peptide was sufficient to reduce brain amyloid burden in the PDAPP Tg mice. In contrast, mAbs 16C11, 21F12 and the control antibody TM2a, directed to other regions of AβP, were inactive. This result is consistent with the inability of these two antibodies to decorate plaques after *in vivo* administration and explains their inability to trigger plaque clearance. These *in vivo* data confirm previous *in vitro* data,[21–23] that only antibodies directed to the 'strategic' epitopes involved in the aggregation process, such as EFRH, exhibit so-called 'chaperone-like' properties in dissolving amyloid plaques.

However, Pfeifer *et al.*[43] using APP23-Tg mice suggested that high concentrations of antibodies against the N-terminus of Aβ, which also recognise amyloid present in the cerebral vasculature, might predispose patients with cerebral amyloid angiopathy to microhemorrhages and hemorrhagic stroke.

6.4.2.2 Antibodies Against Central Region of Aβ Peptide

DeMattos and colleagues used the PDAPP Tg mice model for i.v. injection of anti-Aβ antibody 266.[44] The overall effect of the peripheral administration of the anti-Aβ antibody was to reduce amyloid plaque burden within the brain, presumably by increasing efflux of Aβ peptide outside the brain. Administration of antibody 266 was found to reduce the memory deficits that had previously been reported in the transgenic mice without altering the brain amyloid burden. The authors proposed that the memory improvements might be due to sequestration of soluble Aβ from the brain.

6.4.2.3 Antibodies Against C-terminal of Aβ Peptide

The effect of anti-C-terminal antibodies[45–47] on amyloid levels and cognitive functions was reproduced in a subsequent study in aged APP-transgenic mice, but adverse effects including cerebral microhemorrhages were observed in this study, probably because of redistribution of disaggregated Aβ from brain parenchyma to cerebral vasculature, resulting in increased cerebral amyloid angiopathy.[47]

6.4.2.4 Antibodies Against Conformational Epitopes of Aβ Oligomers

Recently, several anti-Aβ antibodies targeting conformation epitopes have been shown to bind to pathogenic aggregated Aβ forms (protofibrils, oligomers and amyloid plaques) but have no effect on monomeric Aβ and APP. Antibodies raised against synthetic Aβ oligomers detected a 70-fold increase in oligomeric species in AD patients over control brains.[48] Moreover, Kayed *et al.* found that soluble oligomers display a conformation-dependent structure common to all oligomers independent of their sequence, which suggests a shared mechanism of toxicity. Functionally, it has been found that naturally secreted oligomers inhibit hippocampal long-term potentiation *in vivo*.[49] Taken together, these results suggest that strategies aimed at treating amyloid disorders should target oligomers of Aβ. In doing so, the equilibrium between monomers and higher-order aggregates can be disrupted, resulting in neutralisation of soluble, toxic species.

6.4.3 Intravenous Immunoglobulin Treatment

Affinity-purified anti-Aβ antibodies from IVIg increased Aβ levels in blood and decreased Aβ levels in cerebrospinal fluid (CSF) from APP-transgenic mice.[50] Furthermore, IVIg depleted of anti-Aβ antibodies had considerably less effect on Aβ levels in the blood or CSF from these animals. It remains to be determined whether the results with IVIg in APP-transgenic mice reflect the mechanisms active in AD patients.

The molecular basis for the direct and indirect effects of IVIg on Aβ clearance was investigated using the BV-2 cellular microglia line. The data show that IVIg dissolves preformed Aβ fibrils *in vitro*. IVIg increases cellular tolerance to Aβ and enhances microglial migration toward Aβ deposits, mediating phagocytosis of Aβ fibrils.[51] Finally, it is also possible that other activities of IVIg, unrelated to its content of anti-Aβ antibodies, such as the modulation of inflammatory and immune reactions, may complement the effects of anti-Aβ antibodies on cognitive function in AD patients.[52]

6.5 Putative Mechanisms of Amyloid Plaque Removal *via* Immunotherapy

Immunisation against Aβ peptide represents a major new approach to AD therapeutics.[28–31]

Plaque clearance as a result of immunotherapy may depend on multiple mechanisms, one involving direct interaction of antibodies, or F(ab')₂ fragments, with the amyloid deposits resulting in disaggregation. The ability of single-chain antibody 508F(Fv), which lacks the Fc regions of antibodies, to dissolve already formed Aβ fibrils suggests that only the antigen binding site of the antibodies (Fab) was involved in disaggregation of Aβ and not the Fc region.[24] Additional studies with F(ab')₂ fragments of anti-AβP antibodies using *in vivo* topical application demonstrated that the mechanism does not require Fc receptor-mediated cellular activation in plaque clearance by immunotherapy.[53,54]

Another mechanism involving cell-mediated clearance suggests Fc-mediated phagocytosis can be involved in the clearance of Aβ deposits,[42,54,55] possibly involving Fc receptor activation on microglia cells. The two mechanisms may occur independently, or may operate in tandem, with cellular removal of Aβ after disaggregation.

Passive redistribution of soluble Aβ in the CSF or plasma may represent another possible mechanism for reduction of amyloid plaque.[44] Characterisation of these mechanisms may lead to an optimised therapeutic plan for efficient clearance of Aβ deposits from diseased brains. Three main mechanisms suggested so far have been summarised (see ref. 55).

6.6   A Novel Target for Immunotherapy – β-Secretase Cleavage Site on APP

In light of the ongoing debate regarding which β-amyloid species is the villain inducing neuronal and synaptic loss in AD, a logical way to intervene with the amyloid cascade is to interfere with Aβ production and partially restrict it, resulting probably in lower levels of all the aforementioned species. Considering previously described drawbacks involved in BACE1 inhibition,[56–60] the approach reviewed here may serve as an attractive alternative specifically aimed at inhibiting β-secretase activity leaving the native function of BACE1 towards its other substrates unharmed.

The importance of limiting intracellular Aβ formation at early stages of AD development is emphasised in many studies.[61,62] It is likely that the reduction in total Aβ levels need not be complete in order to implicate a beneficial effect if preventive therapy can be initiated. In AD caused by most mutations in APP, PS1, or PS2, the levels of Aβ42 are increased by as little as 30%.[63] Such an elevation can result in the onset of AD 30–40 years earlier than typical late onset AD cases. By inference, it is likely that reducing the total Aβ levels by 30%, or effecting similar selective reductions in the highly pathogenic Aβ42, may delay the development of AD to such an extent that it is no longer a major health-care problem.

6.6.1   Aβ Peptide Production from the Amyloid Precursor Protein (APP)

Aβ peptides are generated by sequential proteolytic processing of a larger amyloid precursor protein (APP) by two protease activities termed β- and γ-secretases. β-secretase (BACE1) initiates APP processing by cleaving its luminal domain to generate a secreted ectodomain (soluble β-APP). The remaining membrane-bound APP carboxy fragment (CTFβ/C99) is subsequently cleaved within the transmembrane domain by γ-secretase to generate mainly 40 or 42 amino acid Aβ peptides.[64]

APP is transported in the secretory pathway from the endoplasmic reticulum (ER) to the cell surface, from where it can be internalised into endosomal/lysosomal compartments.[65,66] During transport and at the cell surface APP can be cleaved by β-secretase.[67] Interestingly, it has been shown that the Swedish mutation that occurs within β-secretase cleavage site of APP and causes early onset AD led to increased β-secretase cleavage through the secretory pathway (*i.e.* ER and trans-Golgi) and increased secretion of Aβ peptides.[68–70] This effect is consistent with the finding that Swedish-mutated APP is cleaved much more efficiently than wild-type APP by the purified BACE1. In contrast, wild-type APP is cleaved by β-secretase preferentially after internalisation from the cell surface in endocytic compartments.[71,72] In neurons, APP metabolism releasing the Aβ peptide was found to occur at all sorting stations.

Aβ production through the endocytic pathway is well established.[71–75] Within the secretory and endocytic pathways, BACE1 shares major trafficking routes with its protein substrate APP. All these experiments confirmed that Aβ is derived through processing of APP endocytosed from the cell surface in addition to the secretory pathway.[67]

The fact that BACE1 initiates the amyloidogenic processing of APP along with its currently known structural and physiological characteristics, have rapidly promoted it as a prime target for drug discovery in AD. Although theoretically promising, BACE1 was shown to be a difficult pharmacologic target. Indeed, only two research groups have succeeded to obtain low molecular weight peptidomimetic inhibitors.[76–77] Although these inhibitors were shown to lower Aβ *in vivo*, their development is still in the preclinical stage.

Accumulating data raise some concerns regarding inhibition of BACE1. The last few years of research have shown that BACE1 is responsible for the proteolytic processing of several other substrates of the enzyme, including P-selectin glycoprotein ligand-1,[56] ST6Gal I,[57] APLP2[58] and Type-III NRG1.[59] While P-selectin glycoprotein ligand-1 and ST6Gal I play a role in leukocytes adhesion and B cells maturation, respectively, Type-III NRG1 is required for glial cells development and myelination. The exact function of the APP homologue, APLP2, is not entirely clear. BACE1 knockout mice display mild deficits in a task that assesses spatial memory,[60] suggesting that chronic reduction in BACE1 activity could impact learning and memory. Other proteases including cysteine proteases were shown to cleave APP at the β-secretase cleavage site,[78–80] insinuating that BACE1 inhibition may not preclude the generation of Aβ peptide.

6.6.2 Antibodies Against β-Secretase Cleavage Site of APP

In order to overcome the limitations presented by BACE1 inhibition methodologies mentioned previously, we proposed a new approach aimed to restrict β-secretase cleavage of APP itself rather than inhibit the enzyme activity. For that purpose, we developed a novel approach to inhibit the β-secretase cleavage of APP mainly through the endocytic pathway.[81] This approach exploits the interaction of APP and BACE1 at the cell surface, prior to their internalisation into the early endosomes,[75] to limit BACE1 activity by blocking β-secretase cleavage site (β-site) of APP using site-directed antibodies.

We have isolated a panel of monoclonal antibodies against APP β-site and have focused on the characterisation of one antibody in particular termed BBS1 (blocking β-site 1).[81] BBS1 antibody showed high-affinity binding to its immunogen which mimic the APP β-site and the generated immunocomplex was found to retain 95% stability within pH 5, the optimal pH for β-secretase activity that occurs in the early endosomes. The antibody ability to bind APP β-site in the context of full-length APP was demonstrated by using CHO cells overexpressing wild-type human APP as well as brain sections from human APP transgenic mice models. Conversely, despite the presence of the first two amino acid residues of Aβ in the immunogen used to generate this monoclonal antibody, BBS1 does not bind any form of Aβ peptides.

6.6.2.1   Antibody Interferes with Aβ Production in Cellular Model

Since most β-secretase activity is localised within intracellular compartments, we tested antibody ability to cointernalise into the cell with APP after binding at the cell surface at different time points. For that purpose, we transiently transfected the wild-type APP expressing cells with fluorescent resident proteins of either the early endosomes (EEA1), lysosomes (Lg120) or the Golgi apparatus (GalT).[82] Twenty-four hours post-transfection, we added the antibody to the condition media at different time points, after which cells were fixed, permeabilised, and antibody presence was detected with fluorescent antimouse IgG antibodies. BBS1 was rapidly internalised into the cell and could be detected in the early endosomes after very short intervals. At longer intervals (from 10 min and above), antibody was visualised in the early endosomes of an increasing number of cells in each field and its levels within the early endosomes were elevated with time. From 15 min incubation and above, BBS1 antibody could be detected in the lysosomes, with a more profound staining after 45 min. Antibody presence in the Golgi apparatus could be detected only after 45 min incubation. Antibody trafficking suggests receptor-mediated internalisation and trafficking through the endocytic pathway, as was expected knowing that APP is internalised into the early endosomes via clathrin-coated pits.[82]

Once administered in the growing media of cells overexpressing wild-type APP, BBS1 reduced the extra- and intracellular Aβ levels. The secreted Aβ levels were reduced by about 20% compared with the basal levels after short intervals (from 3 to 12 h). Interestingly, after five days incubation with the antibody, the intracellular Aβ levels were dramatically reduced (estimated as 50% reduction) in comparison with the basal intracellular Aβ levels. The levels of C99, the direct product of APP cleavage by BACE, were reduced by 20% after a 4-h treatment with BBS1 and were in accordance with the decrease in secreted Aβ levels at short intervals. Notably, antibodies directed to the amino terminal of APP have failed to produce such an effect, both with the extra- and intracellular Aβ pools. Importantly, BBS1 incubation did not harm the cells viability as determined by MTT reduction assay.[81]

6.6.2.2   Antibody Interference with AβP Production in Animal Model of AD

To investigate BBS1 treatment efficacy while administered through the peritoneum, we used female transgenic mice expressing the London-mutated APP (V717I) under the regulation of thy-1 promotor on a FVB x C57BL background (reMYND Inc., Leuven, Belgium).[83] The antibody was administered at two-week intervals for two months after which mice nonspatial memory was assessed prior to sacrifice (submitted for publication). Mice treated with BBS1 showed a positive trend of improved memory throughout training compared with saline-treated mice in the probe test in terms of time and frequencies. Insoluble Aβ levels in treated animals were reduced by 25% and 40% for Aβ42 and Aβ40, respectively. Biochemical evaluation of cerebral Aβ levels revealed 25% and 40% reduction in Aβ42 and Aβ40 levels in the antibody treated animals, respectively. The soluble Aβ levels were not changed between the two treatment groups. Membrane-associated Aβ oligomers were reduced by 24% in the antibody-treated animals. The relevance of the oligomeric Aβ ELISA, used in this study, to AD pathology has just been reported by Xie *et al.*,[84] demonstrating that Aβ species, detected by this assay in plasma samples of AD patients, seems to decline in the period of 1–2 years. Aβ oligomers measurements in these samples were tightly correlated with monomeric Aβ42, presumably reflecting their increasing insolubility in the brain.[84] Aβ oligomers were shown to form calcium-conducting channels in bilayer membranes *in vitro* and in intact neurons. Thus, it has been hypothesised that calcium conducted into the target neurons by the Aβ channel might be responsible for Aβ neurotoxicity. Considering these findings, reducing membrane associated Aβ as a result of antibody treatment, in addition to reduction of plaque burden and insoluble Aβ species, may be even more beneficial in preventing the cellular toxicity.

These results were further supported by a 40% reduction in the area occupied by the amyloid dense core as a result of antibody treatment. Similar to the results in the cellular model, BBS1 treatment led to about 20% reduction in soluble APPβ levels without altering the levels of sAPPα.

Since the BBS1 antibody does not bind Aβ plaques and is rapidly internalised into the cell after APP binding, microglia activation characterising Aβ-specific antibodies was not expected. We evaluated the levels of microglia activation using F4/80 staining in three distinct areas: dentate gyrus, hippocampus hilus and parietal cortex. A significant and dose-dependent decrease in microglia staining was observed in the treated groups compared with the saline-treated group. GFAP staining for reactive astrocytes did not show any difference between the three treatment groups. We further performed CD3 staining of brain sections to evaluate the possibility of T-cells penetrating the brain parenchyma. No T-cells were detected in any of the sections analysed.[85] Upregulation of the neuronal endocytic pathway was reported to occur in AD and Down syndrome (that develop AD pathology early in life) brains and provides some insight into how β-amyloidogenesis might be promoted in sporadic AD, the most prevalent and least understood form of the disease.[86] Anti-β-site antibodies interfere with APP processing through the endocytic pathway suggesting that this approach may be highly valuable to delay AD onset, so it will no longer be considered a major health problem.

6.7 Conclusions

Various active and passive immunisation protocols targeting Aβ have proven to be capable of providing significant therapeutic benefit in various behavioural and cognitive readouts when assessed in transgenic mouse models of AD.

The first clinical trial employing an active immunisation protocol, although halted for safety reasons, was indicative of clinical efficacy of the approach in humans.

The ongoing clinical trials of immunotherapy are summarised in Table 6.1.

A vaccine that is able to clear amyloid plaques in the brains of AD patients would offer a valuable avenue in the search for a solution to the disease: not only would it offer us potentially more effective clinical therapies, but it would also enable us to determine whether the accumulation of $A\beta$ in the brain in AD is a key causal feature of the disease, or a functionally less significant consequence. These are both key goals in the search to find an effective treatment (and possible cure) for AD. If future clinical trials continue successfully to completion, they are therefore likely to offer both the first, direct therapeutic evaluations of the amyloid hypothesis in AD patients, and potentially the first clinically effective treatment of the disease. Improved immunotherapeutic strategies may be used to obtain a beneficial effect without untoward side effects. The modulation of the inhibitory FcR pathway may be an efficient practical therapeutic approach for controlling autoantibody-mediated

Table 6.1 Ongoing human clinical trials based on antiamyloid immunotherapy.

	Company	*Product*	*Description*	*Status*
Active Immunization	Elan/Wyeth	ACC-001	$A\beta$-related immunotherapeutic conjugate	Phase II
	Affiris GmbH/ GlaxoSmith-Kline Plc	Affitope vaccine	Vaccine against $A\beta$	Phase I
	Cytos Bio-technology AG Novartis AG	CAD106	Vaccine with a fragment of the $A\beta$ protein	Phase II
Passive Immunization	Elan Corp. Plc/ Wyeth	Bapineuzumab (AAB-001)	Humanised mAb against N-terminal of $A\beta$	Phase III
	MorphoSys AG Roche	R1450	HuCAL-derived human mAb against $A\beta$	Phase I
	Eli Lilly and Co.	LY2062430	Antibody against mid-region of $A\beta$	Phase II
	Pfizer Inc.	PF-4360365 (RN1219)	Humanised mAb against $A\beta$ C-terminal	Phase II
	Genentech In. AC Immune SA	Anti-$A\beta$	Mab against $A\beta$	Phase I

inflammation induced by self-antigens or antibodies in immunotherapeutic strategies for the treatment of AD.

IVIg treatment may incorporate such immunomodulatory approaches and, associated with immunisation against Aβ peptide, may induce blockade of the Fc receptors on phagocytic cells by saturating, altering or downregulating the affinity of the Fc receptors, a process that may make the sensitised phagocytic cells unable to function. A therapeutic alternative to conventional mAbs may be the development of smaller Aβ-binding biologicals with more rapid terminal-elimination kinetics than IgGs, such as antibody fragments as single-chain variable domain antibody fragment (scFv) antibodies. These fragments contain the full natural antigen-binding pocket of IgGs but lack their constant domains. Such antibody fragments may have rates and mechanisms of transport via the blood/brain barrier entirely different from those of full-length antibodies, and may lead to a redistribution of Aβ peptide from binding to plasma proteins to an antibody-Aβ peptide pool and may have the potential for fast terminal elimination of Aβ from the circulation. In addition, the intrabody approach, a gene-based technology developed to neutralise or modify the function of intracellular and extracellular target antigens, can potentially target all the different isoforms of a misfolding-prone protein, including pathological conformations.

At present, there is no treatment available that can halt the progressive deterioration of cognitive functions in AD patients. The development of novel drugs with strong disease-modifying properties therefore represents one of the biggest unmet medical needs today. Several of the novel and promising therapeutic strategies specifically address the amyloid pathology. Strategies designed to decrease Aβ levels and thus to modify the disease progression, including active and passive Aβ immunisations, use of short peptides with β-sheet breaker elements, secretase inhibitors, cholesterol-lowering drugs, anti-inflammatory drugs and antioxidants, are currently being evaluated in pre-clinical and human clinical trials.

Acknowledgments

Thanks are extended to all those who helped contribute to this research: Dan Frenkel, Eilat Hanan-Aharon, Odelia Katz, Rela Koppel, Rachel Cohen-Kupiec, Vered Lavie, Maria Becker, Michal Arbel and Idan Rakover. Financial support was kindly accorded by ELAN and Wyeth; the Zenith Foundation of Alzheimer's Association; ISOA, and the Dana Foundation.

References

1. C. Mount and C. Downton, *Nature Med.*, 2006, **12**, 780.
2. D. J. Selkoe, *Physiol. Rev.*, 2001, **81**(2), 741.
3. J. Hardy, *J. Trends Neurosci.*, 1997, **20**, 154.

4. J. Hardy, K. Duff, K. G. Hardy, J. Perez-Tur and M. Hutton, *Nature Neurosci.*, 1998, **1**, 355.
5. J. Hardy and D. J. Selkoe, *Science*, 2002, **297**(558), 353.
6. D. J. Selkoe, *Neuron*, 1991, **6**, 487.
7. A. Goate, M. C. Chartier-Harlin, M. Mullan, J. Brown, F. Crawford, L. Fidani, L. Giuffra, A. Haynes, N. Irving and L. James, *et al.*, *Nature*, 1991, **349**(6311), 704.
8. E. Levy-Lahad, W. Wasco, P. Poorkaj, D. M. Romano, J. Oshima, W. H. Pettingell, C. E. Yu, P. D. Jondro, S. D. Schmidt and K. Wang, *et al.*, *Science*, 1995, **269**(5226), 973.
9. R. Sherrington, E. I. Rogaev, Y. Liang, E. A. Rogaeva, G. Levesque, M. Ikeda, H. Chi, C. Lin, G. Li and K. Holman, *et al.*, *Nature*, 1995, **375**(6534), 754.
10. M. N. Sabbagh, D. Galasko and L. J. Thal, *Alz. Dis. Rev.*, 1998, **3**, 1.
11. H. Frauenfelder, G. A. Petsko and D. Tsernoglou, *Nature*, 1979, **280**, 558.
12. M. Karplus and G. A. Petsko, *Nature*, 1990, **347**, 632.
13. M. E. Goldberg, *Trends Biochem.*, 1991, **16**, 358.
14. S. Blond and M. Goldberg, *Proc. Natl. Acad. Sci. USA*, 1987, **84**, 1147.
15. J. D. Carlson and M. L. Yarmush, *Bio/Technology*, 1992, **10**, 86.
16. B. Solomon and F. Schwartz, *J. Mol. Recognit.*, 1995, **8**, 72.
17. B. Solomon, T. Katzav-Gozanski, R. Koppel and E. Hanan-Aharon, in *Progress in Biotechnology 15: Stability and Stabilization of Biocatalysts*, ed. A. Ballesteros, F.J. Plou, J.L. Iborra and P.J. Halling, Elsevier, Amsterdam, 1998, p183.
18. B. Solomon and N. Balas, *Biotechnol. and Appl. Biochem.*, 1991, **14**, 202.
19. T. Katzav, E. Hanan and B. Solomon, *Appl. Biochem. Biotechnol.*, 1996, **2**, 227.
20. J. E. Maggio and P. W. Mantyh, *Brain Pathol.*, 1996, **6**, 147.
21. B. Solomon, R. Koppel, D. Frankel and E. Hanan-Aharon, *Proc. Natl. Acad. Sci. USA*, 1997, **94**, 4109.
22. B. Solomon, R. Koppel, E. Hanan and T. Katzav, *Proc. Natl. Acad. Sci. USA*, 1996, **93**(1), 452.
23. E. Hanan and B. Solomon, *Amyloid: Int. J. Exp. Clin. Invest.*, 1996, **3**, 130.
24. D. Frenkel, B. Solomon and I. Benhar, *J. Neuroimmunol.*, 2000, **106**, 23.
25. D. Frenkel, N. Balass and B. Solomon, *J. Neuroimmunol.*, 1998, **88**, 85.
26. F. Van Leuven, *Prog. Neurobiol.* 2000, **61**(3), 305. Review.
27. D. Games, D. Adams, R. Alessandrini, R. Barbour, P. Berthelette, C. Blackwell, T. Carr, J. Clemens, T. Donaldson and F. Gillespie, *et al.*, *Nature*, 1995, **373**(6514), 523.
28. B. Solomon, *Exp. Opin. Biol. Ther.*, 2002, **2**(8), 907–917.
29. B. P. Imbimbo, *Drug Dev. Res.*, 2002, **56**, 150.
30. R. C. Dodel, H. Hampel and Y. Du, *Lancet Neurol.*, 2003, **2**, 215.
31. A. Monsonego and H. L. Weiner, *Science*, 2003, **302**, 834.
32. D. Schenk, R. Barbour, W. Dunn, G. Gordon, H. Grajeda, T. Guido, K. Hu, J. Huang, K. Johnson-Wood, K. Khan, D. Kholodenko, M. Lee, Z. Liao, I. Lieberburg, R. Motter, L. Mutter, F. Soriano, G. Shopp, N.

Vasquez, C. Vandevert, S. Walker, M. Wogulis, T. Yednock, D. Games and P. Seubert, *Nature*, 1999, **400**(6740), 173.
33. H. L. Weiner, C. A. Lemere, R. Maron, E. T. Spooner, T. J. Grenfell, C. Mori, S. Issazadeh, W. W. Hancock and D. J. Selkoe, *Ann. Neurol.*, 2000, **8**(4), 567.
34. E. M. Sigurdsson, H. Scholtzova, P. D. Mehta, B. Frangione and T. Wisniewski, *Am. J. Pathol.*, 2001, **159**, 439.
35. K. Hsiao, P. Chapman, S. Nilsen, C. Eckman, Y. Harigaya, S. Younkin, F. Yang and G. Cole, *Science.*, 1996, **274**(5284), 99.
36. D. Frenkel, N. Balass, E. Katchalski-Katzir and B. Solomon, *J. Neuroimmunol.*, 1999, **95**, 136.
37. D. Frenkel, O. Katz and B. Solomon, *Proc. Natl. Acad. Sci. USA*, 2000, **97**, 11455.
38. D. Frenkel, N. Kariv and B. Solomon, *Vaccine (Elsevier)*, 2001, **19**, 2615.
39. D. Frenkel, I. Dewachter, F. Van Leuven and B. Solomon, *Vaccine*, 2003, **21**(11–12), 1060.
40. B. Solomon and D. Frenkel, *Drugs of Today*, 2000, **36**(9), 655.
41. V. Lavie, M. Becker, R. Cohen-Kupiec, I. Yacoby, R. Koppel, M. Wedenig, B. Hutter-Paier and B. Solomon, *J. Molec. Neurosc.*, 2004, **24**, 105.
42. F. Bard, C. Cannon, R. Barbour, R. L. Burke, D. Games, H. Grajeda, T. Guido, K. Hu, J. Huang, K. Johnson-Wood, K. Khan, D. Kholodenko, M. Lee, I. Lieberburg, R. Motter, M. Nguyen, F. Soriano, N. Vasquez, K. Weiss, B. Welch, P. Seubert, D. Schenk and T. Yednock, *Nature Med.*, 2000, **6**(8), 916.
43. M. Pfeifer, S. Boncristiano, L. Bondolfi, A. Stalder, T. Delle, M. Staufenbiel, P. M. Mathews and M. Jucker, *Science*, 2002, **298**(5597), 1379.
44. R. B. DeMattos, K. R. Bales, D. J. Cummins, J. C. Dodart, S. M. Paul and D. M. Holtzman, *Proc. Natl. Acad. Sci. USA*, 2001, **17**, 8850.
45. N. C. Carty, D. M. Wilcock, A. Rosenthal, J. Grimm, J. Pons, V. Ronan, P. E. Gottschall, M. N. Gordon and D. Morgan, *J. Neuroinflammation*, 2006, **10**, 3.
46. A. Asami-Odaka, Y. Obayashi-Adachi, Y. Matsumoto, H. Takahashi, H. Fukumoto, T. Horiguchi, N. Suzuki and M. Shoji, *Neurodegenerative Dis.*, 2005, **2**, 36.
47. D. M. Wilcock, A. Rojiani, A. Rosenthal, G. Levkowitz, S. Subbarao, J. Alamed, D. Wilson, N. Wilson, M. J. Freeman, M. N. Gordon and D. Morgan, *J. Neurosci.*, 2004, **24**(27), 6144.
48. R. Kayed, E. Head, J. L. Thompson, T. M. McIntire, S. C. Milton, C. W. Cotman and C. G. Glabe, *Science*, 2003, **300**, 486.
49. R. D. Moir, K. A. Tseitlin, S. Soscia, B. T. Hyman, M. C. Irizarry and R. E. Tanzi, *J. Biol. Chem.*, 2005, **280**(17), 17458.
50. R. C. Dodel, Y. Du, C. Depboylu, H. Hampel, L. Frölich, A. Haag, U. Hemmeter, S. Paulsen, S. J. Teipel, S. Brettschneider, A. Spottke, C. Nölker, H. J. Möller, X. Wei, M. Farlow, N. Sommer and W. H. Oertel, *J. Neurol. Neurosurg. Psychiatry*, 2004, **75**(10), 1472.
51. G. Istrin, E. Bosis and B. Solomon, *J. Neurosc. Res.*, 2006, **84**(2), 434.

52. F. Nimmerjahn and J. V. Ravetch, *J. Exp. Med.*, 2007, **204**, 11.
53. B. J. Bacskai, S. T. Kajdasz, R. H. Christie, C. Carter, D. Games and P. Seubert, *et al., Nature Med.*, 2001, **7**, 369.
54. B. J. Bacskai, S. T. Kajdasz, M. E. McLellan, D. Games, P. Seubert, D. Schenk and B. T. Hyman, *J. Neurosci.*, 2002, **22**(18), 7873.
55. B. Solomon, *Curr. Alz. Dis.*, 2004, **1**, 149.
56. S. F. Lichtenthaler, D. I. Dominguez, G. G. Westmeyer, K. Reiss, C. Haass, P. Saftig, B. De Strooper and B. Seed, *J. Biol. Chem.*, 2003, **278**, 48713.
57. S. Kitazume, Y. Tachida, R. Oka, N. Kotani, K. Ogawa, M. Suzuki, N. Dohmae, K. Takio, T. C. Saido and Y. Hashimoto, *J. Biol. Chem.*, 2003, **278**, 14865.
58. L. Pastorino, A. F. Ikin, S. Lamprianou, N. Vacaresse, J. P. Revelli, K. Platt, P. Paganetti, P. M. Mathews, S. Harroch and J. D. Buxbaum, *Mol. Cell Neurosci.*, 2004, **25**, 642.
59. M. Willem, A. N. Garratt, B. Novak, M. Citron, S. Kaufmann, A. Rittger, B. DeStrooper, P. Saftig, C. Birchmeier and C. Haass, *Science*, 2006, **314**, 664.
60. M. Ohno, E. A. Sametsky, L. H. Younkin, H. Oakley, S. G. Younkin, M. Citron, R. Vassar and J. F. Disterhoft, *Neuron*, 2004, **41**, 27.
61. G. K. Gouras, J. Tsai, J. Naslund, B. Vincent, M. Edgar, F. Checler, J. P. Greenfield, V. Haroutunian, J. D. Buxbaum, H. Xu, P. Greengard and N. R. Relkin, *Am. J. Pathol.*, 2000, **156**(1), 15.
62. F. M. LaFerla, *Nature Rev. Neuro.*, 2007, **8**, 499.
63. D. Scheuner, C. Eckman, M. Jensen, X. Song, M. Citron, N. Suzuki, T. D. Bird, J. Hardy, M. Hutton and W. Kukull, *et al., Nature Med.*, 1996, **2**, 864.
64. S. S. Sisodia and P. H. St George-Hyslop, *Nature Rev. Neurosci.*, 2002, **3**, 281.
65. E. H. Koo, S. L. Squazzo, D. J. Selkoe and C. H. Koo, *J. Cell Sci.*, 1996, **109**(Pt 5), 991.
66. T. Yamazaki, E. H. Koo and D. J. Selkoe, *J. Cell. Sci.*, 1996, **109**(Pt 5), 999.
67. S. S. Sisodia, E. H. Koo, K. Beyreuther, A. Unterbeck and D. L. Price, *Science*, 1990, **248**, 92.
68. R. G. Perez, S. Soriano, J. D. Hayes, B. Ostaszewski, W. Xia, D. J. Selkoe, X. Chen, G. B. Stokin and E. H. Koo, *J. Biol. Chem.*, 1999, **274**, 18851.
69. C. Haass, C. A. Lemere, A. Capell, M. Citron, P. Seubert, D. Schenk, L. Lannfelt and D. J. Selkoe, *Nature Med.*, 1995, **1**, 1291.
70. C. Haass, E. H. Koo, A. Mellon, A. Y. Hung and D. J. Selkoe, *Nature*, 1992, **357**, 500.
71. E. H. Koo and S. L. Squazzo, *J. Biol. Chem.*, 1994, **269**, 17386.
72. J. H. Chyung and D. J. Selkoe, *J. Biol. Chem.*, 2003, **278**, 51035.
73. J. T. Huse, D. S. Pijak, G. J. Leslie, V. M. Lee and R. W. Doms, *J. Biol. Chem.*, 2000, **275**, 33729.
74. J. Walter, R. Fluhrer, B. Hartung, M. Willem, C. Kaether, A. Capell, S. Lammich, G. Multhaup and C. Haass, *J. Biol. Chem.*, 2001, **276**, 14634.

75. A. Kinoshita, H. Fukumoto, T. Shah, C. M. Whelan, M. C. Irizarry and B. T. Hyman, *J. Cell. Sci.*, 2003, **116**, 3339.
76. W. P. Chang, G. Koelsch, S. Wong, D. Downs, H. Da, V. Weerasena, B. Gordon, T. Devasamudram, G. Bilcer and A. K. Ghosh, *et al.*, *J. Neurochem.*, 2004, **89**, 1409.
77. M. Asai, C. Hattori, N. Iwata, T. C. Saido, N. Sasagawa, B. Szabo, Y. Hashimoto, K. Maruyama, S. Tanuma and Y. Kiso, *et al.*, *J. Neurochem.*, 2006, **96**, 533.
78. V. Y. Hook, T. Toneff, W. Aaron, S. Yasothornsrikul, R. Bundey and T. Reisine, *J. Neurochem.*, 2002, **81**, 237.
79. V. Y. Hook and T. D. Reisine, *J. Neurosci. Res.*, 2003, **74**, 393.
80. V. Hook, M. Kindy and G. Hook, *Biol. Chem.*, 2007, **388**, 247.
81. M. Arbel, I. Yacoby and B. Solomon, *Proc. Natl. Acad. Sci. USA*, 2005, **102**, 7718.
82. C. Nordstedt, G. L. Caporaso, J. Thyberg, S. E. Gandy and P. Greengard, *J. Biol. Chem.*, 1993, **268**(1), 608.
83. D. Moechars, M. Gilis, C. Kuiperi, I. Laenen and F. Van Leuven, *Neuroreport*, 1998, **9**, 3561.
84. W. Xia, T. Yang, G. Shankar, I. M. Smith, Y. Shen, D. M. Walsh and D. J. Selkoe, *Arch. Neurol.*, 2009, **66**, 190.
85. I. Rakover, M. Arbel and B. Solomon, *Neurodegen. Dis.*, 2007, **4**(5), 392.
86. R. A. Nixon, P. M. Mathews and A. M. Cataldo, *J. Alz. Dis.*, 2001, **3**(1), 97.

TAU PROTEIN AS TARGET

Alzheimer Neurofibrillary Degeneration: Pivotal Role and Therapeutic Targets

KHALID IQBAL AND INGE GRUNDKE-IQBAL

Department of Neurochemistry, New York State Institute for Basic Research in Developmental Disabilities, 1050 Forest Hill Road, Staten Island, New York, 10314, USA

7.1 Introduction

Neurofibrillary degeneration of abnormally hyperphosphorylated tau is not only a hallmark but also the most pivotal lesion of Alzheimer disease (AD) and related tauopathies. Neurofibrillary degeneration is seen as neurofibrillary tangles, neuropil threads and as dystrophic neurites surrounding the amyloid core in the neuritic (senile) plaques. The major protein subunit of neurofibrillary tangles/paired helical filaments (PHF) was isolated in 1974 from bulk separated tangles from AD brain and identified by Western blots as microtubule-associated protein tau in 1986. The same year, we demonstrated (1) that tau in AD brain was abnormally hyperphosphorylated and, in this state, was polymerised into PHF/neurofibrillary tangles,[1] and (2) that, unlike normal tau, cytosolic abnormally hyperphosphorylated tau in AD brain was unable to promote microtubule assembly.[2] Subsequent studies showed the presence of tau in PHF by amino acid sequencing[3–5] and identified various abnormally hyperphosphorylated sites of tau.[6]

RSC Drug Discovery Series No. 2
Emerging Drugs and Targets for Alzheimer's Disease
Volume 1: Beta-Amyloid, Tau Protein and Glucose Metabolism
Edited by Ana Martinez

Published by the Royal Society of Chemistry, www.rsc.org

AD is multifactorial and heterogeneous. Identification of various etio-pathogenic mechanisms and of various subgroups of the disease is critical for the development of potent disease-modifying drugs. In less than 1% of the AD cases the disease cosegregates with certain mutations in β-amyloid precursor protein, presenilin-1 and presenlin-2.[7] Over 99% of the AD cases are not associated with any known mutations, and the nature of the etiological agent is not yet understood, but might involve metabolic and signal-transduction abnormalities.[8] These different etiological factors, nevertheless, may lead to some common downstream pathogenic events that ultimately produce the disease clinically. Independent of the etiology, AD is histopatholo-gically characterised by the presence of numerous neurofibrillary tangles and neuritic (senile) plaques with neurofibrillary changes in the dystrophic neur-ites.[9] In a large number of tangles, tau is ubiquitinated.[10–12] Several detailed reviews on each of these aspects of AD and therapeutic approaches have recently been published.[8,13–16] Here, we update some of the major findings on neurofibrillary degeneration of the abnormally hyperphosphorylated tau concerning the pivotal role of this lesion in the disease and as a drug target.

7.2 Pivotal Role

Although the amyloid cascade hypothesis postulates Aβ to cause neurofi-brillary degeneration,[17,18] the number of neurofibrillary tangles, and not the plaques, correlates best with the presence and/or the degree of dementia in AD.[19–21] Whereas neurofibrillary degeneration appears to be required for the clinical expression of the disease, the dementia, β-amyloidosis alone in the absence of neurofibrillary degeneration does not produce the disease clinically. In fact, some of the normal aged individuals have as much β-amyloid plaque burden in the brain as typical cases of AD, except that in the former case plaques lack dystrophic neurites with neurofibrillary changes surrounding the beta-amyloid cores.[19–20,22–24] On the other hand, neurofibrillary degeneration of the AD type, but in the absence of β-amyloidosis, is seen in several condi-tions such as Guam Parkinsonism-dementia complex, dementia pugilistica, frontotemporal dementia with Parkinsonism linked to chromosome-17 (FTDP-17), corticobasal degeneration, Pick disease, and progressive supranuclear palsy. All of these neurodegenerative disorders, collectively called tauopathies, are clinically characterised by dementia. Furthermore, in inherited cases of FTDP-17, certain missense mutations in the tau gene, including those that affect the alternate splicing of its mRNA, favouring the 4-repeat tau isoforms, cosegregate with the disease.[25–27] These mutated taus and the 4-repeat tau become more favourable substrates for abnormal hyperphosphorylation.[28]

7.3 Relationship Between β-Amyloidosis and Neurofibrillary Degeneration

Consistent with the amyloid cascade hypothesis, both intracerebral infusion of Aβ in FTDP-17 tau mutation P301L-expressing transgenic mice, as well as crossing these animals with APPTg2576 (APP Swedish plus London mutations), were found to exacerbate neurofibrillary pathology.[29,30] Furthermore, in the triple transgenic mice 3XTgAD (APP_{SWE}-PS1M146V-tau P301L), β-amyloid deposition was found to precede the neurofibrillary pathology and these animals showed more neurofibrillary pathology than the double transgenic Tg2X mice.[31,32] These effects of Aβ on neurofibrillary pathology in transgenic mice could be due to activation of stress-activated protein kinases that are known to abnormally hyperphosphorylate tau.[33,34] However, to date, the data from human conditions apparently do not support the amyloid cascade hypothesis – (1) some of the normal aged individuals show similar level and topography of compact Aβ plaques as typical cases of AD, except that the plaques in the former lack dystrophic neurites with neurofibrillary tau pathology; (2) the plaques and neurofibrillary tangles are seen in disproportionate numbers in AD, especially in the plaque-dominant and tangle-dominant AD subgroups; (3) typically, a considerably higher brain Aβ burden is seen in hereditary cerebral hemorrhage with amyloidosis, Dutch type (HCHWA-D) but without any accompanying neurofibrillary degeneration,[35] and (4) the tauopathies, such as FTDP-17, Pick disease, corticobasal degeneration, dementia pugilistica and Guam Parkinsonism dementia complex, are characterised by dementia associated with neurofibrillary degeneration of abnormally hyperphosphorylated tau in the absence of β-amyloid plaques. Furthermore, recent studies have shown that PS-1 not only promotes or acts as a γ-secretase activity (the cleavage of APP which produces Aβ), but also activates the phosphatidylinositol 3-kinase (PI3K), which downstream through protein kinase B (Akt) inhibits the glycogen synthase kinase-3 (GSK-3), a major tau kinase. Some of the AD-causing mutations in PS-1 result in loss of its ability to activate PI3K pathway, resulting in a sustained activity of GSK-3 and, consequently, abnormal hyperphosphorylation of tau.[36] Finally, several of the AD-causing PS-1 mutations have been reported to produce no change to a decrease in Aβ generation in cultured cells.[37] In our view, AD may be caused by a number of different factors and the amyloid cascade hypothesis is too simplistic and narrow to explain this multifactorial disease. We have proposed[8] that different signal transduction and metabolic factors, through different disease mechanisms, apparently lead to the same two disease characteristic lesions – neurofibrillary degeneration of abnormally hyperphosphorylated tau and β-amyloidosis (Figure 7.1).

7.4   Molecular Mechanism of Neurofibrillary Degeneration

To date, not only in AD but also in every known human tauopathy, the tau pathology is made up of the abnormally hyperphosphorylated protein. Normal tau is highly hydrophilic and is, thus, highly soluble and heat stable. While conformational changes[38–40] and truncation of tau[41–43] have been reported in AD, the most established and the most compelling cause of dysfunctional tau in

Neurofibrillary Degeneration and *Theraputic Targets*

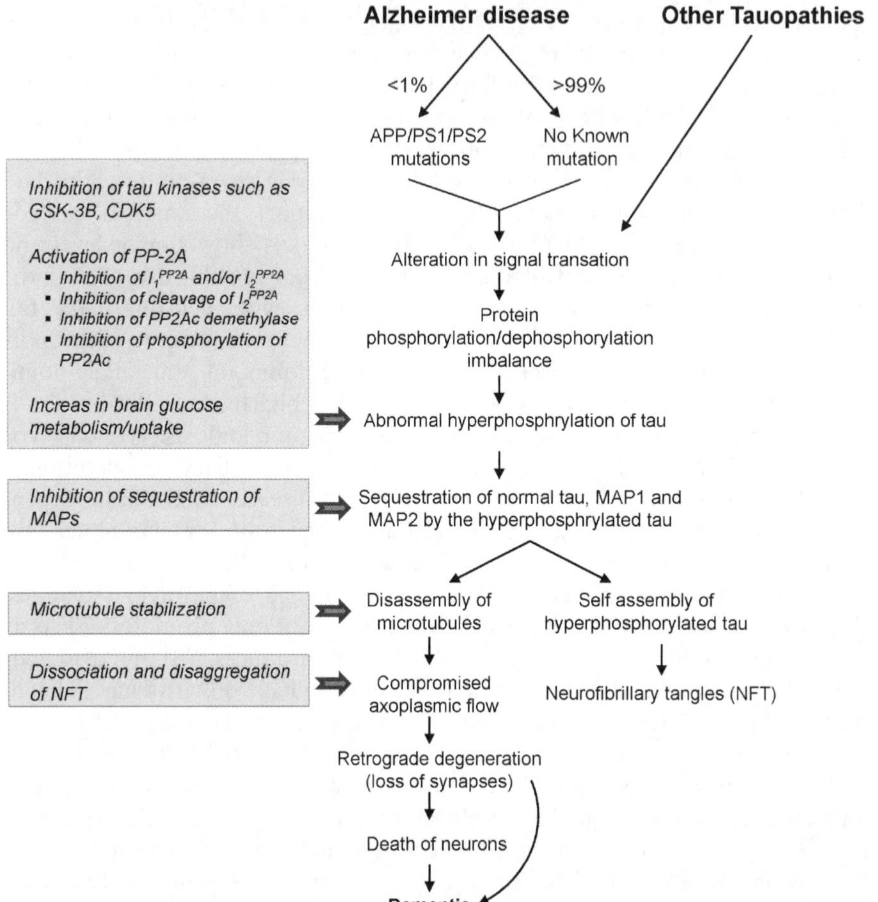

Figure 7.1   Major steps involved in neurofibrillary degeneration of abnormally hyperphosphorylated tau and therapeutic targets that might be considered to inhibit or counteract this lesion at various stages of the pathology. (Modified from Iqbal and Grundke-Iqbal, 2005.)

AD and related tauopathies is the abnormal hyperphosphorylation of this protein.[1,2,44] Tau, a phosphoprotein which normally contains 2–3 mol. of phosphate/mol. of the protein,[45] is abnormally hyperphosphorylated in AD brain[2] and, in this state, is the major protein subunit of the PHF/neurofibrillary tangles.[1,4,46,47] Tau can self-aggregate into PHF/straight filaments (SF) through its microtubule binding repeats R2 and R3, which have the β-structure.[48] However, the flanking regions, especially the amino terminal region to the repeat region, inhibit the self-aggregation of tau. On abnormal hyperphosphorylation as it occurs in AD and related tauopathies, this inhibition is diminished and tau self-assembles into filaments forming the neurofibrillary tangles.[49] Interestingly, some of tau in AD brain is both amino and C-terminally truncated which also promotes its abnormal hyperphosphorylation.[50,51] A transgenic rat model that expresses human brain truncated, tau, $tau_{151-391}$, (found in AD brain) shows neurofibrillary tangles of abnormally hyperphosphorylated tau in three–six months old animals.[52]

Two major known functions of tau are its ability to promote assembly and to maintain the structure of microtubules.[53] These functions of tau are regulated by its degree of phosphorylation.[44,54–56] In AD brain there is as much normal tau as in age-matched control human brain, but, in addition, the diseased brain contains 4–8-fold of abnormally hyperphosphorylated tau.[57,58] As much as 40% of the abnormally hyperphosphorylated tau is present in the cytosol and not polymerised into PHF/neurofibrillary tangles.[45] The tau polymerised into neurofibrillary tangles is apparently inert and neither binds to tubulin nor promotes its assembly into microtubules.[55–56,59] In contrast, the AD cytosolic abnormally hyperphosphorylated tau (AD P-tau) is not only unable to bind to tubulin and promote microtubule assembly, but also inhibits assembly and disrupts microtubules.[44,60] This toxic property of the pathological tau involves the sequestration of normal tau by the diseased protein.[44,61] The AD P-tau also sequesters the other two major neuronal microtubule-associated proteins MAP1 A/B and MAP2.[62] This toxic behaviour of the AD P-tau appears to be solely due to its abnormal hyperphosphorylation because dephosphorylation of diseased tau converts it into a normal-like protein.[44,60,63,64] Furthermore, *in vitro* dephosphorylation of neurofibrillary tangles disaggregates filaments and, as a result, the tau released behaves like normal protein in promoting microtubule assembly.[63] Thus, two characteristics of AD abnormally hyperphosphorylated tau are (1) that it sequesters normal MAPs and disrupts microtubules and (2) that it self-assembles into paired helical and/or SF. On self-aggregation into filaments, the abnormally hyperphosphorylated tau loses its ability to sequester normal tau and inhibit microtubule assembly.[59]

Tau mutations, which cause FTDP-17, result either in an increase in 4-repeat:3-repeat tau ratio or in missense mutations in the protein. Both 4-repeat tau and the mutated protein are more easily abnormally hyperphosphorylated than the normal wild-type protein.[28,65] Thus, inhibition of the abnormal hyperphosphorylation of tau is likely to inhibit neurofibrillary degeneration and consequently the diseases characterised by this lesion.

7.5   Regulation of Tau Phosphorylation

The state of phosphorylation of tau is a function of the balance between the activities of tau protein kinases and the phosphatases and of other post-translational modifications of the serines/threonines involved. Tau, which is phosphorylated at over 38 serine/threonine residues in AD,[6,66] is a substrate for several protein kinases.[67,68] Among these kinases, GSK-3, cyclin-dependent protein kinase-5 (cdk5), protein kinase A (PKA), calcium and calmodulin-dependent protein kinase-II (CaMKII), casein kinase-1 (CK-1), mitogen-activated protein (MAP) kinase ERK 1/2, and stress-activated protein kinases have been most implicated in the abnormal hyperphosphorylation of tau.[69,70] GSK-3β and cdk5 phosphorylate tau at a large number of sites, most of which are common to the two enzymes.[71,72] The expressions of GSK-3β and cdk5 are high in the brain[73–75] and both enzymes have been shown to be associated with all stages of neurofibrillary pathology in AD.[76,77]

In AD brain the activities of protein phosphatase (PP)-2A and PP-1 are compromised by ∼20%.[78,79] The phosphorylation of tau that suppresses its microtubule binding and assembly activities in adult mammalian brain is regulated by PP-2A and not by PP-2B.[80,81] PP-2A accounts for over 70% of all phosphoseryl/phosphothreonyl activity in human brain[82] and also regulates the activities of several tau kinases in brain. Inhibition of PP-2A activity by okadaic acid in cultured cells and in metabolically active rat brain slices results in abnormal hyperphosphorylation of tau at several of the same sites as in AD, not only directly by a decrease in dephosphorylation but also indirectly by promoting the activities of CaM Kinase II,[81] PKA,[83,84] MAP kinase (MEK1/2), extracellular regulated kinase (ERK 1/2) and P70S6 kinase.[70,85]

The intracellular activities of PP2A and PP1 are regulated by endogenous inhibitors. PP-1 activity is regulated mainly by a 18.7-kDa heat-stable protein called inhibitor-1 (I-1).[86,87] In addition, a structurally related protein, DARPP-32 (dopamine and cAMP-regulated phosphoprotein of apparent molecular weight 32 000) is expressed predominantly in the brain.[88] I-1 and DARPP-32 are activated on phosphorylation by protein kinase A and inactivated by calcineurin, and at basal calcium level by PP-2A.[89] Thus, inhibition of PP-2A activity would keep I-1, DARPP-32 in the active form and thereby result in a decrease in PP-1 activity. In AD brain a reduction in PP-2A activity might have decreased the PP-1 activity by allowing the upregulation of the I-1/DARPP-32 activity. In the subgroup of AD cases and/or at moderate to severe stages of the disease, when there is a persistent excitotoxicity and an increase in the intra-neuronal calcium, DARPP-32 is probably mainly dephosphorylated and thereby inactivated as PP-1 inhibitor by calcineurin.

PP-2A is inhibited in the mammalian tissue by two heat-stable proteins: (i) the I_1^{PP2A}, a 30-kDa cytosolic protein[90] that inhibits PP-2A with a K_i of 30 nM and (ii) the I_2^{PP2A}, a 39-kDa nuclear protein that inhibits PP-2A with a K_i of 23 nM.[90] In AD brain there is a shift from nuclear to cytoplasmic localisation of I_2^{PP2A}.[91] Both I_1^{PP2A} and I_2^{PP2A} interact with the catalytic subunit of PP2A.[92,93] The levels of both I_1^{PP2A} and I_2^{PP2A} are upregulated by ∼20% in the affected areas of AD brain.[91] Furthermore, I_2^{PP2A}, which is primarily nuclear, is cleaved into an amino-terminal fragment and a carboxy-terminal fragment selectively in AD brain, resulting in a shift in its intraneuronal localisation from the nucleus to the cytoplasm. In the neuronal cytoplasm the cleaved I_2^{PP2A} colocalises with PP2A and with neurofibrillary tangles. The exact nature of the protease activity that causes the cleavage of I_2^{PP2A} into amino-terminal and carboxy-terminal halves in AD brain is, at present, not known. The upregulation of I_1^{PP2A} and I_2^{PP2A}, and the cleavage and the translocation of the latter from the neuronal nucleus to the cytoplasm are probably causative of the decrease in PP2A activity and the consequent neurofibrillary degeneration of abnormally hyperphosphorylated tau.

Besides I_1^{PP2A} and I_2^{PP2A}, the PP2A activity is also regulated by methylation and phosphorylation of its catalytic subunit and by the nature of its regulatory β subunit that determines both its substrate specificity and subcellular localisation.[94–96] Methylation of the PP2A catalytic subunit at L309 upregulates its activity, whereas the phosphorylation at Y307 decreases the activity. A decrease in methylation[97,98] and an increase in phosphotyrosinylation[99] of PP2A have been observed in AD brain. Thus, these two post-translational modifications of PP2A could also have contributed to the decrease of this enzyme activity in AD brain.

The abnormal hyperphosphorylation of tau could also be due to its decreased O-GlcNAcylation caused by a decreased brain glucose metabolism in AD. In contrast to classical N- or O-glycosylation, O-GlcNAcylation which involves the addition of a single sugar at serine/threonine residues of a protein, dynamically post-translationally modifies cytoplasmic and nuclear proteins in a manner analogous to protein phosphorylation.[100] O-GlcNAcylation and phosphorylation reciprocally regulate each other. Levels of the two major brain glucose transporters, GLUT1 and GLUT3, which are responsible for glucose uptake into neurons, are decreased in AD.[101] This decrease correlates to the downregulation of O-GlcNAcylation, to the hyperphosphorylation of tau, and to the density of neurofibrillary tangles in AD brain.[102] Furthermore, the hypoxia-inducible factor-1, which is a major regulator of GLUT1 and GLUT3 is downregulated in AD brain. Decreased glucose metabolism in cultured cells and in mice, which decreases the O-GlcNAcylation of tau, produces abnormal hyperphosphorylation of this protein.[103]

Based on CSF levels of total tau, $A\beta_{1-42}$, and ubiquitin, we have identified five subgroups of AD.[104] We postulate that more than one disease mechanism and signalling pathway are involved in producing the AD pathology, especially the neurofibrillary degeneration of abnormally hyperphosphorylated tau in these various subgroups of AD.

7.6   Therapeutic Targets

Currently, several hundred drugs for AD are under development by the pharmaceutical industry. Stratification of the test subjects in clinical trials by disease subgroups may increase the chance of success up to several fold. The future of therapeutic drugs for AD may depend on recognition of different subgroups of the disease. Neurofibrillary degeneration of abnormally hyperphosphorylated tau is downstream to alterations in more than one signalling pathway, but pivotally involved in the pathogenesis of AD and related tauopathies (Figure 7.1). Inhibition of this lesion is likely to arrest these diseases, which are products of multiple etiopathogenic mechanisms. The most promising therapeutic approaches to inhibit neurofibrillary degeneration and consequently AD and related tauopathies are (1) to inhibit the abnormal hyperphosphorylation of tau, (2) to inhibit sequestration of normal MAPs by the AD P-tau, (3) to inhibit misfolding of tau, and (4) to directly stabilise microtubules. The inhibition of abnormal hyperphosphorylation of tau can be carried out by inhibiting activities of both GSK-3 and cdk5, by activating the PP-2A activity or by increasing brain glucose uptake/metabolism which could enhance O-GlcNAcylation and consequently the inhibition of the abnormal hyperphosphorylation of tau.

Most inhibitors of GSK-3 also inhibit cdk5 and *vice versa*. To overcome this problem, several highly selective inhibitors of each of these kinases have been recently developed by the pharmaceutical industry, and some of these compounds are at different stages of human clinical trials for the treatment of AD. Our *in vitro* studies on the generation of abnormally hyperphosphorylated tau[105] suggest that compounds that inhibit both GSK-3 and cdk5 might even be more effective than the highly selective inhibitors of one of these enzymes in inhibiting neurofibrillary degeneration.

I_1^{PP2A} and I_2^{PP2A}, as the major regulators of PP2A activity, are very attractive therapeutic targets for inhibition of abnormal hyperphosphorylation of tau. In the case of I_2^{PP2A}, the inhibition of its cleavage into the N-terminal and C-terminal fragments is another therapeutic approach. Inhibition of phosphotyrosinylation of PP2A catalytic subunit and of the activity of PP2A demethylase may also be employed to increase the PP2A activity.

Memantine, a low- to moderate-affinity NMDA receptor antagonist, which improves mental function and the quality of daily living of patients with moderate to severe AD,[106,107] restores the PP-2A activity, and reduces the abnormal hyperphosphorylation of tau at Ser-262 and the associated neurodegeneration in hippocampal slice cultures from adult rats, and PC-12 cells in culture.[83,108] Furthermore, the restoration of the PP-2A activity to normal levels by memantine also results in the restoration of the expression of MAP2 in the neuropil and a reversal of the hyperphosphorylation and the accumulation of neurofilament H and M subunits. Memantine, however, is a positively charged molecule and probably enters a neuron only during excitotoxicity when the NMDA receptor channels are open. Therefore, its therapeutic benefit might be limited to only those patients and/or the advanced states of the disease when there is a persistent excitotoxicity. Generation of cell-permeable memantine-like compounds can help develop potent therapeutic drugs for AD and related tauopathies. The restoration of the PP-2A activity appears to be due to the binding of memantine to I_2^{PP2A} and disinhibition of its activity towards PP-2A.[108] The CSF level of phosphotau is significantly reduced in AD patients after one year treatment with memantine.[109] All these findings taken together suggest that PP-2A is a promising therapeutic target for AD and related tauopathies.

Recent studies suggest inhibition of calpains as another approach for the inhibition of neurofibrillary degeneration.[110] Calpain, the activity of which is upregulated in AD brain, activates cdk5 through cleavage of its activators p39 and p35 to p29 and p25, respectively.[111,112] Calpain also cleaves and activates calcineurine, which regulates the phosphorylation of CREB.[110] Thus, inhibition of calpain can be neuroprotective, both by inhibition of cdk5 activity and by increase in CREB activity.

The abnormally hyperphosphorylated tau causes neurofibrillary degeneration by sequestration of normal MAPS. A competitive inhibition of this sequestration by small molecules can arrest this pathology. Though inhibition of protein–protein interaction is generally a somewhat challenging task, there is a great advantage in developing drugs against such specific targets.

Since O-GlcNAcylation and phosphorylation reciprocally regulate each other, an approach, independent of modulation of tau kinases and phosphatases, is to restore the O-GlcNAcylation of tau, which is compromised in AD[103] to a normal level. This could probably be achieved by increasing brain glucose levels through upregulating the activity of neuronal glucose transporters such as GLUT3 and/or by mediation of the activity of the O-GlcNAcylase with PUGNec- and NAG-AE-like compounds.[113]

The accumulation of abnormally hyperphosphorylated tau in AD and related tauopathies indicates that either the ubiquitin-proteasome system and/or the chaperones in the affected neurons are overwhelmed. Isopeptidase activity might be increased, keeping the hyperphosphorylated tau from poly-ubiquitination for degradation by the ubiquitin proteasome pathway. Thus, isopeptidase inhibitors and drugs that promote heat-shock protein-mediated clearance of tau may also inhibit neurofibrillary degeneration. Recently, HSP90 inhibitors have been shown to result in clearance of hyperphosphorylated tau through increase in the expression of chaperones HSP70-interacting protein (CHIP), a tau ubiquitin ligase, that refolds the misfolded proteins.[114,115] Most recently, immunisation of P301L transgenic mice with a small tau phospho-peptide has been reported to clear the hyperphosphorylated tau.[116] Another approach to overcome the inhibitory activity of the hyperphosphorylated tau is the use of microtubule stabilising drugs like taxol.[84,115,117] Drugs such as methylene blue that can dissociate and disaggregate neurofibrillary tangles/PHF have also been recently reported to ameliorate cognitive impairment of AD patients in a double-blind placebo-controlled clinical trial.[118] The exact molecular mechanism by which this drug apparently seems to benefit the AD patients remains to be understood.

A large majority of therapeutic drugs currently under development for AD are focused on inhibiting β-amyloid. However, there is increasing interest in developing drugs that can inhibit tau pathology. After all, neurofibrillary degeneration of abnormally hyperphosphorylated tau is apparently required for the clinical expression of AD and tau pathology alone in the absence of any β-amyloid causes frontotemporal dementia and other tauopathies. Independent of whether the amyloid cascade hypothesis proves to be true or untrue, inhibition of neurofibrillary degeneration is likely to inhibit the clinical phonotype of AD and related tauopathies. A polytherapy targeting both β-amyloid and neurofibrillary degeneration might be synergistically beneficial in AD patients.

At present, a large majority of the antineurofibrillary degeneration drugs under development are GSK-3 inhibitors; a few target extracellular regulated kinase (ERK)-2, tau phosphatases, and aggregation of the hyperphosphory-lated tau into filaments. Inhibition of the abnormal hyperphosphorylation of tau appears to be the most promising therapeutic target for AD and related tauopathies. Inhibition of GSK-3 activity and modulation of PP-2A, which is the major tau phosphatase and the activity of which is compromised in AD brain, are among the most attractive approaches to inhibit the abnormal hyperphosphorylation of tau. In the case of PP-2A, the restoration of its activity to a normal level should have low risk of any deleterious side effects. GSK-3 is involved in several important signalling pathways and inhibiting its activity carries risks, but then, apparently there is a safe therapeutic window for this enzyme activity; LiCl, a GSK-3 inhibitor, has been successfully prescribed for bipolar disorder for many years.

An advantage of targeting neurofibrillary degeneration for the development of therapeutic drugs is that the efficacy of such neuroprotective drugs can be directly monitored by assaying the CSF level of total tau as a marker of neurodegeneration and of various phosphotaus as markers of inhibition of the abnormal hyperphosphorylation of tau. A retrospective study of 2661 autopsied brains has revealed that neurofibrillary degeneration precedes by several years the clinical expression, *i.e.* dementia, in AD.[119] It will be very important to be able to detect neurofibrillary degeneration, probably by determining the CSF levels of total tau and phosphotaus at the presymptomatic stage of the disease. Inhibition of neurofibrillary degeneration in presymptomatic individuals can most probably prevent AD and related tauopathies. $APOE_4$ carriers who are at ~ 3.5-fold (one $APOE_4$ allele) to ~ 10-fold ($APOE_{4/4}$) higher risk than the noncarriers of this allele for developing late onset AD,[120] individuals with Down syndrome who invariably develop AD histopathology in the fourth decade of life, individuals with a strong family history of AD or related tauopathies and individuals with the risk of familial forms of these diseases are among others who can be employed for prevention clinical trials.

Acknowledgments

We are grateful to Janet Murphy for secretarial assistance. Studies in our laboratories were supported in part by the New York State Office of Mental Retardation and Developmental Disabilities and NIH grants AG019158 and AG028538, and Alzheimer's Association (Chicago, IL) grant IIRG-00-2002.

References

1. I. Grundke-Iqbal, K. Iqbal, Y. C. Tung, M. Quinlan, H. M. Wisniewski and L. I. Binder, *Proc. Natl. Acad. Sci. USA*, 1986, **83**, 4913.
2. K. Iqbal, I. Grundke-Iqbal, T. Zaidi, P. A. Merz, G. Y. Wen, S. S. Shaikh, H. M. Wisniewski, I. Alafuzoff and B. Winblad, *Lancet*, 1986, **2**, 421.

3. C. M. Wischik, M. Novak, H. C. Thogersen, P. C. Edwards, M. J. Runswick, R. Jakes, J. E. Walker, C. Milstein, M. Roth and A. Klug, *Proc. Natl. Acad. Sci. USA*, 1988, **85**, 4506.

4. K. Iqbal, I. Grundke-Iqbal, A. J. Smith, L. George, Y. C. Tung and T. Zaidi, *Proc. Natl. Acad. Sci. USA*, 1989, **86**, 5646.

5. J. Kondo, T. Honda, H. Mori, Y. Hamada, R. Miura, M. Ogawara and Y. Ihara, *Neuron*, 1988, **1**, 827.

6. M. Morishima-Kawashima, M. Hasegawa, K. Takio, M. Suzuki, H. Yoshida, K. Titani and Y. Ihara, *J. Biol. Chem.*, 1995, **270**, 823.

7. D. Campion, C. Dumanchin, D. Hannequin, B. Dubois, S. Belliard, M. Puel, C. Thomas-Anterion, A. Michon, C. Martin, F. Charbonnier, G. Raux, A. Camuzat, C. Penet, V. Mesnage, M. Martinez, F. Clerget-Darpoux, A. Brice and T. Frebourg, *Am. J. Hum. Genet.*, 1999, **65**, 664.

8. K. Iqbal and I. Grundke-Iqbal, *Acta Neuropathol. (Berl.)*, 2005, **109**, 25–31.

9. H. Braak, E. Braak, I. Grundke-Iqbal and K. Iqbal, *Neurosci. Lett.*, 1986, **65**, 351.

10. H. Mori, J. Kondo and Y. Ihara, *Science*, 1987, **235**, 1641.

11. G. Perry, R. Friedman, G. Shaw and V. Chau, *Proc. Natl. Acad. Sci. USA*, 1987, **84**, 3033.

12. G. P. Wang, S. Khatoon, K. Iqbal and I. Grundke-Iqbal, *Brain Res.*, 1991, **566**, 146.

13. D. J. Selkoe, *Arch. Neurol.*, 2005, **62**, 192.

14. K. Iqbal and I. Grundke-Iqbal, *Cell. Mol Life. Sci.*, 2007, **64**, 2234.

15. J. Hardy, *Neuron*, 2006, **52**, 3.

16. K. Iqbal and I. Grundke-Iqbal, *J. Cell. Mol. Med.*, 2008, **12**, 38.

17. J. Hardy and D. J. Selkoe, *Science*, 2002, **297**, 353.

18. J. A. Hardy and G. A. Higgins, *Science*, 1992, **256**, 184.

19. I. Alafuzoff, K. Iqbal, H. Friden, R. Adolfsson and B. Winblad, *Acta Neuropathol. (Berl.)*, 1987, **74**, 209.

20. P. V. Arriagada, J. H. Growdon, E. T. Hedley-Whyte and B. T. Hyman, *Neurology*, 1992, **42**, 631.

21. B. E. Tomlinson, G. Blessed and M. Roth, *J. Neurol. Sci.*, 1970, **11**, 205.

22. D. W. Dickson, J. Farlo, P. Davies, H. Crystal, P. Fuld and S. H. Yen, *Am. J. Pathol.*, 1988, **132**, 86.

23. D. W. Dickson, H. A. Crystal, L. A. Mattiace, D. M. Masur, A. D. Blau, P. Davies, S. H. Yen and M. K. Aronson, *Neurobiol. Aging*, 1992, **13**, 179.

24. R. Katzman, R. Terry, R. DeTeresa, T. Brown, P. Davies, P. Fuld, X. Renbing and A. Peck, *Ann. Neurol.*, 1988, **23**, 138.

25. M. Hutton, C. L. Lendon, P. Rizzu, M. Baker, S. Froelich, H. Houlden, S. Pickering-Brown, S. Chakraverty, A. Isaacs, A. Grover, J. Hackett, J. Adamson, S. Lincoln, D. Dickson, P. Davies, R. C. Petersen, M. Stevens, E. de Graaff, E. Wauters, J. van Baren, M. Hillebrand, M. Joosse, J. M. Kwon, P. Nowotny, L. K. Che, J. Norton, J. C. Morris, L. A. Reed, J. Trojanowski, H. Basun, L. Lannfelt, M. Neystat, S. Fahn, F. Dark, T. Tannenberg, P. R. Dodd, N. Hayward, J. B. Kwok, P. R. Schofield, A.

Andreadis, J. Snowden, D. Craufurd, D. Neary, F. Owen, B. A. Oostra, J. Hardy, A. Goate, J. van Swieten, D. Mann, T. Lynch and P. Heutink, *Nature*, 1998, **393**, 702.

26. P. Poorkaj, T. D. Bird, E. Wijsman, E. Nemens, R. M. Garruto, L. Anderson, A. Andreadis, W. C. Wiederholt, M. Raskind and G. D. Schellenberg, *Ann. Neurol.*, 1998, **43**, 815.

27. M. G. Spillantini, J. R. Murrell, M. Goedert, M. R. Farlow, A. Klug and B. Ghetti, *Proc. Natl. Acad. Sci. USA*, 1998, **95**, 7737.

28. A. D. Alonso, A. Mederlyova, M. Novak, I. Grundke-Iqbal and K. Iqbal, *J. Biol. Chem.*, 2004, **279**, 34873.

29. J. Gotz, F. Chen, J. van Dorpe and R. M. Nitsch, *Science*, 2001, **293**, 1491–1495.

30. J. Lewis, E. McGowan, J. Rockwood, H. Melrose, P. Nacharaju, M. Van Slegtenhorst, K. Gwinn-Hardy, M. Paul Murphy, M. Baker, X. Yu, K. Duff, J. Hardy, A. Corral, W. L. Lin, S. H. Yen, D. W. Dickson, P. Davies and M. Hutton, *Nature Genet.*, 2000, **25**, 402.

31. S. Oddo, A. Caccamo, M. Kitazawa, B. P. Tseng and F. M. LaFerla, *Neurobiol. Aging*, 2003, **24**, 1063.

32. S. Oddo, A. Caccamo, J. D. Shepherd, M. P. Murphy, T. E. Golde, R. Kayed, R. Metherate, M. P. Mattson, Y. Akbari and F. M. LaFerla, *Neuron*, 2003, **39**, 409.

33. S. Kins, P. Kurosinski, R. M. Nitsch and J. Gotz, *Am. J. Pathol.*, 2003, **163**, 833.

34. J. J. Pei, S. Khatoon, W. L. An, M. Nordlinder, T. Tanaka, H. Braak, I. Tsujio, M. Takeda, I. Alafuzoff, B. Winblad, R. F. Cowburn, I. Grundke-Iqbal and K. Iqbal, *Acta Neuropathol. (Berl.)*, 2003, **105**, 381.

35. E. Levy, M. D. Carman, I. J. Fernandez-Madrid, M. D. Power, I. Lieberburg, S. G. van Duinen, G. T. Bots, W. Luyendijk and B. Frangione, *Science*, 1990, **248**, 1124.

36. L. Baki, J. Shioi, P. Wen, Z. Shao, A. Schwarzman, M. Gama-Sosa, R. Neve and N. K. Robakis, *Embo J.*, 2004, **23**, 2586.

37. J. Shioi, A. Georgakopoulos, P. Mehta, Z. Kouchi, C. M. Litterst, L. Baki and N. K. Robakis, *J. Neurochem.*, 2007, **101**, 674.

38. G. A. Jicha, E. Lane, I. Vincent, L. Otvos Jr, R. Hoffmann and P. Davies, *J. Neurochem.*, 1997, **69**, 2087.

39. G. A. Jicha, B. Berenfeld and P. Davies, *J. Neurosci. Res.*, 1999, **55**, 713.

40. G. A. Jicha, J. M. Rockwood, B. Berenfeld, M. Hutton and P. Davies, *Neurosci. Lett.*, 1999, **260**, 153.

41. M. Novak, R. Jakes, P. C. Edwards, C. Milstein and C. M. Wischik, *Proc. Natl. Acad. Sci. USA*, 1991, **88**, 5837.

42. T. C. Gamblin, F. Chen, A. Zambrano, A. Abraha, S. Lagalwar, A. L. Guillozet, M. Lu, Y. Fu, F. Garcia-Sierra, N. LaPointe, R. Miller, R. W. Berry, L. I. Binder and V. L. Cryns, *Proc. Natl. Acad. Sci. USA*, 2003, **100**, 10032.

43. C. W. Cotman, W. W. Poon, R. A. Rissman and M. Blurton-Jones, *J. Neuropathol. Exp. Neurol.*, 2005, **64**, 104.

44. A. D. Alonso, T. Zaidi, I. Grundke-Iqbal and K. Iqbal, *Proc. Natl. Acad. Sci. USA*, 1994, **91**, 5562.
45. E. Kopke, Y. C. Tung, S. Shaikh, A. C. Alonso, K. Iqbal and I. Grundke-Iqbal, *J. Biol. Chem.*, 1993, **268**, 24374.
46. I. Grundke-Iqbal, K. Iqbal, M. Quinlan, Y. C. Tung, M. S. Zaidi and H. M. Wisniewski, *J. Biol. Chem.*, 1986, **261**, 6084.
47. V. M. Lee, B. J. Balin, L. Otvos Jr and J. Q. Trojanowski, *Science*, 1991, **251**, 675.
48. M. Arrasate, M. Perez, R. Armas-Portela and J. Avila, *FEBS Lett.*, 1999, **446**, 199.
49. A. D. Alonso, T. Zaidi, M. Novak, I. Grundke-Iqbal and K. Iqbal, *Proc. Natl. Acad. Sci. USA*, 2001, **98**, 6923.
50. N. Zilka, M. Korenova and M. Novak, *Acta Neuropathol.*, 2009, Epub ahead of print.
51. M. Novak, J. Kabat and C. M. Wischik, *Embo J.*, 1993, **12**, 365.
52. N. Zilka, P. Filipcik, P. Koson, L. Fialova, R. Skrabana, M. Zilkova, G. Rolkova, E. Kontsekova and M. Novak, *FEBS Lett.*, 2006, **580**, 3582.
53. M. D. Weingarten, A. H. Lockwood, S. Y. Hwo and M. W. Kirschner, *Proc. Natl. Acad. Sci. USA*, 1975, **72**, 1858.
54. G. Lindwall and R. D. Cole, *J. Biol. Chem.*, 1984, **259**, 5301.
55. K. Iqbal, T. Zaidi, C. Bancher and I. Grundke-Iqbal, *FEBS Lett.*, 1994, **349**, 104.
56. S. Khatoon, I. Grundke-Iqbal and K. Iqbal, *J. Neurochem.*, 1995, **64**, 777.
57. S. Khatoon, I. Grundke-Iqbal and K. Iqbal, *J. Neurochem.*, 1992, **59**, 750.
58. S. Khatoon, I. Grundke-Iqbal and K. Iqbal, *FEBS Lett.*, 1994, **351**, 80.
59. A. D. Alonso, B. Li, I. Grundke-Iqbal and K. Iqbal, *Proc. Natl. Acad. Sci USA*, 2006, **23**, 8864.
60. B. Li, M. O. Chohan, I. Grundke-Iqbal and K. Iqbal, *Acta Neuropathol. (Berl.)*, 2007, **113**, 501.
61. A. D. Alonso, I. Grundke-Iqbal and K. Iqbal, *Nature Med.*, 1996, **2**, 783.
62. A. D. Alonso, I. Grundke-Iqbal, H. S. Barra and K. Iqbal, *Proc. Natl. Acad. Sci. USA*, 1997, **94**, 298.
63. J. Z. Wang, C. X. Gong, T. Zaidi, I. Grundke-Iqbal and K. Iqbal, *J. Biol. Chem.*, 1995, **270**, 4854.
64. J. Z. Wang, I. Grundke-Iqbal and K. Iqbal, *Brain Res. Mol. Brain Res.*, 1996, **38**, 200.
65. K. Bhaskar, S. H. Yen and G. Lee, *J. Biol. Chem.*, 2005, **280**, 35119.
66. D. P. Hanger, J. C. Betts, T. L. Loviny, W. P. Blackstock and B. H. Anderton, *J. Neurochem.*, 1998, **71**, 2465.
67. T. J. Singh, I. Grundke-Iqbal, B. McDonald and K. Iqbal, *Mol. Cell. Biochem.*, 1994, **131**, 181.
68. G. V. Johnson and J. A. Hartigan, *J. Alzheimers Dis.*, 1999, **1**, 329.
69. K. Iqbal, C. Alonso Adel, S. Chen, M. O. Chohan, E. El-Akkad, C. X. Gong, S. Khatoon, B. Li, F. Liu, A. Rahman, H. Tanimukai and I. Grundke-Iqbal, *Biochim. Biophys. Acta*, 2005, **1739**, 198.

70. J. J. Pei, C. X. Gong, W. L. An, B. Winblad, R. F. Cowburn, I. Grundke-Iqbal and K. Iqbal, *Am. J. Pathol.*, 2003, **163**, 845.
71. J. Z. Wang, Q. Wu, A. Smith, I. Grundke-Iqbal and K. Iqbal, *FEBS Lett.*, 1998, **436**, 28.
72. B. H. Anderton, J. Betts, W. P. Blackstock, J. P. Brion, S. Chapman, J. Connell, R. Dayanandan, J. M. Gallo, G. Gibb, D. P. Hanger, M. Hutton, E. Kardalinou, K. Leroy, S. Lovestone, T. Mack, C. H. Reynolds and M. Van Slegtenhorst, *Biochem. Soc. Symp.*, 2001, **73**.
73. J. R. Woodgett, *Embo J.*, 1990, **9**, 2431.
74. L. H. Tsai, T. Takahashi, V. S. Caviness Jr and E. Harlow, *Development*, 1993, **119**, 1029.
75. J. Lew, Q. Q. Huang, Z. Qi, R. J. Winkfein, R. Aebersold, T. Hunt and J. H. Wang, *Nature*, 1994, **371**, 423.
76. J. J. Pei, I. Grundke-Iqbal, K. Iqbal, N. Bogdanovic, B. Winblad and R. F. Cowburn, *Brain Res.*, 1998, **797**, 267.
77. J. J. Pei, E. Braak, H. Braak, I. Grundke-Iqbal, K. Iqbal, B. Winblad and R. F. Cowburn, *J. Neuropathol. Exp. Neurol.*, 1999, **58**, 1010.
78. C. X. Gong, T. J. Singh, I. Grundke-Iqbal and K. Iqbal, *J. Neurochem.*, 1993, **61**, 921.
79. C. X. Gong, S. Shaikh, J. Z. Wang, T. Zaidi, I. Grundke-Iqbal and K. Iqbal, *J. Neurochem.*, 1995, **65**, 732.
80. C. X. Gong, T. Lidsky, J. Wegiel, L. Zuck, I. Grundke-Iqbal and K. Iqbal, *J. Biol. Chem.*, 2000, **275**, 5535.
81. M. Bennecib, C. X. Gong, I. Grundke-Iqbal and K. Iqbal, *FEBS Lett.*, 2001, **490**, 15.
82. F. Liu, I. Grundke-Iqbal, K. Iqbal and C. X. Gong, *Eur. J. Neurosci.*, 2005, **22**, 1942.
83. L. Li, A. Sengupta, N. Haque, I. Grundke-Iqbal and K. Iqbal, *FEBS Lett.*, 2004, **566**, 261.
84. T. Tanaka, J. Zhong, K. Iqbal, E. Trenkner and I. Grundke-Iqbal, *FEBS Lett.*, 1998, **426**, 248.
85. W. L. An, R. F. Cowburn, L. Li, H. Braak, I. Alafuzoff, K. Iqbal, I. G. Iqbal, B. Winblad and J. J. Pei, *Am. J. Pathol.*, 2003, **163**, 591.
86. P. Cohen, *Ann. Rev. Biochem.*, 1989, **58**, 453.
87. P. Cohen, S. Alemany, B. A. Hemmings, T. J. Resink, P. Stralfors and H. Y. Tung, *Methods Enzymol.*, 1988, **159**, 390.
88. S. I. Walaas and P. Greengard, *Pharmacol. Rev.*, 1991, **43**, 299.
89. A. Nishi, G. L. Snyder, A. C. Nairn and P. Greengard, *J. Neurochem.*, 1999, **72**, 2015.
90. M. Li, H. Guo and Z. Damuni, *Biochemistry*, 1995, **34**, 1988.
91. H. Tanimukai, I. Grundke-Iqbal and K. Iqbal, *Am. J. Pathol.*, 2005, **166**, 1761.
92. S. Chen, I. Grundke-Iqbal and K. Iqbal, *Alzheimer's & Dementia*, 2006, **2**, S471.
93. S. Chen, B. Li, I. Grundke-Iqbal and K. Iqbal, *J. Biol. Chem.*, 2008, **283**, 10513.

94. D. M. Virshup, *Curr. Opin. Cell. Biol.*, 2000, **12**, 180.

95. J. Chen, B. L. Martin and D. L. Brautigan, *Science*, 1992, **257**, 1261.

96. V. Janssens and J. Goris, *Biochem. J.*, 2001, **353**, 417.

97. E. Sontag, A. Luangpirom, C. Hladik, I. Mudrak, E. Ogris, S. Speciale and C. L. White 3rd, *J. Neuropathol. Exp. Neurol.*, 2004, **63**, 287.

98. X. W. Zhou, J. A. Gustafsson, H. Tanila, C. Bjorkdahl, R. Liu, B. Winblad and J. J. Pei, *Neurobiol. Dis.*, 2008.

99. R. Liu, X. W. Zhou, H. Tanila, C. Bjorkdahl, J. Z. Wang, Z. Z. Guan, Y. Cao, J. A. Gustafsson, B. Winblad and J. J. Pei, *J. Cell. Mol. Med.*, 2008, **12**, 241.

100. G. W. Hart, *Ann. Rev. Biochem.*, 1997, **66**, 315.

101. Y. Liu, F. Liu, K. Iqbal, I. Grundke-Iqbal and C. X. Gong, *FEBS Lett.*, 2008, **582**, 359.

102. F. Liu, J. Shi, H. Tanimukai, J.-H. Gu, J.-L. Gu, I. Grundke-Iqbal, K. Iqbal and C. X. Gong, Brain, 2009, E-pub ahead of print.

103. F. Liu, K. Iqbal, I. Grundke-Iqbal, G. W. Hart and C. X. Gong, *Proc. Natl. Acad. Sci. USA*, 2004, **101**, 10804.

104. K. Iqbal, M. Flory, S. Khatoon, H. Soininen, T. Pirttila, M. Lehtovirta, I. Alafuzoff, K. Blennow, N. Andreasen, E. Vanmechelen and I. Grundke-Iqbal, *Ann. Neurol.*, 2005, **58**, 748.

105. J. Z. Wang, I. Grundke-Iqbal and K. Iqbal, *Eur. J. Neurosci.*, 2007, **25**, 59.

106. B. Winblad and N. Poritis, *Int. J. Geriatr. Psychiatry*, 1999, **14**, 135.

107. B. Reisberg, R. Doody, A. Stoffler, F. Schmitt, S. Ferris and H. J. Mobius, *N. Engl. J. Med.*, 2003, **348**, 1333.

108. M. O. Chohan, S. Khatoon, I. G. Iqbal and K. Iqbal, *FEBS Lett.*, 2006, **580**, 3973.

109. M. Degerman Gunnarsson, L. Kilander, H. Basun and L. Lannfelt, *Dement. Geriatr. Cogn. Disord.*, 2007, **24**, 247.

110. F. Liu, I. Grundke-Iqbal, K. Iqbal, Y. Oda, K. Tomizawa and C. X. Gong, *J. Biol. Chem.*, 2005, **280**, 37755.

111. G. Kusakawa, T. Saito, R. Onuki, K. Ishiguro, T. Kishimoto and S. Hisanaga, *J. Biol. Chem.*, 2000, **275**, 17166.

112. H. Patzke and L. H. Tsai, *J. Biol. Chem.*, 2002, **277**, 8054.

113. M. K. Tallent, N. Varghis, Y. Skorobogatko, L. Hernandez-Cuebas, K. Whelan, D. J. Vocadlo and K. Vosseller, *J. Biol. Chem.*, 2009, **284**, 174.

114. C. A. Dickey, A. Kamal, K. Lundgren, N. Klosak, R. M. Bailey, J. Dunmore, P. Ash, S. Shoraka, J. Zlatkovic, C. B. Eckman, C. Patterson, D. W. Dickson, N. S. Nahman Jr, M. Hutton, F. Burrows and L. Petrucelli, *J. Clin. Invest.*, 2007, **117**, 648.

115. B. Zhang, A. Maiti, S. Shively, F. Lakhani, G. McDonald-Jones, J. Bruce, E. B. Lee, S. X. Xie, S. Joyce, C. Li, P. M. Toleikis, V. M. Lee and J. Q. Trojanowski, *Proc. Natl. Acad. Sci. USA*, 2005, **102**, 227.

116. A. A. Asuni, A. Boutajangout, D. Quartermain and E. M. Sigurdsson, *J. Neurosci.*, 2007, **27**, 9115.

117. M. L. Michaelis, S. Ansar, Y. Chen, E. R. Reiff, K. I. Seyb, R. H. Himes, K. L. Audus and G. I. Georg, *J. Pharmacol. Exp. Ther.*, 2005, **312**, 659.
118. C. Wischik and R. Staff, *J. Nutr. Health Aging*, 2009, **13**, 367.
119. H. Braak and E. Braak, *Neurobiol. Aging*, 1997, **18**, 351.
120. A. D. Roses, *J. Alzheimers Dis.*, 2006, **9**, 361.

Multiple Roles of Glycogen Synthase Kinase-3 in Alzheimer's Disease

MATHIEU LESORT AND RICHARD S. JOPE

Department of Psychiatry and Behavioral Neurobiology, University of Alabama at Birmingham, Birmingham, AL 35294-0017, USA

8.1 Introduction

One of the most notable developments in research concerning Alzheimer's disease is the solid evidence directly linking glycogen synthase kinase-3 (GSK-3) to most, if not all, of the major mechanisms known to contribute to the neuropathology of the disease. GSK-3 interacts with multiple components of the plaque-producing amyloid system associated with Alzheimer's disease. GSK-3 participates in phosphorylating the microtubule-binding protein tau, which may contribute to the formation of neurofibrillary tangles (NFTs) in Alzheimer's disease. GSK-3 interacts with presenilin and other Alzheimer's disease-associated proteins. GSK-3 has a central role in neuronal plasticity and memory. GSK-3 promotes both inflammation and apoptosis. These many links between GSK-3 and Alzheimer's disease are discussed in this chapter following an initial brief review of the function and regulation of GSK-3, which have been covered in more detail in recent reviews.[1,2]

RSC Drug Discovery Series No. 2
Emerging Drugs and Targets for Alzheimer's Disease
Volume 1: Beta-Amyloid, Tau Protein and Glucose Metabolism
Edited by Ana Martinez
© Royal Society of Chemistry 2010
Published by the Royal Society of Chemistry, www.rsc.org

8.2   GSK-3 Function and Regulation

GSK-3 is a constitutively partially active serine/threonine kinase. Two GSK-3 mammalian genes encode two highly homologous forms of GSK-3, GSK-3β and GSK-3α, that often, but not always, have overlapping functions.[3,4] GSK-3 knockout mouse models have provided evidence of functional divergence, as disruption of the murine GSK-3β gene results in embryonic lethality caused by severe liver degeneration during mid-gestation, indicating GSK-3α cannot compensate for the loss of GSK-3β.[5] GSK-3α knockout mice are viable but display enhanced glucose and insulin sensitivity accompanied by reduced fat mass.[6] GSK-3 was first identified as enzymes that phosphorylate glycogen synthase in the glycogen synthesis pathway.[7] Subsequently, more than 50 substrates for GSK-3 have been identified, demonstrating that GSK-3 is involved in regulating a large variety of cellular processes, such as neuronal plasticity and gene transcription,[8] cell growth and survival,[8] apoptosis,[9] inflammation,[10] and microtubule stability,[11] Within the central nervous system GSK-3 immunoreactivity is detected in most brain regions although significant local variations of levels occur.[12,13] GSK-3β expression is more prevalent in neurons relative to astrocytes, and in adult neurons GSK-3β is particularly localised in the perikarya and proximal part of dendrites.[13]

The actions of GSK-3 in numerous cellular functions suggests that its activity is tightly regulated. Four fundamental mechanisms by which GSK-3's substrate-specific activity is regulated have been identified, including regulation by phosphorylation of GSK-3 itself, its subcellular localisation, the formation of regulatory protein complexes, and the phosphorylation state of GSK-3 substrates.[1] The most well-defined regulatory mechanism is inhibition of the activity of GSK-3 by phosphorylation of a regulatory serine, serine-9 in GSK-3β or serine-21 in GSK-3α.[14] The phosphatidylinositol 3-kinase (PI3K)/Akt signalling pathway activated in response to insulin and many other growth factors is often a major regulator of GSK-3, because Akt phosphorylates GSK-3 on these inhibitory serine residues.[14] However, several other kinases are also capable of mediating this modification, including P70S6 kinase, P90Rsk, certain isoforms of protein kinase C, and cyclic AMP-dependent protein kinase.[8] In addition to inhibitory serine phosphorylation, GSK-3β and GSK-3α activity can be regulated by tyrosine phosphorylation at residues 216 and 279, respectively.[15] Under physiological conditions GSK-3 is phosphorylated at these residues and increased tyrosine phosphorylation, in contrast to phosphorylation at the serine residue, enhances the enzymatic activity of GSK-3.[15] The kinases mediating this modification have not been clearly identified but could include the src tyrosine kinase Fyn[16] and ZAK1.[17] Alternatively, it has been proposed that during the protein-folding process GSK-3β and GSK-3α autophosphorylate at residues 216 and 279, respectively.[18] In addition to phosphorylation, GSK-3 activity is also regulated by its intracellular localisation. Some substrates of GSK-3 are cytosolic, whereas others, notably several transcription factors including phospho-CREB, are nuclear and others are mitochondrial.[19,20] Thus, it is evident that GSK-3 must be localised and in these cellular compartments, and there is evidence that the regulation of nuclear and mitochondrial levels or phosphorylation of GSK-3 is a dynamic process controlled by signalling cascades.[21–23] GSK-3 activity is also modulated by protein complex formation. Regulation of GSK-3 by protein complexes is most well described in regard to the evolutionary conserved Wnt/wingless signalling pathway.[24,25] GSK-3, along with the product of the adenomatous polyposis coli gene, and the scaffolding protein Axin, functions as a negative regulator of the Wnt signalling pathway. When Wnt signalling is inactive (*i.e.* Wnt ligand is not bound to the Frizzled receptor), the GSK-3/adenomatous polyposis coli gene product/Axin protein complex facilitates the phosphorylation of β-catenin by GSK-3, which targets β-catenin for degradation. Activation of the Wnt signalling pathway disables GSK-3-mediated phosphorylation of β-catenin. In this case, the binding of Wnt to Frizzled activates the protein Disheveled that, in combination with GSK-3-binding proteins (GBP, Frat-1 and Frat-2), binds GSK-3 and blocks its phosphorylation of β-catenin. This allows β-catenin to accumulate and translocate to the nucleus where it binds TCF/LEF proteins, which facilitates the transcription of Wnt target genes. These studies demonstrate both the regulatory effects of protein complex formation in controlling the actions of GSK-3, and the critical role that GSK-3 plays in regulating the β-catenin/TCF/LEF transcription factors. Finally, the action of GSK-3 is usually regulated by the phosphorylation state of its substrate, an indirect mechanism that can regulate how efficiently GSK-3 phosphorylates a substrate. This is because most substrates of GSK-3 require a previous "primed" phosphorylation by a priming kinase on a Ser or Thr residue located four amino acids C-terminal to the Ser or Thr residue to be phosphorylated by GSK-3. This necessitates temporal coordination of the activity of the priming kinase along with GSK-3 activity for the latter to phosphorylate the primed substrate.[24] Resolution of the crystal structure of GSK-3 has provided a model for the binding of prephosphorylated substrates to GSK-3.[26,27] In this model, the primed Ser or Thr sites are recognised by a positively charged binding pocket that facilitates the binding and phosphorylation of the GSK-3 substrate.[26,27] Thus, GSK-3 activity is tightly regulated in several ways, providing important mechanisms to control the actions of an enzyme that can phosphorylate many substrates and regulate numerous cellular functions.

8.3   GSK-3 Status in Alzheimer's Disease

The implication that GSK-3 is involved in the pathology of Alzheimer's disease would suggest there may be alterations in its level, activity or cellular localisation in Alzheimer's disease brains. Increased levels of GSK-3β have been found in Alzheimer's disease, compared with nondiseased, human brain,[28–33] and other tauopathies.[34,35] Expression of GSK-3 is increased in the hippocampus of Alzheimer's disease patients and in the postsynaptosomal supernatant from Alzheimer's disease brains as compared to controls,[32] however, in the latest study no increase in GSK-3 enzyme activity was detected.[32] In an attempt to identify peripheral markers of Alzheimer's disease, total GSK-3 and inhibited serine-phosphorylated GSK-3β were measured in white cells from Alzheimer's disease patients, elderly controls, and patients with mild cognitive impairment, which demonstrated that the GSK-3 level was increased without a compensatory decrease in activity in both subjects with Alzheimer's disease or mild cognitive impairment.[36] These findings are in line with studies showing increased expression levels of GSK-3 in Alzheimer's disease brains, and suggest that peripheral GSK-3 level may constitute a surrogate biomarker of disease. Previous studies examining the cellular localisation in postmortem brains provided evidence that GSK-3β colocalises with neuropathological lesions characteristic of Alzheimer's disease. Immunohistochemical measurements located GSK-3β associated with neurofibrillary tangles in Alzheimer's disease brain[28–33] and other tauopathies.[34,35] Regional and intracellular examinations revealed that GSK-3 initially accumulates in the cytoplasm of pretangle neurons in the prealpha layer of the entorhinal cortex and extended to other brain regions, coincident with the development of the neurofibrillary pathology.[32,33] In line with these findings, a significant increase in the level of the active form of GSK-3β (pTyr216) was observed in the frontal cortex of Alzheimer's disease patients,[30] and interestingly the majority of neurons at the pretangle and tangle stages were also positive for GSK-3β (pTyr216)-positive neurons, suggesting that GSK-3 activation is an early event preceding and accompanying the formation of NFT in AD.[30] However, it is important to note that in these studies the association of NFT with GSK-3 was variable, not all NFT colocalised with GSK-3 and in some cases there was no colocalisation. Altogether and interpreted within the limitations of the immunohistochemical approach in postmortem tissues, these studies provide evidence that GSK-3 levels and activity are increased in Alzheimer's disease, a condition that could contribute to the neuropathological manifestations of the disease.

8.4   GSK-3 and Aβ

According to the amyloid hypothesis for the pathogenesis of Alzheimer's disease, the amyloid β (Aβ) peptide directly affects neurons resulting in tau phosphorylation leading to the production of paired helical filaments in NFTs and neurodegeneration.[37] Evidence from several studies has associated GSK-3 with this hypothesis through its interactions with the amyloid precursor protein, the Aβ peptide product of the proteolysis of amyloid precursor protein, the Aβ-mediated increases in tau phosphorylation, and Aβ-induced neurotoxicity. Early evidence revealed that GSK-3β phosphorylates the amyloid precursor protein.[38,39] Conversely, there is evidence that proteolytic products of the amyloid precursor protein may affect GSK-3 activity. Physiological cleavage of amyloid precursor protein by α-secretase results in the production of a soluble amyloid precursor protein and a C-terminal fragment (CTF). Expression of the CTF of amyloid precursor protein in cell cultures resulted in increased GSK-3 expression and activity, and subsequent increased tau phosphorylation and apoptosis.[40] In pathological conditions proteolysis of amyloid precursor protein by the sequential actions of β-secretase and γ-secretase results in the formation of Aβ and an amyloid precursor protein intracellular domain (AICD), which may function as a transcription factor.[41] *In vitro* studies have provided evidence that AICD translocates into the nucleus resulting in increased GSK-3β expression and activity.[40] Further studies examining the relationship between Aβ and GSK-3 revealed that Aβ leads to increased GSK-3β activity (but not GSK-3α) through inhibition of PI3K, resulting in phosphorylation on several tau residues, which may contribute to the accumulation of hyperphosphorylated tau in Alzheimer's disease (discussed further below).[42–44] Importantly, GSK-3 has been reported to promote the processing of amyloid precursor protein to Aβ, but there is conflicting data about which isoform of GSK-3 mediates this effect.[45,46] Lithium and valproic acid, both known to be GSK-3 inhibitors, regulate amyloid precursor protein processing and inhibit Aβ production in cell-culture systems,[45] and lithium treatment reduces Aβ and the plaque burden in the brains of the APP(V717F) Alzheimer's disease transgenic mouse model.[45] However in this study the effect of lithium on Aβ production was mediated by GSK-3β, as the presence of GSK-3β dominant-negative constructs and GSK-3β antisense mimicked the effect of lithium.[45] Furthermore, Aβ-mediated neurotoxicity was reduced by antisense GSK-3β oligonucleotides[47] and by the GSK-3 inhibitor lithium.[48,49] Altogether, these findings provide evidence that GSK-3 may promote Aβ production, resulting in accumulation of Aβ, phosphorylation of tau, and cell death. Thus, GSK-3 may constitute a key connection between the two neuro-pathological hallmarks in Alzheimer's disease, extracellular Aβ accumulation and the intraneuronal formation of NFTs composed of hyperphosphorylated tau.

8.5　GSK-3 and Tau

The accumulation of hyperphosphorylated tau into paired helical filaments (PHFs) constitutes a major neuropathological hallmark of Alzheimer's disease. Tau is a multifunctional microtubule-associated protein that plays key roles in the assembly of microtubules and stabilisation of microtubules against dynamic instability.[50–52] In the brain, the equilibrium between phosphorylation and dephosphorylation of tau modulates the stability of the cytoskeleton and contributes to axonal remodelling and synaptic plasticity.[53] In Alzheimer's disease, tau is abnormally phosphorylated, resulting in structural and con-formational changes that affect its capacity to bind and stabilise microtubules and facilitate its aggregation into paired helical filaments in NFTs.[53] The abnormally increased tau phosphorylation is likely the result of an imbalance in the activities of kinases and phosphatases that regulate the phosphorylation state of tau. Numerous phosphorylation sites on tau have been identified in PHFs isolated from Alzheimer's disease brains and several kinases have been proposed to contribute to these modifications.[54] There is considerable evidence from numerous *in vitro* studies, but also from animal models, that GSK-3 contributes to the phosphorylation of tau, and there is speculation that GSK-3 may contribute to the abnormal hyperphosphorylation of tau observed in neurofibrillary tangles.[55–61]

In a seminal study, two protein kinases initially named tau protein kinase (TPK) I and II were isolated from the microtubule fractions of rat and bovine brain extracts that were associated with the phosphorylation of tau and were subsequently found to be GSK-3β and cyclin-dependent kinase (CDK) 5, respectively.[31] TPKI/GSK-3β antiserum-labelled intraneuronal NFT in subjects with Alzheimer's disease, and double immunolabelling for TPKI/GSK-3β and tau showed that TPKI/GSK-3β was closely associated with NFTs.[31] These findings, together with the studies noted above about GSK-3β colocalisation with NFTs in Alzheimer's disease brains,[30–33,62] provide tantalising evidence that GSK-3β is a critically important kinase for the *in vivo* generation of hyperphosphorylated tau and subsequent formation of the paired helical filaments in NFT. However, it is important to note that not all NFTs colocalise with GSK-3,[62] and in another study the localisation of GSK-3α and GSK-3β did not correspond to the expected pattern of neuronal vulnerability to NFT formation in Alzheimer's disease, arguing against a major role of GSK-3β in NFT formation in Alzheimer's disease.[63] More direct evidence of GSK-3-mediated tau phosphorylation has been provided from *in vitro* and cell-culture studies, approaches that have also illuminated the implication of these changes on the biophysics and functional properties of tau.[11,55,56] In a cell-free model system, GSK-3β phosphorylates tau within epitopes recognised by Alzheimer's disease-associated antibodies, and phosphorylation of tau by GSK-3β promotes polymer/polymer interactions that result in stable clusters of tau filaments.[55–56,64,65] Numerous studies in various cell-culture systems have provided compelling evidence that GSK-3 can phosphorylate tau *in situ* at many sites, including disease-relevant epitopes.[55,56] Most of these studies are based on approaches that overexpress GSK-3 and have revealed that the GSK-3-mediated phosphorylation of tau decreases tau binding to microtubules and significantly alters the development of neurite outgrowth or processes containing microtubule bundles. In line with these findings, the GSK-3 inhibitor lithium reduces the phosphorylation of tau, enhances the binding of tau to microtubules, and promotes microtubule assembly.[66,67] As mentioned above GSK-3β can phosphorylate substrates at unprimed sites at Ser/Thr-Pro motifs and at primed sites in the ((S/T)XXX(S/T)[P]) motif.[2] Interestingly although GSK-3β phosphorylates tau at unprimed sites, at least a portion of the GSK-3β-mediated phosphorylation of tau may be hierarchical, as prior phosphorylation of tau by CDK5/p25 enhanced the subsequent phosphorylation of tau by GSK-3β at specific sites (*e.g.* Thr-231),[68–70] indicating that tau may be both a primed and unprimed substrate of GSK-3β.[68,69] This hypothesis and the physiological relevance of primed GSK-3β sites on tau was confirmed by studies employing GSK-3β-R96A, a GSK-3β mutant capable of efficiently phosphorylating only unprimed sites.[71,72] Although both GSK-3β and GSK-3β-R96A phosphorylated tau *in situ*, GSK-3β-mediated phosphorylation of tau at primed sites, and in particular at the Thr231, plays a more significant role in regulating the interaction of tau with microtubules than phosphorylation at unprimed epitopes.[71,72] Although additional studies are required to completely elucidate the functional implications of tau phosphorylation at unprimed and primed sites, it clearly appears that a dual kinase system for the regulation of tau phosphorylation by GSK-3β would provide a high degree of control for GSK-3 substrate selectivity and provide the need for the regulation of tau function in the cells. Another process by which GSK-3 can regulate tau phosphorylation is by modulating the activity of other kinases that have been identified to phosphorylate tau. For instance GSK-3 can directly phosphorylate and inactivate the microtubule affinity-regulating kinase (MARK) 2, which has been implicated in the abnormal phosphorylation of tau in Alzheimer's disease.[73] Thus, in this case GSK-3 may serve the role of an inhibitory tau kinase, providing another process by which GSK-3 impacts tau phosphorylation. Further, there is evidence that the glycosylation of tau affects the ability of GSK-3 to phosphorylate tau.[74] Considerable efforts have been dedicated to the development of cellular models that reproduce events associated with the formation of NFTs and contribute to our understanding of the role of specific kinases in this process. For instance, phosphorylated tau aggregated into NFTs, but not normal tau, can be visualised using the AT100 antibody that recognises a conformation-dependent tau epitope. Both *in vitro* and *in vivo* studies have provided pieces of evidence that the AT100 epitope can be generated by the sequential phosphorylation of tau by GSK-3β and other kinases such as protein kinase A or the c-Jun N-terminal kinase-3 (JNK3).[75,76] Similarly, using fluorescence resonance energy transfer microscopy, activation of GSK-3β has been shown to promote the intermolecular association of tau in cell culture.[77] Together, these results suggest that GSK-3, likely in combination with other kinases, is involved in oligomeric tau formation *in situ*.

While it is intriguing to consider the possibility that phosphorylation of tau by GSK-3β is abnormally induced in Alzheimer's disease and contributes to neurofibrillary tangle formation, these relationships remain to be tested in an unequivocal manner. Significant advances on this topic undoubtedly came from approaches manipulating GSK-3 activity in mice. Among other methods, studies of transgenic mice overexpressing GSK-3β have provided strong evidence that GSK-3β is an effective kinase for protein tau *in vivo*. Bigenic mice that overexpress a constitutively active form of the human kinase GSK-3β(S9A) and the human protein tau (htau40) demonstrated an increase in tau phosphorylation that was associated with a reduced association of tau with microtubules.[78] In this model no PHFs or NFTs were detected and the amount of insoluble tau was unaltered when compared with wild-type mice.[78] In line with these observations, transgenic mice overexpressing GSK-3β via a tetracycline-regulated system were used to study the temporal effect of GSK-3β deregulation in Alzheimer's disease.[60] Induction of GSK-3β expression in adult mice resulted in an increased GSK-3β in the hippocampus and frontal cortex that was associated with an increase in tau phosphorylation at several epitopes relevant to Alzheimer's disease.[60] When the tetracycline-regulated GSK-3β mice were crossed with mice overexpressing tau protein carrying a FTDP-17 mutation, the GSK-3β-mediated tau hyperphosphorylation in hippocampal neurons resulted in the formation of NFTs and neurodegeneration.[59] Lithium administration to these mice prevented the progression or reversed early tau pathologies, although NFT-like structures were not affected.[59] Interestingly, hippocampal neuronal subpopulations were found to be differentially affected by increased GSK-3 activity, with dentate gyrus granular neurons undergoing apoptotic death, while CA1 pyramidal neurons accumulated hyperphosphorylated tau both in the axonal and in the somatodendritic compartments.[59] Others studies have exploited GSK-3 inhibitors to examine the effect of endogenous GSK-3 activity in mouse models of Alzheimer's disease. For instance, treatment of mice overexpressing mutant human tau (P301L) with the GSK-3β inhibitors lithium or AR-A014418 resulted in significantly reduced levels of tau phosphorylation, insoluble tau, and aggregated tau when compared to nontreated mice.[79] Further, in a recent study administration of the GSK-3 inhibitor SB216763 corrected the elevation in levels of phosphorylated tau, aberrant dendritic morphology, and induction of apoptosis mediated by the intracerebroventricular infusion of Aβ oligomers.[80] Altogether these studies are particularly significant as they provide strong evidence that tau is an *in vivo* substrate of GSK-3 and suggest that the action of GSK-3 plays an important role in Alzheimer's disease.[80]

Although most studies linking GSK-3 and tau point towards a role of GSK-3 as an enzyme capable of altering tau phosphorylation, there is also evidence suggesting that GSK-3 plays a role in regulating tau exon 10 splicing.[81] In adult central nervous system, six different tau isoforms are expressed that differ by the presence or absence of exons 2, 3 and 10.[82,83] Exon 10 encodes one of the four repeat sequences that form the microtubule binding domain.[54,82,83] Thus, the presence of exon 10 results in tau isoforms (4R) with increased microtubule binding capacity, whereas the alternative splicing results in tau isoforms (3R) with a reduced microtubule-binding capacity.[54,82,83] The tau pre-mRNA splicing generating the 3R isoform is catalysed by a multimolecular complex including SC35, a member of the serine/arginine-rich (SR) family of splicing factor.[81] Interestingly recombinant GSK-3 phosphorylates SC35, and GSK-3 activity affects the intranuclear localisation of SC35.[81] Furthermore, inhibition of GSK-3 by lithium or AR-18 in cortical neuronal cultures results in increased 4R tau transcription,[81] providing evidence that GSK-3 is able to modulate tau-mRNA splicing *in situ*.[81] These findings reveal an additional aspect of GSK-3 regulation of the microtubule-binding capacity of tau and raise the possibility that conditions associated with increased GSK-3 activity may result in an alteration in the tau4R/tau3R proportion of isoforms.

8.6 GSK-3 and Presenilin

Very intriguing, but still incompletely understood, associations between GSK-3β and presenilin proteins have been identified by investigators interested in the neurodegenerative mechanisms that cause Alzheimer's disease. Mutations in the genes for presenilin 1 and 2 account for most of the cases of familial Alzheimer's disease, although the large majority of Alzheimer's disease patients are classified as sporadic with no known genetic cause.[84] Mutations in the presenilin 1 gene cause an autosomal dominant inheritance of early onset Alzheimer's disease with 100% penetrance in causing Alzheimer's disease.[85] Thus, there is widespread interest in identifying the functions of presenilins and how mutations contribute to neuronal dysfunction and degeneration. Since their identification, presenilins have emerged as the catalytic component of the multiprotein γ-secretase enzyme complexes that promote the proteolysis of the amyloid precursor protein and several other Type-I transmembrane proteins.[86,87] The presenilins are synthesised as inactive holoproteins, which undergo endoproteolysis to generate an active heterodimer made up of the amino- and carboxyl-terminal fragments (NTF/CTF).[88] Presenilin-dependent cleavage of amyloid precursor protein contributes to the generation of Aβ peptides and the formation of pathological Aβ senile plaques. Mutations in presenilins 1 and 2, which contribute to the onset of familial form of Alzheimer's disease, lead to an increase in the generation of the more amyloidogenic 42 amino acid Aβ peptide. From biochemical and genetic analyses it has become evident that presenilins have biological roles beyond the γ-secretase cleavage of amyloid precursor protein, some of which are independent of its proteolytic activity.[89] One possibility is suggested by the finding that Alzheimer's disease-associated mutations in presenilin 1 facilitate apoptosis in several model systems, such as apoptosis caused by Aβ[90–92] (but for a contrasting viewpoint see ref. 93). However, the mechanisms by which mutations in presenilin genes facilitate apoptosis and contribute to Alzheimer's disease remain elusive.

Several investigators found that presenilin 1 binds to GSK-3β.[94–96] Takashima *et al.*[94] first reported a direct interaction between presenilin 1 and GSK-3β by showing that GSK-3β was detectable in presenilin 1 immunoprecipitates from human brain. The same immunoprecipitates also contained tau, suggesting that presenilin 1 may facilitate the colocalisation of GSK-3β and tau, and thus may facilitate GSK-3β-mediated phosphorylation of tau.[94] In contrast to the consistent findings that presenilin 1 binds GSK-3β, contradictory results have been published concerning how Alzheimer's disease-associated mutations of presenilin 1 affect this interaction. Some studies indicate that mutations in presenilin 1 increase its association with GSK-3β and, correspondingly, increase its tau-directed kinase activity.[94] Takashima *et al.* reported that expression of the C263R and P264L mutants of presenilin 1 in COS-7 cells caused a 3-fold increase in the amount of GSK-3β that coimmunoprecipitated with presenilin 1 compared with wild-type presenilin 1.[94] An opposite finding was reported by Kang *et al.*, who demonstrated that the M146L and ΔX9 presenilin 1 mutants bound GSK-3β less than wild-type presenilin 1.[95] Gantier *et al.* further showed that the L392V mutation of presenilin 1 decreased its affinity for GSK-3β relative to wild-type presenilin 1.[96] The contradictory findings in these studies could be ascribed to the different presenilin 1 mutants and different methods used in each study, or they may indicate that any perturbation of the interactions between presenilin 1 and GSK-3β, whether it is increased or decreased association, impairs the normal control or function of the proteins. These conflicting data leave open the question of how altered association of GSK-3β with mutant presenilin 1 contributes to the early onset of Alzheimer's disease associated with mutations in the presenilin 1 gene, but all of the studies are in agreement that wild-type presenilin 1 binds GSK-3β and thus may be an important modulator of the actions of GSK-3β. Further evidence revealed that presenilin may be a substrate for GSK-3.[97–99] Subsequently, the possibility that GSK-3 may phosphorylate presenilin 1 was suggested by the presence of three highly conserved GSK-3 consensus phosphorylation sites within the hydrophilic loop domain of presenilin 1.[97,98] Importantly, phosphorylation of presenilin 1 at one of these GSK-3 consensus sites is critical for the interaction between presenilin 1 and β-catenin,[98] suggesting that presenilin 1 regulates β-catenin signalling through a mechanism that involve phosphorylation by GSK-3 and subsequent degradation.[95] However, in another study presenilin 1 was found to regulate β-catenin in a manner independent of GSK-3 activity.[100] Furthermore, GSK-3 phosphorylation at a different consensus motif in the presenilin 1 loop domain alters the turnover of presenilin 1 C-terminal fragment.[97]

As noted previously,[94,95] presenilin 1 also binds other substrates of GSK-3β in addition to tau, including δ-catenin,[101] β-catenin,[95,102] amyloid precursor protein,[39] and Notch.[103] These studies suggest that presenilin 1 may act as a scaffold for bringing GSK-3β into close proximity with these substrates that have critical roles in Alzheimer's disease and cell-survival/apoptotic mechanisms.

In addition to directly binding GSK-3β, there is also evidence that presenilin 1 is capable of modulating its activity by regulating the PI3K/Akt signalling in a γ-secretase-independent manner.[104] In fibroblasts from presenilin 1 knockout mice, phosphorylation on both Ser21-GSK-3α and Ser9-GSK-3β were significantly decreased compared to wild-type cells, indicative of increased GSK-3 activity.[104] Reintroduction of presenilin 1 in these cells increased phosphorylation at both sites and this increase was prevented by pharmacological inhibition of either PI3K or Akt. Importantly, ablation of presenilin 1 was also associated with increased caspase-3 activity and was prevented by the reintroduction of presenilin 1 and by PI3K or Akt inhibitors.[104] Transient transfection of mutant presenilin 1 was associated with decreased Akt activity compared with wild-type presenilin 1.[104,105] Cells expressing mutant presenilin 1 also contained lower levels of phosphorylated Ser9-GSK-3β, in accordance with the lower levels of active Akt, which phosphorylates this site,[104,105] and higher levels of phosphorylated Tyr216-GSK-3β, the active form of the kinase.[105] Mutant presenilin 1 expression was associated with increased caspase-3 activation, raising the possibility that the increased vulnerability to apoptosis rendered by mutant presenilin 1 may be linked to the impaired inhibitory control of GSK-3, which itself has been shown to promote apoptosis.[106,107] In line with these findings, brains from PS1[A246E] transgenic mice revealed an inhibition of the PI3K/Akt signalling, resulting in an increased GSK-3β activity and tau phosphorylation at Alzheimer's disease relevant epitopes.[108] Altogether, these findings provide evidence that wild-type presenilin 1 is capable of inhibiting GSK-3 and promotes cell survival by activating PI3K/Akt signalling, but mutant presenilin 1 is associated with inhibition of PI3K/Akt signalling.[104–105,108]

Overall, these studies provide substantial evidence of interactions between presenilin 1 and GSK-3 and demonstrate that wild-type presenilin 1 can activate the PI3K/Akt signalling and thus modify the activity and function of GSK-3. In contrast, mutant presenilin 1 inhibits the PI3K/Akt signalling and thus promotes GSK-3 activity. Such an action could contribute to increased deleterious actions of GSK-3, such as increased phosphorylation of tau and neuronal apoptosis, conditions associated with Alzheimer's disease. Further, the finding that GSK-3 activity stimulates Aβ production in a γ-secretase independent manner[46] raises the intriguing possibility that the mutant presenilin 1-mediated GSK-3 activity may contribute to the increased Aβ production independently of the presenilin activity associated with the γ-secretase cleavage.

8.7 GSK-3: Neuronal Plasticity and Memory

As mentioned previously, numerous substrates for GSK-3 have been identified and GSK-3 has been implicated in a large variety of cellular functions.[8] Accumulating evidence has revealed that GSK-3 has central roles in neuronal plasticity, including mechanisms involved in memory processes, in the ability of the cells to respond to external stimuli, and in regulating events that participate in cellular death.

The central role for GSK-3 in neuronal plasticity is clearly evident from its ability to regulate the activity of a remarkable array of transcription factors (for a review see ref. 8). GSK-3 is known to directly phosphorylate and thereby regulate the function of several transcription factors known to play a key role in neuronal plasticity and cell survival, such as NF-κB, HSF-1, NFAT, MYC, CREB, AP-1, CEBPα and the well-recognised GSK-3-linked transcriptional coactivator, β-catenin.[8] Interestingly in most of the cases GSK-3 is inhibitory towards the activities of transcription factors.[8] The significance of GSK-3-mediated regulation of transcription factors is exemplified by the protein CREB. CREB regulates many cellular processes, including formation of long-term memory, maintenance of synaptic plasticity and apoptosis.[109] CREB activity is regulated by complex phosphorylation mechanisms that result either in its activation or inhibition. CREB phosphorylation at Ser133 within its activation domain is required for CREB to be transcriptionally active. Phosphorylation of CREB at Ser133 provides a consensus priming site for phosphorylation by GSK-3 at Ser129,[20] and the GSK-3-mediated phosphorylation of CREB reduces its DNA-binding activity.[20] Thus, the inhibitory influence of GSK-3 on CREB activity may result in impairments in neuronal plasticity, formation of long-term memory, and neuronal survival. Altogether, the GSK-3-dependent regulation of a broad range of transcription factors indicates the strong influence that GSK-3 has on cell plasticity and cell death. Since GSK-3 is usually inhibitory towards transcription factors, conditions where GSK-3 activity in not properly controlled by inhibitory mechanisms may result in impaired activation of transcription factors.

The contribution of GSK-3 on the formation of long-term memory has been more directly examined by determining the role of GSK-3 on the learning/cognitive performance of mouse models and on the induction of the NMDA-receptor dependent long-term potentiation (LTP) and long-term depression (LDP).[110–112,58] Behavioral testing of tetracycline-inducible GSK-3β mice showed that an increase in GSK-3 activity in the hippocampus is sufficient to cause a deficit in spatial learning in the Morris water maze, which reflects impairment in hippocampal-dependent spatial learning and memory.[58] Importantly, in this mouse model the significant increase in GSK-3 activity in the hippocampus was not associated with the presence of tau fibrils.[58] Further, the role of GSK-3β in the formation of long-term memory was examined in GSK-3β heterozygote (GSK + /–) mice.[112] Genetic reduction of GSK-3β resulted in about 50% reduction in the brain level of GSK-3β when compared to wild-type mice and the relative activity was about 70% of that in wild-type mice.[112] Examination of spatial memory in the Morris water maze test of GSK + /– mice revealed a retrograde amnesia. Further analysis indicated that GSK + /– mice had impaired memory reconsolidation but normal memory consolidation, suggesting that GSK-3β is necessary for memory reconsolidation in adult mouse brains.[112] This is consistent with the finding that a modest GSK-3 inhibition interfered with the working memory in rats.[80] Further studies have provided evidence that implicates a role for GSK-3β in the control of the NMDA-receptor-dependent synaptic plasticity.[110,111] The NMDA-receptor-dependent LTP is impaired in transgenic mice conditionally overexpressing GSK-3β,[110] and the LTP deficits can be prevented by chronic treatment with lithium, suggesting that inhibition of GSK-3 facilitates the induction of LTP.[110] Furthermore, in normal brain the activation of GSK-3β is essential for NMDA-receptor-dependent LTD. In rat hippocampal slices, GSK-3β inhibitors block the induction of LTD[111] and induction of LTD is associated with inhibition of Akt, probably via the activation of protein phosphatase 1, resulting in activation of GSK-3β.[111] Importantly, induction of LTP is associated with activation of the PI3K/Akt pathway resulting in increased phosphorylation at the inhibitory Ser9 site on GSK-3β, and disrupts the ability of synapses to undergo LTD.[110,111] Thus, the regulation of GSK-3β activity provides a powerful mechanism to coordinate the interaction between the two major forms of synaptic plasticity in the brain. Thus, conditions where the inhibitory control of GSK-3 has been compromised may affect the balance and interplay between NMDA-receptor-dependent LTP and LTD,[111] and explain some of the negative effects of hyperactive GSK-3 on learning and memory observed in mice.

Taking into account the multiple potential links between hyperactive GSK-3 and Alzheimer's disease, consideration should be given to the possibility that transcription factors affected by GSK-3 may be abnormally regulated in Alzheimer's disease. As noted above, evidence for hyperactive GSK-3β associated with Alzheimer's disease has been obtained in assessments of GSK-3β localisation and activity and the interactions of GSK-3β with Aβ, tau, and presenilin 1. Together, this evidence raises the possibility that inhibitory control of GSK-3 has been compromised in afflicted neurons, a condition predicted to result in GSK-3β-mediated impaired activation of transcription factors resulting in impaired neuronal plasticity. In addition, these effects may be further accentuated by the major role that GSK-3 exercises to coordinate the balance and interplay between NMDA-receptor-dependent LTP and LTD. Thus, not only is GSK-3 associated with the major neuropathological markers of Alzheimer's disease, but dysfunctional control of GSK-3β may further contribute to learning and memory defects in Alzheimer's disease through GSK-3β's influences on transcription factors and on the coordination between the two major forms of synaptic plasticity in the brain.

8.8 GSK-3: Diabetes and Alzheimer's Disease

The regulation of glucose utilisation by insulin is one of the most fundamental processes necessary for providing sufficient nutrition to maintain the vitality of mammalian organisms.[113–116] Impaired regulation of glucose production and utilisation is associated with insulin resistance and diabetes, including both Type-I insulin-dependent and Type-II insulin-independent diabetes.[114,116,117] There is a rapidly increasing prevalence of individuals developing insulin resistance, conditions in which the ability of insulin to properly regulate glucose is impaired.[114,116,117] Insulin controls blood glucose levels by promoting glucose transport into peripheral tissues and enhancing formation of glycogen. On the other hand, glycogen formation in resting cells is suppressed by phosphorylation and inactivation of the rate-limiting enzyme glycogen synthase (GS). Insulin indirectly relieves GS inhibition through a signalling cascade beginning with phosphorylation of substrates, including insulin receptor substrate 1 (IRS1), by the tyrosine kinase activity of the activated insulin receptor.[118] Tyrosine phosphorylated IRS-1 initiates additional events, including activation of Akt that phosphorylates and inhibits GSK-3 and dephosphorylation of GS.[14,118] Several enzymes have been implicated in the regulation of GS phosphorylation, including GSK-3, and there is convincing evidence that GSK-3 inactivation and GS activation are causally related, as GSK-3 phosphorylates GS at inactivating sites *in vitro* and overexpression of GSK-3 in cells suppresses GS function.[6,119,120] Since GSK-3 contributes to the regulation of glucose homeostasis, increasing attention has been focused on diabetes-associated changes in GSK-3 and the possibility that GSK-3 may be a useful therapeutic target in diabetes to facilitate control of glucose levels.

Not only do insulin-resistant conditions like diabetes affect peripheral tissues, such as muscle and fat cells, but brain function also can be significantly impaired.[121–123] Although little is understood about the diabetes-associated changes in the brain, it has been demonstrated that insulin activates the same signalling pathway leading to phosphorylation of Akt and GSK-3 in brain as in peripheral tissues. For instance, injection of insulin increased GSK-3 phosphorylation in mouse brain[124] and brain GSK-3 phosphorylation was reduced by blockade of insulin signalling by knockout of the insulin receptor[125] and in mice with diet-induced insulin resistance.[126] The understanding of diabetes-associated changes in the brain has taken on additional importance with the identification of significant links between diabetes and Alzheimer's disease.[127–131] The first troubling signs suggesting that diabetes plays a role in the pathogenesis of Alzheimer's disease emerged from epidemiological studies revealing that patients with diabetes have almost a two-fold increased risk of developing Alzheimer's disease.[131] As mentioned above one of the hallmark pathological characteristics of Alzheimer's disease is the presence of neurofibrillary tangles, which contain as their primary component aggregates of the protein tau in a hyperphosphorylated state.[54,132] Insulin normally regulates signalling cascades that contribute to the regulation of tau phosphorylation,[16,133] raising the possibility that one factor linking diabetes and Alzheimer's disease may involve changes in the phosphorylation state of tau consequent to deficient actions of insulin. When testing this hypothesis and examining whether insulin depletion caused by administration of streptozotocin altered site-specific tau phosphorylation in mouse brain[134] we found a massive average 5-fold increase in the phosphorylation of tau in mouse cerebral cortex and hippocampus that was widespread at multiple residues.[134] This condition could prime tau for the neuropathology of Alzheimer's disease, thereby contributing to the increased susceptibility to Alzheimer's disease caused by diabetes.[134] Surprisingly, the massive increase in tau phosphorylation was associated with a robust increased phosphorylation of Ser9-GSK-3β, indicating decreased activity of GSK-3. The insulin-depletion-induced increases in the phosphorylation of GSK-3 in mouse brain were found to result from the hyperglycemia associated with insulin depletion.[134]

8.9   Conclusions

GSK-3 is a central point of convergence of many signalling pathways, integrating information to influence cellular responses to external and internal stimuli. Thus, GSK-3 is responsive to stimulation of many neurotransmitter and neurotrophin receptors, and also to internal signals, for example those generated by the accumulation of misfolded proteins or DNA damage. Therefore, it is not surprising that GSK-3 is an important component of many of the basic processes that are important in Alzheimer's disease, such as neuronal plasticity, learning and memory, and apoptosis. However, somewhat more surprising is the association of GSK-3 with all of the key neuropathological hallmarks of Alzheimer's disease, the amyloid system and presenilins, and tau, the major component of NFTs. Furthermore, GSK-3 is involved in diabetes and inflammation, conditions that likely influence the onset and progression of Alzheimer's disease. These links to Alzheimer's-disease-associated events, as well as to fundamental functions that are impaired in the disease and conditions that contribute to progression, raise two possibilities. First, that disruption in the regulation and/or actions of GSK-3 may be a critical factor in the development of Alzheimer's disease, Secondly, that GSK-3 may be a valid target for therapeutic intervention to slow the progression of the disease. Ongoing research should soon shed more light on the roles of GSK-3 in the onset, progression, and treatment of Alzheimer's disease.

References

1. R. S. Jope and G. V. Johnson, *Trends Biochem. Sci.*, 2004, **29**, 95.
2. J. R. Woodgett, *Curr. Drug Targets Immune Endocr. Metabol. Disord.*, 2003, **3**, 281.
3. S. Patel and J. R. Woodgett, *Cancer Cell*, 2008, **14**, 351.
4. T. Force and J. R. Woodgett, *J. Biol. Chem.*, 2009, **284**, 9643.
5. K. P. Hoeflich, J. Luo, E. A. Rubie, M. S. Tsao, O. Jin and J. R. Woodgett, *Nature*, 2000, **406**, 86.
6. K. Macaulay, B. W. Doble, S. Patel, T. Hansotia, E. M. Sinclair, D. J. Drucker, A. Nagy and J. R. Woodgett, *Cell Metab.*, 2007, **6**, 329.
7. N. Embi, D. B. Rylatt and P. Cohen, *Eur. J. Biochem.*, 1980, **107**, 519.
8. C. A. Grimes and R. S. Jope, *Prog. Neurobiol.*, 2001, **65**, 391.
9. E. Beurel and R. S. Jope, *Prog. Neurobiol.*, 2006, **79**, 173.
10. R. S. Jope, C. J. Yuskaitis and E. Beurel, *Neurochem. Res.*, 2007, **32**, 577.
11. M. A. Utton, A. Vandecandelaere, U. Wagner, C. H. Reynolds, G. M. Gibb, C. C. Miller, P. M. Bayley and B. H. Anderton, *Biochem. J.*, 1997, **323**(Pt 3), 741.
12. J. R. Woodgett, *Embo. J.*, 1990, **9**, 2431.
13. K. Leroy and J. P. Brion, *J. Chem. Neuroanat.*, 1999, **16**, 279.
14. D. A. Cross, D. R. Alessi, P. Cohen, M. andjelkovich and B. A. Hemmings, *Nature*, 1995, **378**, 785.
15. K. Hughes, E. Nikolakaki, S. E. Plyte, N. F. Totty and J. R. Woodgett, *Embo. J.*, 1993, **12**, 803.
16. M. Lesort, R. S. Jope and G. V. Johnson, *J. Neurochem.*, 1999, **72**, 576.
17. L. Kim, J. Liu and A. R. Kimmel, *Cell*, 1999, **99**, 399.
18. A. Cole, S. Frame and P. Cohen, *Biochem. J.*, 2004, **377**, 249.
19. M. Hoshi, A. Takashima, K. Noguchi, M. Murayama, M. Sato, S. Kondo, Y. Saitoh, K. Ishiguro, T. Hoshino and K. Imahori, *Proc. Natl. Acad. Sci. USA*, 1996, **93**, 2719.
20. C. A. Grimes and R. S. Jope, *J. Neurochem.*, 2001, **78**, 1219.
21. G. N. Bijur and R. S. Jope, *Neuroreport*, 2003, **14**, 2415.
22. T. D. King, G. N. Bijur and R. S. Jope, *Brain Res.*, 2001, **919**, 106.
23. G. N. Bijur and R. S. Jope, *J. Biol. Chem.*, 2001, **276**, 37436.
24. B. W. Doble and J. R. Woodgett, *J. Cell Sci.*, 2003, **116**, 1175.
25. W. J. Nelson and R. Nusse, *Science*, 2004, **303**, 1483.

26. R. Dajani, E. Fraser, S. M. Roe, M. Yeo, V. M. Good, V. Thompson, T. C. Dale and L. H. Pearl, *Embo. J.*, 2003, **22**, 494.

27. R. Dajani, E. Fraser, S. M. Roe, N. Young, V. Good, T. C. Dale and L. H. Pearl, *Cell*, 2001, **105**, 721.

28. K. Imahori and T. Uchida, *J. Biochem.*, 1997, **121**, 179.

29. K. Imahori, M. Hoshi, K. Ishiguro, K. Sato, M. Takahashi, R. Shiurba, H. Yamaguchi, A. Takashima and T. Uchida, *Neurobiol. Aging*, 1998, **19**, S93.

30. K. Leroy, Z. Yilmaz and J. P. Brion, *Neuropathol. Appl. Neurobiol.*, 2007, **33**, 43.

31. H. Yamaguchi, K. Ishiguro, T. Uchida, A. Takashima, C. A. Lemere and K. Imahori, *Acta Neuropathol.*, 1996, **92**, 232.

32. J. J. Pei, T. Tanaka, Y. C. Tung, E. Braak, K. Iqbal and I. Grundke-Iqbal, *J. Neuropathol. Exp. Neurol.*, 1997, **56**, 70.

33. J. J. Pei, E. Braak, H. Braak, I. Grundke-Iqbal, K. Iqbal, B. Winblad and R. F. Cowburn, *J. Neuropathol. Exp. Neurol.*, 1999, **58**, 1010.

34. I. Ferrer, P. Pastor, M. J. Rey, E. Munoz, B. Puig, E. Pastor, R. Oliva and E. Tolosa, *Neuropathol. Appl. Neurobiol.*, 2003, **29**, 23.

35. I. Ferrer, M. Barrachina, M. Tolnay, M. J. Rey, N. Vidal, M. Carmona, R. Blanco and B. Puig, *Brain Pathol.*, 2003, **13**, 62.

36. A. Hye, F. Kerr, N. Archer, C. Foy, M. Poppe, R. Brown, G. Hamilton, J. Powell, B. anderton and S. Lovestone, *Neurosci. Lett.*, 2005, **373**, 1.

37. J. Hardy and D. J. Selkoe, *Science*, 2002, **297**, 353.

38. A. E. Aplin, J. S. Jacobsen, B. H. Anderton and J. M. Gallo, *Neuroreport*, 1997, **8**, 639.

39. A. E. Aplin, G. M. Gibb, J. S. Jacobsen, J. M. Gallo and B. H. Anderton, *J. Neurochem.*, 1996, **67**, 699.

40. H. S. Kim, E. M. Kim, J. P. Lee, C. H. Park, S. Kim, J. H. Seo, K. A. Chang, E. Yu, S. J. Jeong, Y. H. Chong and Y. H. Suh, *Faseb. J.*, 2003, **17**, 1951.

41. S. H. Kim, T. Ikeuchi, C. Yu and S. S. Sisodia, *J. Biol. Chem.*, 2003, **278**, 33992.

42. A. Takashima, K. Noguchi, G. Michel, M. Mercken, M. Hoshi, K. Ishiguro and K. Imahori, *Neurosci. Lett.*, 1996, **203**, 33.

43. A. Takashima, K. Noguchi, K. Sato, T. Hoshino and K. Imahori, *Proc. Natl. Acad. Sci. USA*, 1993, **90**, 7789.

44. A. Takashima, M. Sato, M. Mercken, S. Tanaka, S. Kondo, T. Honda, K. Sato, M. Murayama, K. Noguchi, Y. Nakazato and H. Takahashi, *Biochem. Biophys. Res. Commun.*, 1996, **227**, 423.

45. Y. Su, J. Ryder, B. Li, X. Wu, N. Fox, P. Solenberg, K. Brune, S. Paul, Y. Zhou, F. Liu and B. Ni, *Biochemistry*, 2004, **43**, 6899.

46. C. J. Phiel, C. A. Wilson, V. M. Lee and P. S. Klein, *Nature*, 2003, **423**, 435.

47. A. Takashima, H. Yamaguchi, K. Noguchi, G. Michel, K. Ishiguro, K. Sato, T. Hoshino, M. Hoshi and K. Imahori, *Neurosci. Lett.*, 1995, **198**, 83.

48. H. Wei, P. R. Leeds, Y. Qian, W. Wei, R. Chen and D. Chuang, *Eur. J. Pharmacol.*, 2000, **392**, 117.
49. G. Alvarez, J. R. Munoz-Montano, J. Satrustegui, J. Avila, E. Bogonez and J. Diaz-Nido, *FEBS Lett.*, 1999, **453**, 260.
50. D. W. Cleveland, S. Y. Hwo and M. W. Kirschner, *J. Mol. Biol.*, 1977, **116**, 207.
51. D. W. Cleveland, S. Y. Hwo and M. W. Kirschner, *J. Mol. Biol.*, 1977, **116**, 227.
52. M. D. Weingarten, A. H. Lockwood, S. Y. Hwo and M. W. Kirschner, *Proc. Natl. Acad. Sci. USA*, 1975, **72**, 1858.
53. K. Iqbal, I. Grundke-Iqbal, A. J. Smith, L. George, Y. C. Tung and T. Zaidi, *Proc. Natl. Acad. Sci. USA*, 1989, **86**, 5646.
54. M. Goedert, *Trends Neurosci.*, 1993, **16**, 460.
55. G. V. Johnson and S. M. Jenkins, *J. Alzheimers Dis.*, 1999, **1**, 307.
56. G. V. Johnson and J. A. Hartigan, *J. Alzheimers Dis.*, 1999, **1**, 329.
57. J. J. Lucas, F. Hernandez, P. Gomez-Ramos, M. A. Moran, R. Hen and J. Avila, *Embo. J.*, 2001, **20**, 27.
58. F. Hernandez, J. Borrell, C. Guaza, J. Avila and J. J. Lucas, *J. Neurochem.*, 2002, **83**, 1529.
59. T. Engel, J. J. Lucas, P. Gomez-Ramos, M. A. Moran, J. Avila and F. Hernandez, *Neurobiol. Aging*, 2006, **27**, 1258.
60. T. Engel, F. Hernandez, J. Avila and J. J. Lucas, *J. Neurosci.*, 2006, **26**, 5083.
61. D. Terwel, D. Muyllaert, I. Dewachter, P. Borghgraef, S. Croes, H. Devijver and F. Van Leuven, *Am. J. Pathol.*, 2008, **172**, 786.
62. K. Leroy, A. Boutajangout, M. Authelet, J. R. Woodgett, B. H. Anderton and J. P. Brion, *Acta Neuropathol.*, 2002, **103**, 91.
63. S. D. Harr, R. D. Hollister and B. T. Hyman, *Neurobiol. Aging*, 1996, **17**, 343.
64. C. A. Rankin, Q. Sun and T. C. Gamblin, *Neurobiol. Dis.*, 2008, **31**, 368.
65. C. A. Rankin, Q. Sun and T. C. Gamblin, *Mol. Neurodegener.*, 2007, **2**, 12.
66. S. Lovestone, D. R. Davis, M. T. Webster, S. Kaech, J. P. Brion, A. Matus and B. H. Anderton, *Biol. Psychiatry*, 1999, **45**, 995.
67. M. Hong, D. C. Chen, P. S. Klein and V. M. Lee, *J. Biol. Chem.*, 1997, **272**, 25326.
68. A. Sengupta, J. Kabat, M. Novak, Q. Wu, I. Grundke-Iqbal and K. Iqbal, *Arch. Biochem. Biophys.*, 1998, **357**, 299.
69. A. Sengupta, Q. Wu, I. Grundke-Iqbal, K. Iqbal and T. J. Singh, *Mol. Cell Biochem.*, 1997, **167**, 99.
70. K. Ishiguro, A. Shiratsuchi, S. Sato, A. Omori, M. Arioka, S. Kobayashi, T. Uchida and K. Imahori, *FEBS Lett.*, 1993, **325**, 167.
71. J. H. Cho and G. V. Johnson, *J. Biol. Chem.*, 2004, **279**, 54716.
72. J. H. Cho and G. V. Johnson, *J. Neurochem.*, 2004, **88**, 349.
73. T. Timm, K. Balusamy, X. Li, J. Biernat, E. Mandelkow and E. M. Mandelkow, *J. Biol. Chem.*, 2008, **283**, 18873.

74. F. Liu, K. Iqbal, I. Grundke-Iqbal and C. X. Gong, *FEBS Lett.*, 2002, **530**, 209.
75. F. Liu, Z. Liang, J. Shi, D. Yin, E. El-Akkad, I. Grundke-Iqbal, K. Iqbal and C. X. Gong, *FEBS Lett.*, 2006, **580**, 6269.
76. S. Sato, Y. Tatebayashi, T. Akagi, D. H. Chui, M. Murayama, T. Miyasaka, E. Planel, K. Tanemura, X. Sun, T. Hashikawa, K. Yoshioka, K. Ishiguro and A. Takashima, *J. Biol. Chem.*, 2002, **277**, 42060.
77. W. Chun and G. V. Johnson, *J. Biol. Chem.*, 2007, **282**, 23410.
78. K. Spittaels, C. Van Den Haute, J. Van Dorpe, H. Geerts, M. Mercken, K. Bruynseels, R. Lasrado, K. Vandezande, I. Laenen, T. Boon, J. Van Lint, J. Vandenheede, D. Moechars, R. Loos and F. Van Leuven, *J. Biol. Chem.*, 2000, **275**, 41340.
79. W. Noble, E. Planel, C. Zehr, V. Olm, J. Meyerson, F. Suleman, K. Gaynor, L. Wang, J. Lafrancois, B. Feinstein, M. Burns, P. Krishna-murthy, Y. Wen, R. Bhat, J. Lewis, D. Dickson and K. Duff, *Proc. Natl. Acad. Sci. USA*, 2005, **102**, 6990.
80. S. Hu, A. N. Begum, M. R. Jones, M. S. Oh, W. K. Beech, B. H. Beech, F. Yang, P. Chen, O. J. Ubeda, P. C. Kim, P. Davies, Q. Ma, G. M. Cole and S. A. Frautschy, *Neurobiol. Dis.*, 2009, **33**, 193.
81. F. Hernandez, M. Perez, J. J. Lucas, A. M. Mata, R. Bhat and J. Avila, *J. Biol. Chem.*, 2004, **279**, 3801.
82. M. Goedert, M. G. Spillantini, R. Jakes, D. Rutherford and R. A. Crowther, *Neuron*, 1989, **3**, 519.
83. M. Goedert and R. Jakes, *Embo. J.*, 1990, **9**, 25.
84. J. Hardy, *Trends Neurosci.*, 1997, **20**, 154.
85. C. Haass and B. De Strooper, *Science*, 1999, **286**, 916.
86. C. Kaether, C. Haass and H. Steiner, *Neurodegener. Dis.*, 2006, **3**, 275.
87. H. Steiner, R. Fluhrer and C. Haass, *J. Biol. Chem.*, 2008, **283**, 29627.
88. R. Kopan and M. X. Ilagan, *Nature Rev. Mol. Cell. Biol.*, 2004, **5**, 499.
89. D. J. Selkoe and M. S. Wolfe, *Cell*, 2007, **131**, 215.
90. D. M. Kovacs, R. Mancini, J. Henderson, S. J. Na, S. D. Schmidt, T. W. Kim and R. E. Tanzi, *J. Neurochem.*, 1999, **73**, 2278.
91. Q. Guo, B. L. Sopher, K. Furukawa, D. G. Pham, N. Robinson, G. M. Martin and M. P. Mattson, *J. Neurosci.*, 1997, **17**, 4212.
92. H. Tanii, M. Ankarcrona, F. Flood, C. Nilsberth, N. D. Mehta, J. Perez-Tur, B. Winblad, E. Benedikz and R. F. Cowburn, *Neuroscience*, 2000, **95**, 593.
93. S. Bursztajn, R. Desouza, D. L. Mcphie, S. A. Berman, J. Shioi, N. K. Robakis and R. L. Neve, *J. Neurosci.*, 1998, **18**, 9790.
94. A. Takashima, M. Murayama, O. Murayama, T. Kohno, T. Honda, K. Yasutake, N. Nihonmatsu, M. Mercken, H. Yamaguchi, S. Sugihara and B. Wolozin, *Proc. Natl. Acad. Sci. USA*, 1998, **95**, 9637.
95. D. E. Kang, S. Soriano, M. P. Frosch, T. Collins, S. Naruse, S. S. Sisodia, G. Leibowitz, F. Levine and E. H. Koo, *J. Neurosci.*, 1999, **19**, 4229.
96. R. Gantier, D. Gilbert, C. Dumanchin, D. Campion, D. Davoust, F. Toma and T. Frebourg, *Neurosci. Lett.*, 2000, **283**, 217.

97. F. Kirschenbaum, S. C. Hsu, B. Cordell and J. V. Mccarthy, *J. Biol. Chem.*, 2001, **276**, 30701.
98. F. Kirschenbaum, S. C. Hsu, B. Cordell and J. V. Mccarthy, *J. Biol. Chem.*, 2001, **276**, 7366.
99. C. Twomey and J. V. Mccarthy, *FEBS Lett.*, 2006, **580**, 4015.
100. J. J. Palacino, M. P. Murphy, O. Murayama, K. Iwasaki, M. Fujiwara, A. Takashima, T. E. Golde and B. Wolozin, *J. Biol. Chem.*, 2001, **276**, 38563.
101. J. Zhou, U. Liyanage, M. Medina, C. Ho, A. D. Simmons, M. Lovett and K. S. Kosik, *Neuroreport*, 1997, **8**, 1489.
102. G. Yu, F. Chen, G. Levesque, M. Nishimura, D. M. Zhang, L. Levesque, E. Rogaeva, D. Xu, Y. Liang, M. Duthie, P. H. St. George-Hyslop and P. E. Fraser, *J. Biol. Chem.*, 1998, **273**, 16470.
103. B. De Strooper, W. Annaert, P. Cupers, P. Saftig, K. Craessaerts, J. S. Mumm, E. H. Schroeter, V. Schrijvers, M. S. Wolfe, W. J. Ray, A. Goate and R. Kopan, *Nature*, 1999, **398**, 518.
104. L. Baki, J. Shioi, P. Wen, Z. Shao, A. Schwarzman, M. Gama-Sosa, R. Neve and N. K. Robakis, *Embo. J.*, 2004, **23**, 2586.
105. C. C. Weihl, G. D. Ghadge, S. G. Kennedy, N. Hay, R. J. Miller and R. P. Roos, *J. Neurosci.*, 1999, **19**, 5360.
106. G. N. Bijur and R. S. Jope, *J. Neurochem.*, 2000, **75**, 2401.
107. M. Pap and G. M. Cooper, *J. Biol. Chem.*, 1998, **273**, 19929.
108. T. Robakis, B. Bak, S. H. Lin, D. J. Bernard and P. Scheiffele, *J. Biol. Chem.*, 2008, **283**, 36369.
109. K. Deisseroth, H. Bito and R. W. Tsien, *Neuron*, 1996, **16**, 89.
110. C. Hooper, V. Markevich, F. Plattner, R. Killick, E. Schofield, T. Engel, F. Hernandez, B. anderton, K. Rosenblum, T. Bliss, S. F. Cooke, J. Avila, J. J. Lucas, K. P. Giese, J. Stephenson and S. Lovestone, *Eur. J. Neurosci.*, 2007, **25**, 81.
111. S. Peineau, C. Taghibiglou, C. Bradley, T. P. Wong, L. Liu, J. Lu, E. Lo, D. Wu, E. Saule, T. Bouschet, P. Matthews, J. T. Isaac, Z. A. Bortolotto, Y. T. Wang and G. L. Collingridge, *Neuron*, 2007, **53**, 703.
112. T. Kimura, S. Yamashita, S. Nakao, J. M. Park, M. Murayama, T. Mizoroki, Y. Yoshiike, N. Sahara and A. Takashima, *Plos ONE*, 2008, **3**, E3540.
113. R. J. Schulingkamp, T. C. Pagano, D. Hung and R. B. Raffa, *Neurosci. Biobehav. Rev.*, 2000, **24**, 855.
114. A. R. Saltiel and C. R. Kahn, *Nature*, 2001, **414**, 799.
115. K. Gerozissis, *Cell. Mol. Neurobiol.*, 2003, **23**, 1.
116. M. F. White, *Science*, 2003, **302**, 1710.
117. P. Zimmet, K. G. Alberti and J. Shaw, *Nature*, 2001, **414**, 782.
118. J. P. Whitehead, S. F. Clark, B. Urso and D. E. James, *Curr. Opin. Cell Biol.*, 2000, **12**, 222.
119. S. Patel, B. W. Doble, K. Macaulay, E. M. Sinclair, D. J. Drucker and J. R. Woodgett, *Mol. Cell. Biol.*, 2008, **28**, 6314.
120. H. Eldar-Finkelman, G. M. Argast, O. Foord, E. H. Fischer and E. G. Krebs, *Proc. Natl. Acad. Sci. USA*, 1996, **93**, 10228.

121. A. B. Smit, R. E. Van Kesteren, K. W. Li, J. Van Minnen, S. Spijker, H. Van Heerikhuizen and W. P. Geraerts, *Prog. Neurobiol.*, 1998, **54**, 35.

122. L. Plum, F. T. Wunderlich, S. Baudler, W. Krone and J. C. Bruning, *Physiology (Bethesda)*, 2005, **20**, 152.

123. W. H. Gispen and G. J. Biessels, *Trends Neurosci.*, 2000, **23**, 542.

124. E. Planel, T. Miyasaka, T. Launey, D. H. Chui, K. Tanemura, S. Sato, O. Murayama, K. Ishiguro, Y. Tatebayashi and A. Takashima, *J. Neurosci.*, 2004, **24**, 2401.

125. M. Schubert, D. Gautam, D. Surjo, K. Ueki, S. Baudler, D. Schubert, T. Kondo, J. Alber, N. Galldiks, E. Kustermann, S. Arndt, A. H. Jacobs, W. Krone, C. R. Kahn and J. C. Bruning, *Proc. Natl. Acad. Sci. USA.*, 2004, **101**, 3100.

126. L. Ho, W. Qin, P. N. Pompl, Z. Xiang, J. Wang, Z. Zhao, Y. Peng, G. Cambareri, A. Rocher, C. V. Mobbs, P. R. Hof and G. M. Pasinetti, *Faseb. J.*, 2004, **18**, 902.

127. L. Gasparini, W. J. Netzer, P. Greengard and H. Xu, *Trends Pharmacol. Sci.*, 2002, **23**, 288.

128. M. Schnaider Beeri, U. Goldbourt, J. M. Silverman, S. Noy, J. Schmeidler, R. Ravona-Springer, A. Sverdlick and M. Davidson, *Neurology*, 2004, **63**, 1902.

129. J. Janson, T. Laedtke, J. E. Parisi, P. O'Brien, R. C. Petersen and P. C. Butler, *Diabetes*, 2004, **53**, 474.

130. S. Hoyer, *J. Neural. Transm.*, 1998, **105**, 415.

131. A. Ott, R. P. Stolk, F. Van Harskamp, H. A. Pols, A. Hofman and M. M. Breteler, *Neurology*, 1999, **53**, 1937.

132. K. S. Kosik, C. L. Joachim and D. J. Selkoe, *Proc. Natl. Acad. Sci. USA*, 1986, **83**, 4044.

133. M. Hong and V. M. Lee, *J. Biol. Chem.*, 1997, **272**, 19547.

134. B. J. Clodfelder-Miller, A. A. Zmijewska, G. V. Johnson and R. S. Jope, *Diabetes*, 2006, **55**, 3320.

Tau Protein Kinases Inhibitors: From the Bench to the Clinical Trials

DANIEL I. PEREZ, CARMEN GIL AND ANA MARTINEZ

Instituto de Química Médica-CSIC, Juan de la Cierva 3, 28006, Madrid, Spain

9.1 Introduction

Tau protein belongs to the family of microtubule-associated proteins (MAPs) whose biological function is to promote microtubule assembly and to stabilise microtubules permitting neurites extension and stabilisation. This protein is integral to the pathogenesis of Alzheimer's disease (AD), as well as a range of neurodegenerative disorders, called tauopathies, in which tau is deposited as intracellular inclusions in affected brain regions. This is disease dependent but all distinctive inclusions, such as flame-shaped or globular neurofibrillary tangles and/or neurophil threads, are comprised of insoluble, highly phosphorylated forms of tau and they play a key role in the neuronal loss evident in tauopathies.[1]

In the adult human brain, the equilibrium between tau phosphorylation and dephosphorylation modulates the stability of the cytoskeleton and thereby the axonal morphology. Moreover, alternative splicing generates two sets of six tau isoforms, each containing either three (3R) or four (4R) microtubule-binding repeats with differential affinity to microtubules.[2] The 4R:3R tau ratio is

RSC Drug Discovery Series No. 2
Emerging Drugs and Targets for Alzheimer's Disease
Volume 1: Beta-Amyloid, Tau Protein and Glucose Metabolism
Edited by Ana Martinez

Published by the Royal Society of Chemistry, www.rsc.org

approximately equal in normal brain, but usually this ratio increases in most of the neurodegenerative tauopathies. In tauopathies, tau abnormalities due to mutations in the tau gene or an altered 4R:3R ratio, cause deposition in the brain of highly phosphorylated tau in an aberrant conformation. In fact, one of the earliest modifications found in Alzheimer brain is tau phosphorylation.[3] These filaments are caused by anomalous processing of tau protein and they have also been described in different neurodegenerative diseases such as Pick's disease,[4] progressive supranuclear palsy (PSP), corticobasal degeneration,[5] frontotemporal dementia with Parkinsonism linked to chromosome 17 (FTDP-17)[6] and amyotrophic lateral sclerosis.[7]

9.2   Relevant Tau Protein Kinases

Recent evidence suggests that targeting tau phosphorylation, mainly through inhibition of specific protein kinases, is one of the most promising strategies for the therapeutic intervention on AD and the above-mentioned neurodegenerative disorders.[8,9] Although some transgenic animal studies have shown that reducing tau phosphorylation is a valid strategy,[10] many kinases have been shown to phosphorylate tau protein *in vitro* and in cells, and more than 40 phosphorylation sites have been identified from AD brain. Therefore, it is important to identify which tau kinase to target.

Based on different studies aimed to identify phosphorylation tau sites in both physiological and pathological states, two protein kinases were first identified as responsible for more than 95% of *in vivo* tau phosphorylation. These were called initially tau protein kinases I and II (TPK-I and TPK-II).[11] After isolation and cloning, TPK-I was found to be identical to the well-known glycogen synthase kinase 3 (GSK-3),[12] while TPK-II consisted of a novel 23-kDa protein activator (p35) and a catalytic subunit that was identical with cyclin-dependent kinase 5 (CDK5).[13] The initial phosphorylation of tau with TPK-II appears to stimulate tau modifications via the TPK-I/GSK-3 kinase system.

More than ten years after this discovery, the most promising candidate kinases for tau phosphorylation include not only GSK-3 and CDK5, but also casein kinase 1delta (CK1δ),[14] mitogen-activated protein kinases (MAPKs), cAMP-dependent protein kinase (PKA), calcium/calmodulin-dependent protein kinase II (CaMKII), c-Jun N-terminal kinase (JNK) families and dual-specificity tyrosine-regulated kinase 1A (Dyrk1A).[15] Recent studies reported that some of these tau kinases, such as GSK3β, CDK5 or CK1 are over-expressed and/or overactivated on AD patients brains, increasing phosphorylation of tau protein in the disease.[16] Moreover, it is demonstrated that the overexpression of GSK-3 affected the ability of spatial memory and accelerated NFT formation in a mouse model.[17] This suggests that activation of protein tau kinases in general and of GSK-3 in particular, is involved in NFT formation and neuronal loss.

Based on this assumption, tau protein kinases inhibitors emerged as a promising therapeutic strategy for effective AD treatment. Currently, the first

GSK-3 inhibitor named **NP-12** is being clinically developed as disease-modifying drugs against AD and other neurodegenerative disease such as PSP.

9.3 GSK-3 Inhibitors Reach Clinical Trials

GSK-3 is a ubiquitous serine/threonine kinase that is present in mammals in two isoforms, α and β sharing over 90% homology in the catalytic site.[18] More recently, another isoform of GSK-3β with an additional 13 amino acids in the catalytic domain was discovered.[19] Although GSK-3 was initially identified as an enzyme that phosphorylates glycogen synthase to inhibit glycogen synthesis,[20] today it is well known that GSK-3 modulates many fundamental cellular processes, including replication, structural control, metabolism, gene expression, and apoptosis. Consequently, dysregulation of GSK-3 can contribute to a variety of pathologies, and has been linked to mood disorders, neurodegenerative diseases such as AD, cancer, inflammation, cardiovascular disease and diabetes. The use of GSK-3 inhibitors is claimed to be one of the most promising approaches for the treatment of these severe diseases.[21,22]

As has been mentioned in Chapter 8, GSK-3 has been linked to all the primary abnormalities associated with AD and today there is no doubt of the central role that GSK-3 plays in this pathology.[23,24] Therefore, drugs that target GSK-3 could interfere with AD pathogenesis via multiple mechanisms of action and can be considered as potential disease-modifying drugs for therapeutic intervention in AD neurodegenerative progression.[25,26]

Over the last five years, significant progress has been made in the search for GSK-3 inhibitors[27] being one candidate on clinical trials for AD.[28] The most frequent approach until now for identifying GSK-3 inhibitors has exploited screening programs specifically aimed to find new 'hits' in compounds that exhibit other pharmacological profiles. This approach identified GSK-3 inhibitors both from marine natural compounds sources, like **hymenialdisine**,[29] **manzamines**[30] and **palinurin**,[31] and synthetic small molecules from organic synthesis programs, like the **paullones**,[32] **indirubines**,[33] **maleimides**[34] and 2,4-disubstituted thiadiazolidindiones (TDZDs)[35] (Figure 9.1). However, the availability of X-ray crystallographic data of GSK-3β and several of its complexes with different inhibitors has enabled the application of rational drug optimisation to discover new lead compounds.[36]

Several recent reviews have been published regarding the preclinical efficacy of GSK-3 inhibitors and their potential for AD,[37,38] so we have focused this section only on two inhibitors that have reached clinical trials: the well-known lithium chloride, a standard treatment for bipolar disorder, and the unique ATP non-competitive GSK-3 inhibitors like the TDZD small-molecule family.[39]

9.3.1 Lessons Learnt from Clinical Use of Lithium

It is worth mentioning that lithium was the first GSK-3 inhibitor discovered[40] and has been widely used to test the putative role of this enzyme in multiple

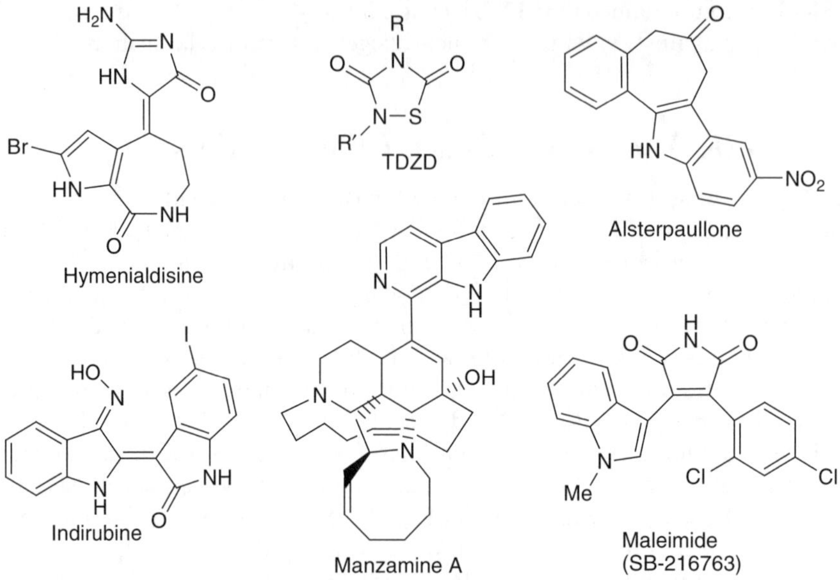

Figure 9.1 Chemical structures of known GSK-3 inhibitors.

experimental settings as well as in different AD animal models. Preclinical studies demonstrate that lithium chloride significantly decreases Aβ production and amyloid plaque burden through the inhibition of GSK-3 activity in mice.[41,42] Moreover, reduction of tau phosphorylation was also observed *in vivo* after lithium treatment in different animal models[43,44] further suggesting that lithium may also be useful for treating tauopathies other than AD.

Different human studies have been performed on AD patients under lithium treatment reporting a reduction of mRNA levels of GSK-3 and therefore proposing a new regulatory effect for GKS-3 inhibitors.[45] A recent Japanese clinical study designed to measure the cognitive performance using mini-mental state examination (MMSE) found that psychiatric patients who were treated with lithium have significantly better scores than control patients.[46] However, clinical trial results from mild to moderate AD patients treated with lithium are controversial.[47,48] The mixed lithium mechanism of action and the high discontinuation rates observed in these studies could be underlying all the studies. Thus, specific GSK-3 inhibitors are expected to achieve a clinical efficacy similar to lithium in AD being a field of high interest. To date only one candidate from the TDZDs family named **NP-12** is on Phase II of clinical development for AD and a Phase-II study for PSP is pending approval.

9.3.2 TDZDs, a Work Case on Clinical Development

Among the wide range of chemical structures that exhibit GSK-3 inhibition, TDZDs were identified as the first ATP non-competitive GSK-3 inhibitors.[35]

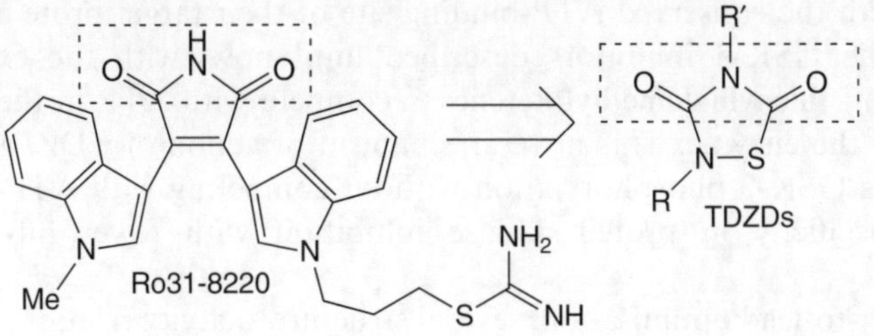

Figure 9.2   Common features between Ro31-8220 and TDZDs.

This unique mechanism of action for the TDZDs has enabled their use as a tool to resolve the complex signalling pathways where GSK-3β is implicated. The commercially available candidate TDZD-8 has been widely used both in deciphering the role of GSK-3 in cellular physiology and in confirming the involvement of GSK-3 in different pathologies in various preclinical studies.[37]

The kinase inhibitory activity of TDZDs was discovered in a GSK-3 screen-directed program initiated in the late 1990s. The activity was confirmed in a radiometric assay using the rabbit recombinant GSK-3β enzyme with the peptide derived from the human glycogen synthase GS-1 sequence as substrate.

The previously characterised PKC inhibitor **Ro31-8220** was reported to inhibit GSK-3.[49] Based on similarities in the chemical structure between TDZDs and **Ro31-8220,** *i.e.* the 1,3-dicarbonyl moiety in a five-member ring with a nitrogen atom between both carbonyls groups (Figure 9.2), we included several of the side products obtained in our laboratory from the synthesis of biologically active compounds, such as potassium channel openers,[50] muscarinic agonists,[51] or acetylcholinesterase inhibitors,[52] in our screening program. Three out of the four tested compounds had an IC_{50} (the compound concentration that inhibits 50% of the enzyme activity) in the low micromolar range.[53] After the discovery of TDZDs as GSK-3 inhibitors, other compounds with the maleimide core were reported as inhibitors of the GSK-3 enzyme,[34,54] highlighting the relevance of the 1,3-dicarbonyl substitution joined by a nitrogen atom in a pentagonal ring for the interaction with GSK-3.

The TDZD synthesis pathway is based on the reactivity of *N*-alkyl-*S*-[*N'*-(chlorocarbonyl)amino]isothiocarbamoyl chlorides with isocyanates.[55] These intermediate heterocyclic salts are exceptionally reactive, and in the presence of moist air and via hydrolysis, it was possible to obtain the 1,2,4-thiadiazolidine-3,5-diones as white crystalline solids after evolution of hydrogen chloride. Recently, some efforts to optimise the synthesis have been made due to the pharmacological relevance of the TDZDs.[56]

To investigate their mechanism of action on GSK-3β, several kinetic experiments were performed by varying concentrations of ATP, the substrate GS-1 and the inhibitor. The double-reciprocal Lineweaver–Burk plot of enzyme kinetics suggests that TDZDs act as non-competitive inhibitors of ATP nor the substrate binding. These results define the first ATP non-competitive inhibitors of GSK-3β. Most of the small-molecule protein kinase inhibitors interact with the conserved ATP-binding site of their target protein kinase. In fact, all the GSK-3 inhibitors described until now, with the exception of TDZDs and **thienylhalomethyl ketones**,[57] compete with ATP in their mode of binding to the enzyme. This novel mechanism of action of TDZDs inhibitors that blocks GSK-3 phosphorylation without competing with ATP may allow greater specificity in protein kinase inhibition with fewer off-target side effects.[58]

In the hit-to-lead optimisation, several structure–activity relationships (SAR) studies have been done to define crucial chemical features required for GSK-3 inhibition and, more importantly, to identify the best candidate for further therapeutic development. Different modifications both on the type of TDZD substituents[59] and on the nature of heteroatoms in the heterocyclic moiety,[60] have revealed the TDZDs as a privilege scaffold for GSK-3 inhibition and points to the sulfur atom in the heterocyclic moiety as a key feature for enzyme inhibition.

Drug-like properties were considered for the design of the second generation of TDZDs in the lead selection step. The selection of a candidate for pharmaceutical development should be based primarily in the specific activity of the compound, but some pharmacokinetic properties are also essential for effective therapeutic application. For example, if Alzheimer's disease is the planned therapeutic target, the selected candidate should not only be able to cross the blood/brain barrier (BBB), but should also be orally bioavailable for convenient administration to the patients. Safety properties should also be taken into account prior to regulatory toxicological studies to ensure the success of the candidate.

The Lipinski "rule of five" is a good approach to estimate the oral bioavailability or permeability of a new drug. Accordingly, the calculation of molecular weight, octanol/water partition coefficient, and the number of acceptors and donors of hydrogen atoms for several selected TDZDs indicated that they were within limits for a promising bioavailable drug. Additionally, we also applied CODES to the TDZD study case, an *in silico* model based on a neural network previously used in our group to successfully predict both oral absorption and BBB permeability of new drugs.[61] Results confirmed that several of the new compounds could cross the BBB and be orally administered. Using the experimental *in vitro* artificial membrane methodology, both pharmaceutical properties were further verified.[62]

Additionally, analysis with human microsomes, as well as both Cyp2D6 and Cyp3A4, showed that a number of the chemicals were stable with no inhibitory effects on the selected cytochromes.[63] Moreover, a screening approach based on a reduced Ames assay and an *in vitro* micronucleus test was performed using **NP00111**. The results were negative in these assays,[64] pointing to a safe genotoxicity profile for TDZDs. Taken together, these results confirm that TDZDs exhibit a promising drug profile for further development.

The candidates presenting the best properties were further explored *in vivo* in rodents to test oral absorption and BBB permeability. Several TDZDs showed acceptable drug levels in plasma and were able to cross the BBB, confirming these as good candidates for further development. However, the elimination half-lifetime for the TDZDs was not compatible with a single daily administration in humans. Considering that pharmacological treatment of AD patients must be chronic, a second generation of TDZDs was synthesised. Different stability studies were run in biologically simulated environments to define the best substituents to be introduced in the TDZD scaffold. Several compounds were obtained with increased elimination half-lifetime, a suitable GSK-3 inhibitory activity and drug-like properties, thus they were confirmed as candidates for starting the regulatory toxicological studies before entering clinical trials.[65]

To further assess the potential of TDZDs in the treatment of Alzheimer's disease, we determined the target selectivity of this heterocyclic family by investigating their effect on other kinases. More than twenty TDZDs were tested against a wide panel of related kinases with no resulting significant inhibitory effect, indicating a remarkable specificity of TDZDs with respect to GSK-3 inhibition. We also examined their biological profile in cell-based assays, testing in parallel for cell permeability and potential effects on cell viability. The cell-based assays generally applied for specifically assessing GSK-3 activity use either GSK3-dependent phosphorylation events or their cellular consequences as readouts. Two assays were selected for the evaluation of TDZDs: the first assesses GSK-3 inhibition in human neuroblastoma cell cultures using the phosphatidyl inositol 3 kinase (PI3K) pathway, and the second assays for GSK-3-mediated tau phosphorylation in the same cell culture. In both cases, we found a dose-dependent protection from cell death induced by blocking the PI3K/PKB pathway in cultured human neuroblastoma cells,[66] and a decrease of 15 to 25% of endogenous tau phosphorylation after the treatment with different TDZDs.[67]

Finally, the efficacy of several candidates was tested as proof-of-concept for these new drugs in different transgenic AD animal models. As GSK-3 functions in AD pathogenesis involve not only tau phosphorylation but also amyloid pathology and neuronal death, these three effects were investigated *in vivo* using different animal models and the administration of a TDZD compound.

First, a conditional double-transgenic mouse that overexpressed GSK-3 specifically in hippocampus (the tet/GSK-3 model) was used.[68] In this case, we demonstrated that chronic oral treatment with TDZDs decreased tau hyperphosphorylation in a dose- and time-dependent manner. These animals were treated for 3 and 6 weeks with a single, daily oral administration. Furthermore, in a double-transgenic model that overexpressed hAPP (with London mutation) and presenilin-1,[69] chronic oral treatment of inhibitors for 3 weeks decreased brain-soluble $A\beta_{1-40}$ and $A\beta_{1-42}$ levels in a dose-dependent manner and had beneficial effects on nonspatial memory function and curiosity, whilst longer treatment for 2 months also reduced the amyloid plaque load on the cortex and hippocampus.[70] TDZDs have also been proven to be effective in a kainate excitotoxicity model, in which acute i.c.v. treatment produced a marked reduction in clinical symptoms, a decrease in the number and intensity of inflammatory lesions (both measured by NMR and inmunohistochemistry) and a reduction in neuronal loss.[71] GFAP showed a significant increase in gliosis observed after kainate injection, which is prevented in the animals treated with **NP031112.** Moreover, a decrease in the production of TNF-α both in neurons and astrocytes was observed. Finally, neuronal cell loss was more prominent in kainate-injected animals by day 9, whereas animals treated with the TDZD compound showed no cell loss.

A double-transgenic mouse line based on overexpression of human mutant APP and tau has been recently developed and characterised,[72] showing all the features present on AD. APP^{sw}-tau^{vlw} mice developed enhanced amyloid deposition, plaque-associated glial proliferation, neuronal loss and neurofibrillary tangle-like formation in selectively vulnerable regions in AD, like the entorhinal cortex (EC) and the CA1 region of the hippocampus, and progressive memory impairment. This model offers a unique pharmacological tool to test our hypothesis based on GSK-3 inhibition. Thus, treatment with a TDZD compound might prevent/arrest the clinicopathological phenotype that characterises the APP^{sw}-tau^{vlw} transgenic mouse line. The novel non-ATP-competitive GSK-3 inhibitor **NP-12** was administered for 3 consecutive months by oral gavages. Treatment with this thiadiazolidinone compound resulted in lower levels of tau phosphorylation, decreased amyloid deposition and plaque-associated astrocytic proliferation, protection of neurons in the entorhinal cortex and CA1 hippocampal subfield against cell death, and prevention of memory deficits in this transgenic mouse model.[73] These results show that this novel GSK-3 inhibitor has a dual impact on amyloid and tau alterations and, perhaps even more important, on neuronal survival *in vivo* stressing the important role of GSK-3 as AD therapeutic target.

These successful data need to be confirmed in clinical trials. Phase-I studies to assess safety and tolerability were completed some months ago. Now, the recruitment of patients with mild to moderate AD for Phase-IIa studies is completed and the results are expected to be disclosed in the next three months. Only the results obtained from clinical trial Phase III will demonstrate the real efficacy of the new therapy.

9.4   CDK5 in Tau Phosphorylation

Cyclin-dependent kinase 5 (CDK5) is a unique member of the cyclin-dependent kinase family of proline-directed serine/threonine kinases. Unlike its family members, CDK5 has no function in the cell-cycle control although it plays an important role in neurotransmission and neuronal development.[74] CDK5 has common features with other members of the CDK family (CDK1 and CDK2). They present 60% identity in the protein sequence together with their crystal-structure homology.

CDK5 is expressed in all tissues, its kinase activity is observed in neurons. In its absence, neuronal migration and axonal path finding are deranged. Conversely, excessive and mislocalised CDK5 activity appears to be detrimental to neuronal function and pahological hallmarks of AD. Thus, both β-amyloid peptide (Aβ) aggregates and neurofibrillary tangles have been linked to CDK5-mediated neuronal death.[75,76] As for other members of the CDK family, which need cyclins to activate their kinase activity, CDK5 is activated by p35 and p39.[77] These activators are mainly expressed in neurons and regulate the spatial and temporal expression of active CDK5 to restrict its activity primarily to postmitotic neurons.

CDK5 is, together with GSK-3, one of the major kinases implicated in the abnormal tau phosphorylation in AD brain and it was firstly recognised as TPK-II. Elevated levels of p25 were found in AD brain tissues, suggesting that production of p25 is responsible for the increased CDK5 activity observed in AD.[78] Upon toxic stimulation such as Aβ, oxidative stress, ischemic brain damage and excitotoxicity, a disruption of the intracellular calcium homeostasis is caused, leading to the activation of calpains, a family of calcium-dependent proteases.[79] p35 activator is cleaved by calpain giving the fragment p25, which possesses a longer half-life, lacks the myristoylation sequence for membrane targeting, but retains the capacity to activate CDK5.[80] Deregulation of CDK5 activity targets tau as a substrate for hyperphosphorylation, which is a prerequisite of paired helical filament (PHF) formation.[81] *In vitro* experiments have shown that inhibition of this pathway, either by inhibiting CDK5 or calpain, can significantly rescue the neuronal death caused by Aβ accumulation. Thus, the development of small molecules able to reduce the increased CDK5 activity both by direct inhibition of CDK5 and indirectly by inhibition of calpain, represents one strategy that could halt the progression of AD decreasing tau phosphorylation.[82]

9.4.1 CDK5 Inhibitors

CDK5 has been an attractive target to prevent tau phosphorylation and neurofibrillary pathology in AD and other tauopathies for a number of years. Several classes of potent chemical inhibitors for CDK5 have been described, although none of them have reached clinical trials for these CNS diseases yet. One of the main challenges to overcome with these compounds is the lack of specificity among other protein kinases and specifically among cyclin-dependent kinases. This is not only due to ATP competition but also to the high homology of the primary amino acid sequence and 3D structure of CDK5 compared to other CDKs in their kinase domain.[83]

Among the CDK5 inhibitors described, only some of them have been pre-clinically studied for AD or other tauopathies. Purine derivatives are one of these cases. (*R*)-roscovitine, **Olomoucine** and **Purvalanol A** and **B** are 2,6,9-tri-substituted purine derivatives with CDK5 inhibitory properties (Figure 9.3). They are nonselective kinase inhibitors and have residual inhibition on other CDKs. **Roscovitine** ability to cross the blood/brain barrier and to reduce the activity of hyperphosphorylated tau *in vivo* was described after intracerebroventricular administration of this compound in a Niemann–Pick disease type C[84] and in a p25-transgenic mouse.[85] Clinical trials in various cancer types and inflammatory diseases are currently being carried out with this

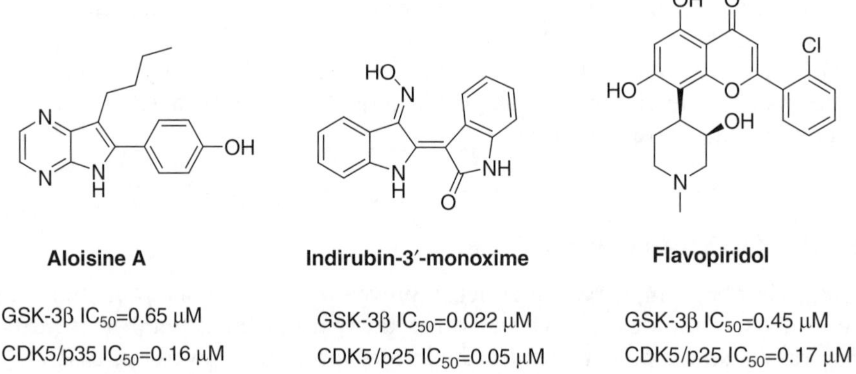

Inhibitor	R^1	R^2	R^3	CDK5 IC$_{50}$(µM)
(*R*)-Roscovitine	^{i}Pr	CH$_2$Ph	NHCH(Et)CH$_2$OH	0.16
Olomoucine	Me	CH$_2$Ph	NH(CH$_2$)$_2$OH	3.0
Purvanalol A	^{i}Pr	3-Cl-Ph	NHCH(^{i}Pr)CH$_2$OH	0.075
Purvanalol B	^{i}Pr	3-Cl- 4-CO$_2$H-Ph	NHCH(^{i}Pr)CH$_2$OH	0.006

Figure 9.3 2,6,9- Trisubstituted purines derivatives as CDK5 inhibitors.

Aloisine A

GSK-3β IC$_{50}$=0.65 µM

CDK5/p35 IC$_{50}$=0.16 µM

Indirubin-3′-monoxime

GSK-3β IC$_{50}$=0.022 µM

CDK5/p25 IC$_{50}$=0.05 µM

Flavopiridol

GSK-3β IC$_{50}$=0.45 µM

CDK5/p25 IC$_{50}$=0.17 µM

Figure 9.4 Dual CDK5/GSK-3β inhibitors.

compound,[86] but no clinical studies for neurodegenerative disorders have been reported yet.

Several dual CDK5/GSK-3 inhibitors such as **aloisines**,[87] **indirubines** and **flavopiridol** are described in the literature (Figure 9.4). Most of them have been studied as antiproliferative agents and **flavopiridol**, which inhibit the proliferation of mammalian cell lines at nanomolar concentrations, undergoes clinical trials against a variety of cancers.[88] The potential of these dual CDK5/GSK-3 inhibitors for neurodegenerative diseases decreasing tau phosphorylation are shown in the bis-indolyl compounds. Thus, **indirubin-3-monoxime** showed nanomolar inhibition against GSK-3β by competing with ATP for binding to the catalytic site and CDK5 and inhibited tau phosphorylation *in vitro* and *in vivo* at AD specific sites.[33] It has been reported that this compound showed potent antineurotoxicity induced by aged Aβ$_{25-35}$ in SH-SY5Y cells.[89]

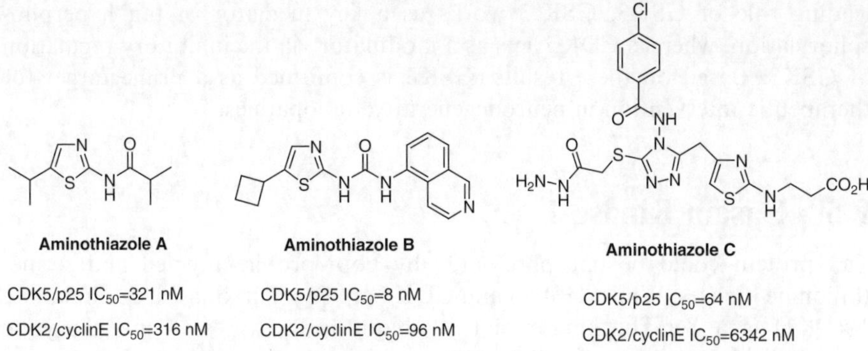

Figure 9.5　Aminothiazoles inhibitors of CDK5.

Once the X-ray crystallographic information on the CDK5/p25 complex was available[90] the design of different small molecules and peptides to inhibit this enzyme has been rationally developed. Thus, the CDK5 inhibitory peptide (CIP) was designed and synthesised. This compound was able to specifically inhibit CDK5/p25 without inhibiting other members of the CDKs family and decreased tau phosphorylation in cell cultures,[91] but the poor pharmacokinetic properties associated to the peptide structure prevents its use in animal models.

Rational design is also used in the optimisation of the aminothiazole family discovered as CDK5/p25 inhibitors using a high-throughput screening (HTS) technique (Figure 9.5). The first compound described, the *N*-(5-isopropyl-thiazol-2-yl) isobutyramide (**aminothiazole A**), had an IC_{50} value regarding CDK5/p25 of 321 nM, being competitive at the ATP binding site and with no selectivity regarding CDK2. With the aim of improving the selectivity of CDK5 towards CDK2, different chemical modifications of the lead compound were carried out. From the new series of synthesised compounds, **aminothiazole B** showed an inhibitory activity over CDK5/p25 (IC_{50} = 8 nM) and CDK2/cyclinE (IC_{50} = 96 nM) which has an improvement in selectivity of CDK5 towards CDK2 of 12 times.[92] Moreover, **aminothiazole C** showed CDK5/p25 (IC_{50} = 64 nM) and CDK2/cyclinE (IC_{50} = 6342 nM) respectively, which means a 100-fold selectivity between CDK5 towards CDK2. This compound was obtained under microwave irradiation using acidic alumina and its potential for AD treatment has been studied.[93]

Finally, the design approach to use calpain inhibitors as an indirect way of inhibiting CDK5 is worth mentioning. As we previously mentioned, calpain is the cysteine protease responsible for the cleavage of p35 to the pathogenic fragment p25, which is responsible for the increased activity of the catalytic subunit CDK5. Different calpain inhibitors have been described[94] showing positive results in AD models although their drawback is the poor ability to pass the blood/brain barrier.[95]

However, despite the initial pieces of evidence showing the pivotal role of CDK5 in AD and the development of specific CDK5 inhibitors as an interesting approach for the treatment of tauophaties, *in vivo* studies reveal the leading role of GSK3. GSK-3 works as a key mediator of tau hyperphosphorylation, whereas CDK5 acts as a modulator via the inhibitory regulation of GSK3. Based on these results, GSK3 is confirmed as a prime target for therapeutic intervention in neurodegenerative tauopathies.[96,97]

9.5   Casein Kinase 1

Tau protein could be phosphorylated by both proline-directed and serine/ threonine kinases. Whilst GSK-3 and CDK5 are Pro/Thr kinases, casein kinase 1 (CK1) is a Ser/Thr kinase that is ubiquitously expressed in eukaryotic organisms.[98] Even though CK1 is involved in diverse biological functions, during recent years different studies highlighted the importance of CK1 in neurodegenerative diseases, especially in tauopathies like AD.

So far, at least seven mammalian isoforms (α, β, γ1–3, δ and ϵ)[99] and their various splice variants[100] have been characterised in mammals. All CK1 isoforms are highly homologous within their kinase domains and they use exclusively ATP as a phosphate donor being, in general, cofactor independent.[101]

Results from a distribution study of the three CK1 isoforms (CK1α, CK1δ, and CK1ϵ) in AD and control brains showed that CK1 was found not only with elements of the fibrillar pathology but also within the matrix of granulovacuolar degeneration bodies.[102] Moreover, levels of all CK1 isoforms are high in the CA1 region of AD hippocampus relative to controls being CK1δ isoform elevated more than 30-fold. Following studies confirmed a sharp upregulation of CK1δ mRNA in AD brain, this upregulation being directly correlated with the degree of regional pathology.[103] It is worth pointing out that the sites directly phosphorylated by CK1δ on tau protein are those involved on its binding to tubulin, which illustrates the important role of CK1 in tau aggregation.[104] The presence of Aβ on brain stimulates CK1 activities contributing to abnormal tau protein phosphorylation in AD.[105]

9.5.1   CK1 Inhibitors

As the CK1 role in physiology and pathology is quite recent, not much work has been reported concerning inhibitors of this protein kinase. However, different chemical families of ATP competitive inhibitors both from synthetic and natural origin have been discovered and/or designed to target more or less specifically CK1 (Figure 9.6). Different methodologies including structure-based virtual-screening approaches,[106] and crystallographic complex studies have been applied to describe CK1 inhibitors and their specific interactions.[107]

The first CK1 inhibitor described in the micromolar range (IC$_{50}$ = 9.5 μM), was the compound (*N*-(2-aminoethyl)-5-chloroisoquinoline-8-sulfonamide) known as **CK1-7**.[108] It did not show any specificity against the different CK1 isoforms, inhibiting both α and δ isoforms but did not inhibit the gamma one. Although, **CK1-7** inhibited CK1 competitively with ATP, some kinase

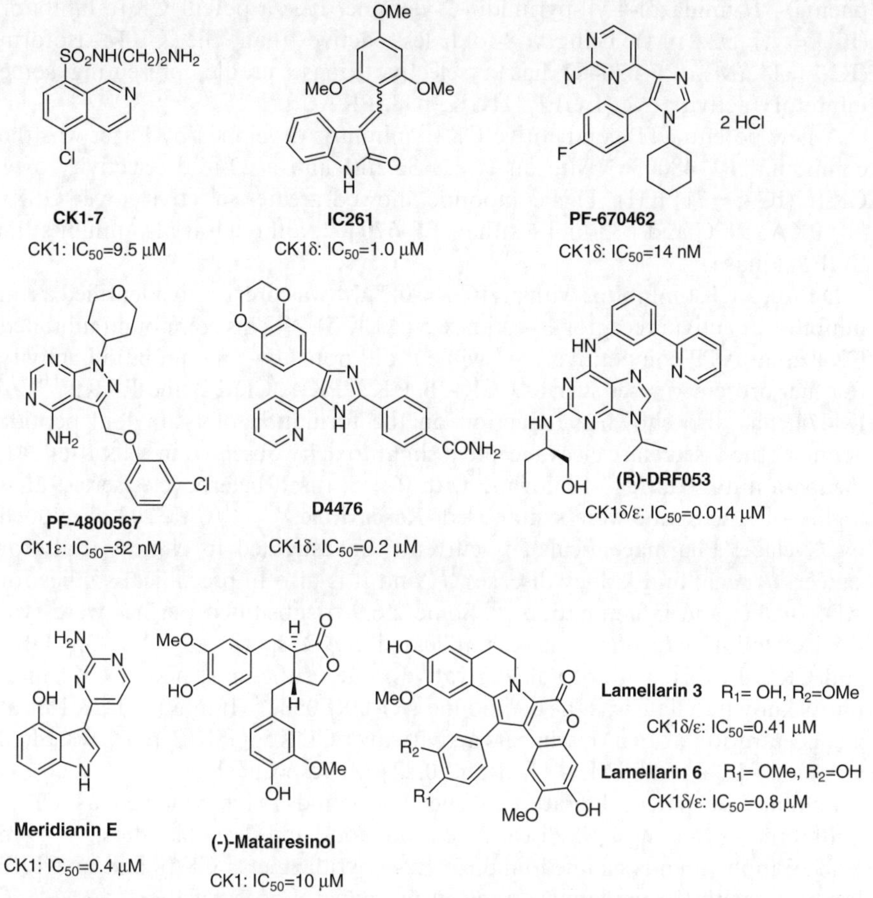

Figure 9.6 CK1 inhibitors.

specificity profile was found and it did not inhibit CK2, PKC, PKA or CaM kinase II. Its potential for AD has been recently found on cell cultures and a decrease on Aβ formation is described by the action of **CK1-7** at the γ-secretase cleavage pathway.[109]

The oxoindole heterocyclic compound **IC261** showed inhibitory activity over three CK1 isoforms (CK1α, IC$_{50}$ = 16 μM; CK1δ and CK1ε, IC$_{50}$ = 1.0 μM) without inhibiting other protein such as PKA, p34^{cdc2} and p55^{fyn}. Competition with ATP was also shown in its binding mode. One advantage of **IC261** compared to **CK1-7** relies on its ability to cross the cell membranes.[107] Compound **IC261** is also able to reduce the formation of Aβ acting at the γ-secretase cleavage.[109] and still reduces AKt phosphorylation,[110] which is a very important pathway for several development processes including cell metabolism, cell-cycle control, and cancer.[111]

Two imidazole derivatives have been reported as potent CK1 inhibitors too. The first one is **PF-670462,** chemically named as 4-[3-cyclohexyl-5-(4-fluorophenyl)-3*H*-imidazol-4-yl]-pyrimidin-2-ylamine. It is a potent CK1ε inhibitor ($IC_{50} = 7.7 \pm 2.2$ nM) being 1.8-fold less active than the CK1ε isoform ($IC_{50} = 14$ nM). **PF-670462** has a selective kinase profile, presenting some inhibitory activity over EGFR, HGK, p38. PKACα.[112]

A new potent ATP competitive CK1ε inhibitor developed by Pfizer was the compound **PF-4800567** with an $IC_{50} = 32$ nM and a 22-fold selectivity over CK1ε ($IC_{50} = 711$ nM). This compound showed greater selectivity over GSK-3β, PKA, PKC and p38 kinases than **PF-670462,** although it also inhibits the EGFR kinase.

D4476, a CK1 inhibitor with an $IC_{50} = 0.2$ μM was previously identified as an inhibitor of activin receptor-like kinase 5 (ALK 5).[113] This compound inhibited CK1 in an ATP competitive way, while it did not show any inhibition activity to other protein kinases such as GSK-3β, CK2, PKA, CDK2, and ERK2.[114,115] **D4476**, has also shown a reduction on the formation of β-amyloid peptide acting at the γ-secretase cleavage and a slight toxicity observed in cells after 24 h incubation at 50 μM.[109] Following with [6 + 5] fused heterocycles series, analogues of roscovitine were synthesised. **Roscovitine,**[116,117] (CYC202 developed by Cyclacel Pharmaceuticals), is currently investigated in clinical trials for cancer,[118] renal and kidney diseases,[119] and it is also in preclinical studies for AD, stroke, and inflammation.[120] Some 2,6,9-trisubsituted purines were synthesised to inhibit protein kinases implicated in AD, specifically GSK-3, CDK5 and CK1. In this series of new derivatives, one of the most potent CK1 inhibitors known to date was the compound **(*R*)-DRF053**.[121] It inhibited CK1δ/ε at a concentration of 14 nM, while its activity over CDK5, GSK-3 and CDK1 had a value of IC_{50} of 80 nM, 4.1 μM and 0.22 μM, respectively.

Among the patent literature, some compounds were reported as CK1ε inhibitors for treating CNS disease focus on mood and sleep disorders.[122–125] In 2009, Sanofi-Aventis claimed imidazo[1,2-*b*]pyridazine as CK1ε and/or CK1δ inhibitors with the potential to prevent tau hyperphosphorylation.[126]

Marine organisms represent a very promising source of new bioactive molecules in the pharmaceutical field.[127] Nowadays, different molecules obtained from marine organisms are either commercialised, in clinical trials or they are used as lead compounds for new derivatives. For the inhibition of CK1, different marine compounds have been reported such as **Meridianins,**[128] **(–)-Matairenisol,**[129] **Lamellarins 3** and **6** and **Hymenialdisine,**[29] where different analogues programs are currently in progress in order to design specific inhibitors for CK1 and to investigate the molecular mechanisms of action on neurodegenerative diseases.[130]

9.6 Inhibitors of Other Tau Protein Kinases

In addition to GSK-3β, CDK5 and CK1δ, other protein kinases considered as targets for inhibiting tau hyperphosphorylation are the mitogen-activated protein kinases (MAPKs), cAMP-dependent protein kinase (PKA) or calcium/calmodulin-dependent protein kinase II (CaMKII) among others.[131]

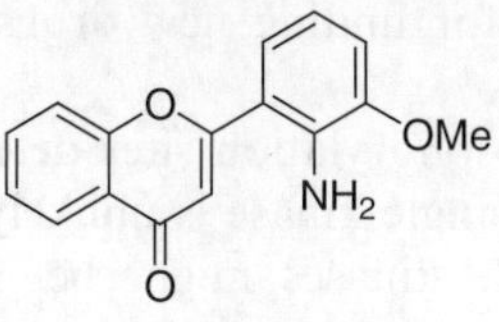

PD98059

Figure 9.7   MAPK inhibitor chemical structure.

Extracellular signal-regulated kinases (ERKs) and c-Jun N-terminal kinases (JNK) are the major brain MAPKs and play important roles in neuronal differentiation and development as well as in tau phosphorylation, especially under stress condition. Although the common problem of kinase inhibitors is their insufficient kinase selectivity, a specific MAPK inhibitor **PD98059** that prevents the phosphorylation of tau at Ser199/Ser202 induced by fibrillar Aβ has been reported.[132] Moreover, **PD98059** was used to show that the inhibition of MAPK activation partially prevented neurite degeneration (Figure 9.7). Taken together, these results suggest that the activation of the MAPK signal-transduction pathway induced by fibrillar Aβ may lead to the abnormal phosphorylation of tau and the neuritic degeneration observed in AD. For that reason, its inhibitors may represent a valid therapeutic approach.[133]

9.7   Conclusions

Tau protein is undoubtedly involved in AD and related tauopathies, with neurofibrillary degeneration of abnormally hyperphosphorylated tau not only a hallmark but also the most pivotal lesion of these neurodegenerative diseases. Reducing tau phosphorylation is regarded by many as the preferred tau-based therapeutic strategy, although whether phosphorylation of tau precedes or follows tau aggregation remains a subject of debate. Thus, targeting tau kinases is a rational strategy that is currently undergoing clinical investigation by pharmaceutical companies, with some GSK-3 inhibitors having entered clinical trials for AD.[28]

Restoring tau phosphorylation to normal, physiological levels involves not only a decrease in the activity of tau kinases, but also increases directly or indirectly the activity of protein phosphatases. The Ser/Thr phosphatase PP2A is one of the key players that catalyse the removal of phosphate moieties from tau protein.[134] The activity of PP2A is significantly suppressed in the brain of AD patients and its activation as a therapeutic intervention strategy for AD will be described in the next chapter. Recently, the activation of certain kinases, specifically GSK-3, is reported as one of the suppressors of PP2A activity in this pathology. High levels of GSK-3 activity, as is the case of AD brains, induced an accumulation of inhibitor-2 of PP2A. Inhibitor-2 blocked the activity of PP2A increasing tau phosphorylation.[135] Targeting kinases therefore remains the most rational strategy for finding new drugs to ameliorate tau-induced neurodegeneration.

The large number of phosphorylation sites detected on tau extracted from AD brain suggested that a single kinase is unlikely to direct activity at all the residues and hence multiple kinases might be involved.[136] Until now, the existing therapeutic strategies are aimed at specific and complete inhibition of individual tau kinases. In some way, off-target kinase inhibition is regarded as disadvantageous to drug discovery. An alternative therapeutic strategy to this specific and potent inhibition of an individual kinase might be to target multiple tau kinases with the aim of reducing the overall level of tau phosphorylation. Such a strategy has shown promising results in a preclinical model of AD,[137] and a subtle modulation of overall tau kinase activity might be a suitable approach, particularly when targeting diseases that primarily affect elderly patients.

Finally, several protein kinases have been recently discovered to play an important role in Aβ hypothesis. CDK5, GSK-3 and CK1 can alter APP processing and neurotoxicity of Aβ depositions. Thus, inhibitors of such kinases may have the potential not only to reduce neurofibrillary degeneration in the tauopathies, but also to afford neuroprotection in AD by reducing the amyloidogenesis. This hypothesis has been recently confirmed in a double-transgenic mice model of AD. Treatment with a moderate GSK-3 inhibitor produces a dual impact on amyloid and tau alterations and, perhaps even more important, on neuronal survival *in vivo* suggesting that GSK-3 is a relevant therapeutic target in AD.[73] Nowadays, compound **NP12**, recently called **NYPTA**, is in Phase II of clinical trials. Ongoing basic and clinical research should shed more light soon on the potential of tau kinases inhibitors for the effective treatment of Alzheimer's disease in the next decades.

References

1. I. Grundke-Iqbal, K. Iqbal, M. Quinlan, Y. C. Tung, M. S. Zaidi and H. M. Wisniewski, *J. Biol. Chem.*, 1986, **261**, 6084.
2. M. Goedert, M. G. Spillantini, R. Jakes, D. Rutherford and R. A. Crowther, *Neuron*, 1989, **3**, 519.
3. M. Goedert, A. Klug and R. A. Crowther, *J. Alzheimers Dis.*, 2006, **9**, 195.
4. B. Puig, F. Vinals and I. Ferrer, *Acta Neuropathol.*, 2004, **107**, 185.
5. A. Webb, B. Miller, S. Bonasera, A. Boxer, A. Karydas and K. C. Wilhelmsen, *Arch. Neurol.*, 2008, **65**, 1473.
6. L. Gasparini, B. Terni and M. G. Spillantini, *Neurodegener. Dis.*, 2007, **4**, 236.
7. M. J. Strong, W. Yang, W. L. Strong, C. Leystra-Lantz, H. Jaffe and H. C. Pant, *Neurology*, 2006, **66**, 1770.
8. A. Castro and A. Martinez, *Exp. Opin. Ther. Patents*, 2000, **10**, 1519.

9. D. P. Hanger, B. H. Anderton and W. Noble, *Trends Mol. Med.*, 2009, **15**, 112.

10. M. P. Mazanetz and P. M. Fischer, *Nature Rev. Drug Discov.*, 2007, **6**, 464.

11. H. Yamaguchi, K. Ishiguro, T. Uchida, A. Takashima, C. A. Lemere and K. Imahori, *Acta Neuropathol.*, 1996, **92**, 232.

12. K. Ishiguro, A. Shiratsuchi, S. Sato, A. Omori, M. Arioka, S. Kobayashi, T. Uchida and K. Imahori, *FEBS Lett.*, 1993, **325**, 167.

13. K. Ishiguro, S. Kobayashi, A. Omori, M. Takamatsu, S. Yonekura, K. Anzai, K. Imahori and T. Uchida, *FEBS Lett.*, 1994, **342**, 203.

14. D. P. Hanger, H. L. Byers, S. Wray, K. Y. Leung, M. J. Saxton, A. Seereeram, C. H. Reynolds, M. A. Ward and B. H. Anderton, *J. Biol. Chem.*, 2007, **282**, 23645.

15. I. Ferrer, T. Gomez-Isla, B. Puig, M. Freixes, E. Ribe, E. Dalfo and J. Avila, *Curr. Alzheimer Res.*, 2005, **2**, 3.

16. R. B. Maccioni, C. Otth, Concha II and J. P. Munoz, *Eur. J. Biochem.*, 2001, **268**, 1518.

17. T. Engel, J. J. Lucas, P. Gomez-Ramos, M. A. Moran, J. Avila and F. Hernandez, *Neurobiol. Aging*, 2006, **27**, 1258.

18. J. R. Woodgett, *EMBO J.*, 1990, **9**, 2431.

19. F. Mukai, K. Ishiguro, Y. Sano and S. C. Fujita, *J. Neurochem.*, 2002, **81**, 1073.

20. N. Embi, D. B. Rylatt and P. Cohen, *Eur. J. Biochem.*, 1980, **107**, 519.

21. A. Martinez, A. Castro, I. Dorronsoro and M. Alonso, *Med. Res. Rev.*, 2002, **22**, 373.

22. I. Dorronsoro, A. Castro and A. Martinez, *Exp. Opin. Ther. Patents*, 2002, **12**, 1527.

23. C. Hooper, R. Killick and S. Lovestone, *J. Neurochem.*, 2008, **104**, 1433.

24. H. C. Huang and P. S. Klein, *Curr. Drug Targets*, 2006, **7**, 1389.

25. R. M. Kypta, *Exp. Opin. Ther. Patents*, 2005, **15**, 1315.

26. A. Martinez and D. I. Perez, *J. Alzheimers Dis.*, 2008, **15**, 181.

27. M. Alonso and A. Martinez, *Curr. Med. Chem.*, 2004, **11**, 755.

28. M. Medina and A. Castro, *Curr. Opin. Drug Discov. Devel.*, 2008, **11**, 533.

29. L. Meijer, A. M. Thunnissen, A. W. White, M. Garnier, M. Nikolic, L. H. Tsai, J. Walter, K. E. Cleverley, P. C. Salinas, Y. Z. Wu, J. Biernat, E. M. Mandelkow, S. H. Kim and G. R. Pettit, *Chem. Biol.*, 2000, **7**, 51.

30. M. Hamann, D. Alonso, E. Martin-Aparicio, A. Fuertes, M. J. Perez-Puerto, A. Castro, S. Morales, M. L. Navarro, M. Del Monte-Millan, M. Medina, H. Pennaka, A. Balaiah, J. Peng, J. Cook, S. Wahyuono and A. Martinez, *J. Nature Prod.*, 2007, **70**, 1397.

31. D. Alonso and A. Martinez, in *Glycogen synthase kinase 3 (GSK-3) and its inhibitors,* ed. A. Martinez, A. Castro and M. Medina, Wiley-Interscience, New Jersey, 2006, p. 307.

32. M. Leost, C. Schultz, A. Link, Y. Z. Wu, J. Biernat, E. M. Mandelkow, J. A. Bibb, G. L. Snyder, P. Greengard, D. W. Zaharevitz, R. Gussio, A. M.

Senderowicz, E. A. Sausville, C. Kunick and L. Meijer, *Eur. J. Biochem.*, 2000, **267**, 5983.

33. S. Leclerc, M. Garnier, R. Hoessel, D. Marko, J. A. Bibb, G. L. Snyder, P. Greengard, J. Biernat, Y. Z. Wu, E. M. Mandelkow, G. Eisenbrand and L. Meijer, *J. Biol. Chem.*, 2001, **276**, 251.

34. D. G. Smith, M. Buffet, A. E. Fenwick, D. Haigh, R. J. Ife, M. Saunders, B. P. Slingsby, R. Stacey and R. W. Ward, *Bioorg. Med. Chem. Lett.*, 2001, **11**, 635.

35. A. Martinez, M. Alonso, A. Castro, C. Perez and F. J. Moreno, *J. Med. Chem.*, 2002, **45**, 1292.

36. E. ter Haar, in *Glycogen Synthase Kinase 3 (GSK-3) and its Inhibitors,* ed. A. Martinez, A. Castro and M. Medina, Wiley-Interscience, New Jersey, 2006, p. 61.

37. A. Martinez, *Med. Res. Rev.*, 2008, **28**, 773.

38. J. Avila and F. Hernandez, *Expert Rev. Neurother.*, 2007, **7**, 1527.

39. A. Martinez, M. Alonso, A. Castro and I. Dorronsoro, in *Medicinal Chemistry of Alzheimer's Disease,* ed. A. Martinez, Transworld Research Network, Kerala, 2008, p. 225.

40. V. Stambolic, L. Ruel and J. R. Woodgett, *Curr. Biol.*, 1996, **6**, 1664.

41. C. J. Phiel, C. A. Wilson, V. M. Lee and P. S. Klein, *Nature*, 2003, **423**, 435.

42. Y. Su, J. Ryder, B. Li, X. Wu, N. Fox, P. Solenberg, K. Brune, S. Paul, Y. Zhou, F. Liu and B. Ni, *Biochemistry*, 2004, **43**, 6899.

43. A. Caccamo, S. Oddo, L. X. Tran and F. M. LaFerla, *Am. J. Pathol.*, 2007, **170**, 1669.

44. M. Tajes, J. Gutierrez-Cuesta, J. Folch, I. Ferrer, B. Caballero, M. A. Smith, G. Casadesus, A. Camins and M. Pallas, *J. Neuropathol. Exp. Neurol.*, 2008, **67**, 612.

45. C. T. Mendes, F. B. Mury, E. de Sa Moreira, F. L. Alberto, O. V. Forlenza, E. Dias-Neto and W. F. Gattaz, *Eur. Arch. Psychiatry Clin. Neurosci.*, 2009, **259**, 16.

46. T. Terao, H. Nakano, Y. Inoue, T. Okamoto, J. Nakamura and N. Iwata, *Prog. Neuropsychopharmacol. Biol. Psychiatry*, 2006, **30**, 1125.

47. A. Macdonald, K. Briggs, M. Poppe, A. Higgins, L. Velayudhan and S. Lovestone, *Int. J. Geriatr. Psychiatry*, 2008, **23**, 704.

48. H. Hampel, M. Ewers, K. Burger, P. Annas, A. Mortberg, A. Bogstedt, L. Frolich, J. Schroder, P. Schonknecht, M. W. Riepe, I. Kraft, T. Gasser, T. Leyhe, H. J. Moller, A. Kurz and H. Basun, *J. Clin. Psychiatry*, 2009, **70**, 922.

49. I. Hers, J. M. Tavare and R. M. Denton, *FEBS Lett.*, 1999, **460**, 433.

50. A. Martinez, A. Castro, I. Cardelus, J. Llenas and J. M. Palacios, *Bioorg. Med. Chem.*, 1997, **5**, 1275.

51. A. Martinez, D. Alonso, A. Castro, V. J. Aran, I. Cardelus, J. E. Banos and A. Badia, *Arch. Pharm.*, 1999, **332**, 191.

52. A. Martinez, E. Fernandez, A. Castro, S. Conde, I. Rodriguez-Franco, J. E. Banos and A. Badia, *Eur. J. Med. Chem.*, 2000, **35**, 913.

53. A. Martinez, A. Castro, M. C. Pérez, M. Alonso, I. Dorronsoro, F. J. Muñoz, F. Wandosell, 2001, Patent No. WO0185685.
54. M. P. Coghlan, A. A. Culbert, D. A. Cross, S. L. Corcoran, J. W. Yates, N. J. Pearce, O. L. Rausch, G. J. Murphy, P. S. Carter, L. Roxbee Cox, D. Mills, M. J. Brown, D. Haigh, R. W. Ward, D. G. Smith, K. J. Murray, A. D. Reith and J. C. Holder, *Chem. Biol.*, 2000, **7**, 793.
55. G. Ottman and H. Hooks, *Angew. Chem. Int. Ed.*, 1966, **5**, 672.
56. S. Nasim and P. A. Crooks, *Tet. Lett.*, 2009, **50**, 257.
57. S. Conde, D. I. Perez, A. Martinez, C. Perez and F. J. Moreno, *J. Med. Chem.*, 2003, **46**, 4631.
58. M. A. Bogoyevitch and D. P. Fairlie, *Drug Discov. Today*, 2007, **12**, 622.
59. A. Castro, A. Encinas, C. Gil, S. Brase, W. Porcal, C. Perez, F. J. Moreno and A. Martinez, *Bioorg. Med. Chem.*, 2008, **16**, 495.
60. A. Martinez, M. Alonso, A. Castro, I. Dorronsoro, J. L. Gelpi, F. J. Luque, C. Perez and F. J. Moreno, *J. Med. Chem.*, 2005, **48**, 7103.
61. I. Dorronsoro, A. Chana, I. Abasolo, A. Castro, C. Gil, M. Stud and A. Martinez, *QSAR Comb. Sci.*, 2004, **23**, 89.
62. M. Alonso, I. Dorronsoro, A. Castro, M. I. Rodriguez-Franco, G. Abellán, M. Boiani, J. A. Vericat and A. Martínez, XVIIIth International Symposium on Medicinal Chemistry, Copenhagen, Denmark & Malmö, Sweden, 2004.
63. M. Alonso, I. Dorronsoro, A. Martinez and J. A. Vericat, 8th International Montreal/Springfield Symposium on Advances in Alzheimer Therapy, Montreal, Canada, 2004.
64. I. Anglade, N. Fabre, M. Alonso, A. Martinez, J. A. Vericat, 35th Annual Meeting of the European Environmental Mutagen Society (EEMS2005), Kos Island, Greece, 2005.
65. A. Martinez, I. Dorronsoro, M. Alonso, G. Panizo, M. Medina, A. Fuertes, M. J. Pérez Puerto, 2005, Patent No. PCT/EP2005/003613.
66. A. Fuertes, M. Alonso, M. J. Pérez-Puerto, D. Pérez-Navarro, E. Martín-Aparicio, A. Martinez, M. Medina, Cell Signalling Symposium, Dundee, UK, 2004.
67. E. Martín-Aparicio, A. Fuertes, M. J. Pérez-Puerto, M. Alonso, A. Martínez, M. Medina, IXth International Conference of Alzheimer's Disease and Related Disorders, Philadelphia, USA, 2004.
68. J. J. Lucas, F. Hernandez, P. Gomez-Ramos, M. A. Moran, R. Hen and J. Avila, *EMBO J.*, 2001, **20**, 27.
69. I. Dewachter, J. Van Dorpe, L. Smeijers, M. Gilis, C. Kuiperi, I. Laenen, N. Caluwaerts, D. Moechars, F. Checler, H. Vanderstichele and F. Van Leuven, *J. Neurosci.*, 2000, **20**, 6452.
70. J. A. Vericat, M. Alonso, S. Wera, I. Van der Auwera, A. Martínez, 7th International Conference AD/PD, Sorrento, Italy, 2005.
71. R. Luna-Medina, M. Cortes-Canteli, S. Sanchez-Galiano, J. A. Morales-Garcia, A. Martinez, A. Santos and A. Perez-Castillo, *J. Neurosci.*, 2007, **27**, 5766.
72. E. M. Ribe, M. Perez, B. Puig, I. Gich, F. Lim, M. Cuadrado, T. Sesma, S. Catena, B. Sanchez, M. Nieto, P. Gomez-Ramos, M. A. Moran,

F. Cabodevilla, L. Samaranch, L. Ortiz, A. Perez, I. Ferrer, J. Avila and T. Gomez-Isla, *Neurobiol. Dis.*, 2005, **20**, 814.

73. L. Sereno, M. Coma, M. Rodriguez, P. Sanchez-Ferrer, M. B. Sanchez, I. Gich, J. M. Agullo, M. Perez, J. Avila, C. Guardia-Laguarta, J. Clarimon, A. Lleo and T. Gomez-Isla, *Neurobiol. Dis.*, 2009, **35**, 359.

74. T. Tanaka, F. F. Serneo, H. C. Tseng, A. B. Kulkarni, L. H. Tsai and J. G. Gleeson, *Neuron*, 2004, **41**, 215.

75. R. Dhavan and L. H. Tsai, *Nature Rev. Mol. Cell. Biol.*, 2001, **2**, 749.

76. E. A. Monaco, 3rd, *Curr. Alzheimer Res.*, 2004, **1**, 33.

77. J. Lew, Q. Q. Huang, Z. Qi, R. J. Winkfein, R. Aebersold, T. Hunt and J. H. Wang, *Nature*, 1994, **371**, 423.

78. H. C. Tseng, Y. Zhou, Y. Shen and L. H. Tsai, *FEBS Lett.*, 2002, **523**, 58.

79. M. S. Lee, Y. T. Kwon, M. Li, J. Peng, R. M. Friedlander and L. H. Tsai, *Nature*, 2000, **405**, 360.

80. L. H. Tsai, M. S. Lee and J. Cruz, *Biochim. Biophys. Acta*, 2004, **1697**, 137.

81. F. Y. Wei and K. Tomizawa, *Mini Rev. Med. Chem.*, 2007, **7**, 1070.

82. M. S. Lee and L. H. Tsai, *J. Alzheimers Dis.*, 2003, **5**, 127.

83. M. A. Glicksman, G. D. Cuny, M. Liu, B. Dobson, K. Auerbach, R. L. Stein and K. S. Kosik, *Curr. Alzheimer Res.*, 2007, **4**, 547.

84. M. Zhang, J. Li, P. Chakrabarty, B. Bu and I. Vincent, *Am. J. Pathol.*, 2004, **165**, 843.

85. L.-H. Tsai, Alzheimer's Disease: From molecular mechanisms to drug discovery, Cancun, Mexico, 2004.

86. T. Guzi, *Curr. Opin. Investig. Drugs*, 2004, **5**, 1311.

87. Y. Mettey, M. Gompel, V. Thomas, M. Garnier, M. Leost, I. Ceballos-Picot, M. Noble, J. Endicott, J. M. Vierfond and L. Meijer, *J. Med. Chem.*, 2003, **46**, 222.

88. G. Kaur, M. Stetler-Stevenson, S. Sebers, P. Worland, H. Sedlacek, C. Myers, J. Czech, R. Naik and E. Sausville, *J. Natl. Cancer Inst.*, 1992, **84**, 1736.

89. S. Zhang, Y. Zhang, L. Xu, X. Lin, J. Lu, Q. Di, J. Shi and J. Xu, *Neurosci. Lett.*, 2009, **450**, 142.

90. C. Tarricone, R. Dhavan, J. Peng, L. B. Areces, L. H. Tsai and A. Musacchio, *Mol. Cell.*, 2001, **8**, 657.

91. Y. L. Zheng, B. S. Li, N. D. Amin, W. Albers and H. C. Pant, *Eur. J. Biochem.*, 2002, **269**, 4427.

92. C. J. Helal, M. A. Sanner, C. B. Cooper, T. Gant, M. Adam, J. C. Lucas, Z. Kang, S. Kupchinsky, M. K. Ahlijanian, B. Tate, F. S. Menniti, K. Kelly and M. Peterson, *Bioorg. Med. Chem. Lett.*, 2004, **14**, 5521.

93. M. R. Shiradkar, K. C. Akula, V. Dasari, V. Baru, B. Chiningiri, S. Gandhi and R. Kaur, *Bioorg. Med. Chem. Lett.*, 2007, **15**, 2601.

94. P. B. DePetrillo, *IDrugs*, 2002, **5**, 568.

95. M. E. Saez, R. Ramirez-Lorca, F. J. Moron and A. Ruiz, *Drug Discov. Today*, 2006, **11**, 917.

96. F. Plattner, M. Angelo and K. P. Giese, *J. Biol. Chem.*, 2006, **281**, 25457.

97. Y. Wen, E. Planel, M. Herman, H. Y. Figueroa, L. Wang, L. Liu, L. F. Lau, W. H. Yu and K. E. Duff, *J. Neurosci.*, 2008, **28**, 2624.

98. J. M. Peters, R. M. McKay, J. P. McKay and J. M. Graff, *Nature*, 1999, **401**, 345.

99. S. D. Gross and R. A. Anderson, *Cell Signal.*, 1998, **10**, 699.

100. J. Rowles, C. Slaughter, C. Moomaw, J. Hsu and M. H. Cobb, *Proc. Natl. Acad. Sci. USA*, 1991, **88**, 9548.

101. H. Flotow, P. R. Graves, A. Q. Wang, C. J. Fiol, R. W. Roeske and P. J. Roach, *J. Biol. Chem.*, 1990, **265**, 14264.

102. N. Ghoshal, J. F. Smiley, A. J. DeMaggio, M. F. Hoekstra, E. J. Cochran, L. I. Binder and J. Kuret, *Am. J. Pathol.*, 1999, **155**, 1163.

103. K. Yasojima, J. Kuret, A. J. DeMaggio, E. McGeer and P. L. McGeer, *Brain Res.*, 2000, **865**, 116.

104. G. Li, H. Yin and J. Kuret, *J. Biol. Chem.*, 2004, **279**, 15938.

105. A. Chauhan, V. P. Chauhan, N. Murakami, H. Brockerhoff and H. M. Wisniewski, *Brain Res.*, 1993, **629**, 47.

106. G. Cozza, A. Gianoncelli, M. Montopoli, L. Caparrotta, A. Venerando, F. Meggio, L. A. Pinna, G. Zagotto and S. Moro, *Bioorg. Med. Chem. Lett.*, 2008, **18**, 5672.

107. N. Mashhoon, A. J. DeMaggio, V. Tereshko, S. C. Bergmeier, M. Egli, M. F. Hoekstra and J. Kuret, *J. Biol. Chem.*, 2000, **275**, 20052.

108. T. Chijiwa, M. Hagiwara and H. Hidaka, *J. Biol. Chem.*, 1989, **264**, 4924.

109. M. Flajolet, G. He, M. Heiman, A. Lin, A. C. Nairn and P. Greengard, *Proc. Natl. Acad. Sci. USA*, 2007, **104**, 4159.

110. C. Modak and P. Bryant, *Biochem. Biophys. Res. Commun.*, 2008, **368**, 801.

111. J. A. Engelman, J. Luo and L. C. Cantley, *Nature Rev. Genet.*, 2006, **7**, 606.

112. L. Badura, T. Swanson, W. Adamowicz, J. Adams, J. Cianfrogna, K. Fisher, J. Holland, R. Kleiman, F. Nelson, L. Reynolds, K. St Germain, E. Schaeffer, B. Tate and J. Sprouse, *J. Pharmacol. Exp. Ther.*, 2007, **322**, 730.

113. J. F. Callahan, J. L. Burgess, J. A. Fornwald, L. M. Gaster, J. D. Harling, F. P. Harrington, J. Heer, C. Kwon, R. Lehr, A. Mathur, B. A. Olson, J. Weinstock and N. J. Laping, *J. Med. Chem.*, 2002, **45**, 999.

114. G. Rena, J. Bain, M. Elliott and P. Cohen, *EMBO Rep.*, 2004, **5**, 60.

115. D. M. Aud, S. L. Peng, K. Wha, 2009, US20090099237 A1.

116. L. Meijer, A. Borgne, O. Mulner, J. P. Chong, J. J. Blow, N. Inagaki, M. Inagaki, J. G. Delcros and J. P. Moulinoux, *Eur. J. Biochem.*, 1997, **243**, 527.

117. L. Meijer and E. Raymond, *Acc. Chem. Res.*, 2003, **36**, 417.

118. I. N. Hahntow, F. Schneller, M. Oelsner, K. Weick, I. Ringshausen, F. Fend, C. Peschel and T. Decker, *Leukemia*, 2004, **18**, 747.

119. N. O. Bukanov, L. A. Smith, K. W. Klinger, S. R. Ledbetter and O. Ibraghimov-Beskrovnaya, *Nature*, 2006, **444**, 949.

120. A. G. Rossi, D. A. Sawatzky, A. Walker, C. Ward, T. A. Sheldrake, N. A. Riley, A. Caldicott, M. Martinez-Losa, T. R. Walker, R. Duffin, M. Gray, E. Crescenzi, M. C. Martin, H. J. Brady, J. S. Savill, I. Dransfield and C. Haslett, *Nature Med.*, 2006, **12**, 1056.

121. N. Oumata, K. Bettayeb, Y. Ferandin, L. Demange, A. Lopez-Giral, M. L. Goddard, V. Myrianthopoulos, E. Mikros, M. Flajolet, P. Greengard, L. Meijer and H. Galons, *J. Med. Chem.*, 2008, **51**, 5229.

122. A. Metz, F. Halley, G. Dutruc-Rosset, Y. M. Choi-Sledeski, G. B. Poli, D. M. Fink, G. Doerflinger, B.-G. Huang, A. M. Gelormini, J. A. Gamboa, A. Giovanni, J. E. Roehr, J. T. Tsay, F. Camacho, W. J. Hurst, S. W. Harnish, Y. Chiang, 2005, Patent No. WO2005061498 A1.

123. D. M. Fink, Y. Chiang, N. D. Collar, 2006, Patent No. WO2006021000 A2.

124. A. Metz, F.-X. Ding, 2006, Patent No. WO2006023467 A1.

125. A. Metz, 2007, Patent No. US20070142454.

126. P. Burnier, Y. Chiang, S. Cote-des-Combes, A. K. Li, F. Puech, 2009, WO2009037394 A2.

127. M. L. Bourguet-Kondracki and J. M. Kornprobst, *Adv. Biochem. Eng. Biotechnol.*, 2005, **97**, 105.

128. M. Gompel, M. Leost, E. B. De Kier Joffe, L. Puricelli, L. H. Franco, J. Palermo and L. Meijer, *Bioorg. Med. Chem. Lett.*, 2004, **14**, 1703.

129. T. Yokoyama, M. Okano, T. Noshita, S. Funayama and K. Ohtsuki, *Biol. Pharm. Bull.*, 2003, **26**, 371.

130. D. Baunbaek, N. Trinkler, Y. Ferandin, O. Lozach, P. Ploypradith, S. Rucirawat, F. Ishibashi, M. Iwao and L. Meijer, *Mar. Drugs*, 2008, **6**, 514.

131. C. X. Gong and K. Iqbal, *Curr. Med. Chem.*, 2008, **15**, 2321.

132. I. Churcher, *Curr. Top Med. Chem.*, 2006, **6**, 579.

133. M. Rapoport and A. Ferreira, *J. Neurochem.*, 2000, **74**, 125.

134. M. Goedert, R. Jakes, Z. Qi, J. H. Wang and P. Cohen, *J. Neurochem.*, 1995, **65**, 2804.

135. G. P. Liu, Y. Zhang, X. Q. Yao, C. E. Zhang, J. Fang, Q. Wang and J. Z. Wang, *Neurobiol. Aging*, 2008, **29**, 1348.

136. Y. Yu, X. Run, Z. Liang, Y. Li, F. Liu, Y. Liu, K. Iqbal, I. Grundke-Iqbal and C. X. Gong, *J. Neurochem.*, 2009, **108**, 1480.

137. S. Le Corre, H. W. Klafki, N. Plesnila, G. Hubinger, A. Obermeier, H. Sahagun, B. Monse, P. Seneci, J. Lewis, J. Eriksen, C. Zehr, M. Yue, E. McGowan, D. W. Dickson, M. Hutton and H. M. Roder, *Proc. Natl. Acad. Sci. USA*, 2006, **103**, 9673.

CHAPTER 10

Activating PP2A as a Therapeutic Intervention Strategy in Alzheimer's Disease

NIALL M. CORCORAN AND CHRISTOPHER M. HOVENS

Department of Surgery, Royal Melbourne Hospital, University of Melbourne, 5th Floor Clinical Sciences Bldg, Parkville, VIC 3050, Australia

10.1 Maintenance of Signal-Transduction Networks

In order to maintain normal cellular homeostasis, it is vital for the cell to balance the complex network of signal-transduction pathways that coordinate cellular responses to external stimuli. These stimuli can induce complex multistep processes that result in proliferation, differentiation or even cell death, all of which must be tightly controlled in order for the cell to maintain its correct place and function in the broader multicellular organism.

One of the key mechanisms, by which a normal cell coordinates rapid responses to often competing external stimuli, is through the process of reversible protein phosphorylation. Signals emanating from the external milieu, can be rapidly processed by pre-existing protein signal-transduction networks within the cell by switching target proteins into either active or inactive confirmations through the reversible addition of phosphate epitopes on specific serine, threonine or tyrosine residues. In its simplest form, growth factor or hormone signals from the microenvironment are detected by cell-surface receptors, which

RSC Drug Discovery Series No. 2
Emerging Drugs and Targets for Alzheimer's Disease
Volume 1: Beta-Amyloid, Tau Protein and Glucose Metabolism
Edited by Ana Martinez

Published by the Royal Society of Chemistry, www.rsc.org

in turn activate protein kinases, which transduce the original signal by phosphorylating specific substrate proteins, thereby packaging and amplifying the original stimulus into a complex physiological response by the cell.

It should come as no surprise then that the reversal of this process, namely the specific removal of the same phosphoepitopes from specific serine, threonine and tyrosine residues, is also necessary to coordinate and indeed re-establish the ability of the cell to respond to external stimuli.

10.1.1 Protein Phosphatases: The Forgotten Half of Signal Transduction

At any one time in a normal eukaryotic cell, up to one third of all proteins present are phosphorylated, with two amino acid residues, namely serine and threonine containing by far the bulk of the phosphoepitopes. This predilection for phosphorylated serine and threonine residues is similarly maintained in the numbers of known kinase genes in the human genome with over 80% of the known 500 kinases modifying serine/threonine residues specifically. Given that the world of protein-signalling dynamics must similarly obey Newton's third law of motion that 'for every action there is an equal and opposite reaction' there must be a reciprocating enzymatic reaction to specifically remove those same serine and threonine phosphoepitopes from target proteins. Simplistically, one might expect that a similar broad number of phosphatase genes would therefore be encoded in the genome, but this is in fact not the case, with only 13 human phosphatase catalytic genes identified to date, and in fact the two most abundant serine/threonine phosphatases, namely PP1 and PP2A, are encoded by only five catalytic subunit genes in humans. This apparent paucity of phosphatase genes, failing to correspond to the known number of specific kinases, suggested that phosphatases *per se*, were generally promiscuous, a notion borne out by early studies using the isolated catalytic active phosphatase subunits that were shown to be fairly nondiscriminating in their removal of phosphate epitopes from serine/threonine phosphate peptides. This concept, however, has now been shown to be erroneous, with the required specificity and control of the function of phosphatases imparted by their ability to form multimeric subunit structures, composed of a small number of catalytically active subunits, but that in turn can complex with some hundreds of potential alternative regulatory and stabilising/scaffolding subunits. The diversity and complexity of the myriad phosphatase structures possible in a cell, with an estimated two hundred variants of a single serine threonine phosphatase, PP2A alone possible, rebalances the number of distinct kinase and phosphatase entities and provides the obligated precision and control for the removal of phosphate moieties from specific proteins, permitting the rapid modulation of protein function in eukaryotic cells.

Serine/threonine phosphatases belong to the family of phosphoprotein phosphatases, PPPs, (www.genenames.org) of which PP1 and PP2A are the most abundant. PP2A, like all members of the broader PPP family, is a multimeric enzyme. Given that only two genes (PP2CA and PP2BA) encode the catalytic C subunit, which as discussed above, can complex into over two hundred different and distinct multimeric structures, a large amount of this subunit needs to be available to the cell at any particular moment, and in fact over 0.1% of the total cellular protein is composed of the $PP2A_C$ subunit, making it one of the most abundant cellular proteins known. Typically, this catalytic C subunit, is complexed with a stabilising/scaffolding a subunit, $PP2A_A$ and one of an expanding array of regulatory subunits, $PP2A_B$. The importance of PP2A for normal cellular homoeostasis is underscored by the fact that loss of PP2A function results in apoptosis for the individual cell and lethality for the developing organism.[1-3]

Up to twenty different B subunit genes have been identified so far, based on sequence homology, but new unrelated regulatory domains continue to be identified, many of which have multiple splice variants.[4] The formation of different heterotrimeric forms are apparently tightly controlled in a temporal and spatial manner indicated by the tissue restricted formation of certain PP2A complexes.[5-9] The multiplicity of different PP2A complexes, each with unique structural, and topographical interfaces, permits the precise positioning of serine/threonine phosphoepitopes in the extended catalytic groove of the enzyme complex. This enables the specific dephosphorylation of unique phosphate residues even within the same target protein.[10] Given the ubiquitous nature of the extended PP2A heteroprotein family, and their key functional role in modulating intracellular signalling, it is perhaps not surprising that the PP2A enzyme complex has been linked to the regulation of the cell cycle, DNA replication, transcription and translation as well as vital signal-transduction processes.[8,11]

The complexity of PP2A heterotrimeric forms, needs must engender a complex nomenclature. The enzymatic core catalytic domain of 36 kilodaltons termed $PP2A_C$ having α and β isoforms, is ubiquitously complexed with a structural subunit, a 65-kDa $PP2A_A$ or PR65, having α and β isoforms. This heterodimeric $PP2A_{A-C}$, otherwise known as $PP2A_D$, then provides a scaffold for interaction with regulatory B subunits.

The regulatory B subunits are further divided into four families of proteins each with multiple isoforms; the B/PR55 (four isoforms), the B'/PR61 (11 isoforms), the B''/PR72 (four isoforms) and B'''/PR93 or PR110 (two unrelated proteins).[9]

Probing the functional roles of distinct PP2A species has already yielded insights into the nature of the cellular processes certain PP2A complexes control, but much more remains to be done. For instance, the B/PR55 family containing heterotrimers regulate cytokinesis in yeast,[12] regulate the ordered segregation of chromatids[13] and the differentiation of neurons in humans.[14] In contrast, the B'/PR61 family members are implicated in stabilising p53 via dephosphorylation of the p53 regulator mdm2, through complexing with cyclin G.[15]

10.1.1.1 *Regulation of PP2A Function: Post-Translational Modifications*

Biochemical evidence had already implicated post-translational modification of distinct subunits as a further means of regulating the catalytic function of the

PP2A complex and with the recent revealing of the crystal structure of a specific PP2A BαC complex, the structural basis for some of these modifications is starting to be elucidated.[16]

Tyrosine307 at the carboxy terminus of the catalytic PP2A$_C$ subunit can be phosphorylated, rendering the PP2A enzyme complex interactive.[17] The same subunit can also be inactivated by phosphorylation on threonine residues (the exact positioning of which remains to be determined) and analogous to many autophosphorylation reactions amongst kinase enzymes, PP2A can reactivate itself by autodephosphorylation of these same threonine residues.[11]

Another important post-translational modification is the reversible methylation of the Leucine309 also in the carboxy terminus of the PP2A$_C$ catalytic subunit, which is vital for the structural assembly of the complete PP2A heterotrimeric complex.

Interestingly, PP2A is one of a very few number of proteins that appear to be post-translationally regulated by both reversible phosphorylation and carboxy methylation. The regulation of PP2A, so important for its function, is in itself highly regulated, with a specific carboxyl methyltransferase, LCMT1, regulating the PP2A$_C$ subunit, and a specific phosphatase methylesterase, PME1, performing the reverse reaction.[17]

The reversible methylation of the catalytic subunit of PP2A seems to promote the formation of specific PP2A heterotrimers at the expense of others, with the B/PR55 form, promoted by PP2A$_C$ methylation, whilst B'/PR61 and B''/PR72 are only able to formate when this subunit is free of a methylated leucine309 residue.[17]

As if this level of regulation were not enough, a separate family of negative regulators also exists to add further complexity to both the nuances of PP2A regulation as well as its associated nomenclature.

These inhibitors, variously dubbed as I$_1$-PP2A or PHAP-1 and I$_2$-PP2A or the better-known SET protein have been clearly implicated in binding to and abrogating PP2A catalytic activity.[18,19] More is known concerning the function of I$_2$-PP2A or SET, which has been implicated in chromatin remodelling, and the development of acute myeloid leukemia[20,21] and BCR/ABL oncogene-driven chronic myeloid leukemia and acute lymphocytic leukemia, adding credence to the notion that PP2A hypofunction, at least in certain cells, promotes and drives cancer formation.[22,23]

Relatively less is known concerning the physiological role of the I$_1$-PP2A / PHAP-1 inhibitor, however, a recent body of work, firmly implicates this protein in the abnormal phosphorylation of tau, a pathological process germaine to the tauopathies in particular, and neurodegeneration in general.[19,24,25]

10.2   Tau, Phosphorylation and Alzheimer's Disease

Tau is a microtubule-binding protein, a family that also includes the proteins MAP1A/1B and MAP2. It is encoded by a single gene on chromosome 17, with alternative mRNA splicing resulting in 6 isoforms that differ in the number of microtubule binding repeats and/or N-terminal inserts.[26] The lack of hydrophobic regions mean that unbound tau exists predominantly as a random coil without significant secondary structure, and is water soluble and heat stable. Tau is localised primarily to neuronal axons within the central nervous system where it binds to multiple tubulin monomers via its microtubule binding domains. This binding promotes assembly into microtubules and stabilises their structure in filaments, counterbalancing the intrinsic propensity of microtubules to disassemble.[27] This function is important in maintaining neuronal outgrowth and axonal flow, although there appears to be significant redundancy within the family as tau knockout mice demonstrate no major neurological deficit.[28]

Tau is a phosphoprotein in that its microtubule binding activity is determined by its phosphorylation status. Tau contains over 40 serine/threonine residues that may undergo reversible phosphorylation.[29] In normal brain specimens 2–3 moles of phosphate are detected per mole of tau protein, which appears optimal for normal function.[30] Hyperphosphorylation of tau leads to a loss of its tubulin binding activity, causing the microtubule system to disintegrate. It also causes relocation of tau to the somatodendritic compartment of the neuron and also makes it resistant to proteolysis, leading to its accumulation.[31] Numerous kinases have been indentified that phosphorylate tau, including proline-directed kinases such as GSK3β, CDK-5, ERK1.2 and Dryk1A that phosphorylate at multiple canonical sites. In addition, nonproline-directed kinases have a more limited repertoire of phosphorylation sites, but prime tau to phosphorylation by proline-directed kinases.[31]

The aberrant accumulation of hyperphosphorylated tau has been demonstrated to significantly alter microtubule dynamics. Hyperphosphorylated tau dissolved in the cytosol does not bind tubulin, but instead sequesters normal tau and other MAP family members including MAP 1 A/B and MAP2, an action that appears critical to its neurotoxicity.[32] This leads to an inhibition of microtubule assembly and function, ultimately manifesting in neuronal cell loss. Further phosphorylation permits polymerisation into fibrillary tangles, which are characteristic of some tauopathies, particularly Alzheimer's disease. Fibrillar tau is not active, in that it is unable to bind and sequester MAP proteins or tubulin, whereas dephosphorylation leads to disaggregation and disassembly of the tangles and return of normal protein function.[33] In this way fibrillary tangles may be seen as a protective attempt to isolate neurotoxic soluble hyperphosphorylated tau. This primary pathogenic effect of cytosolic tau is supported by evidence from animal models in which cognitive decline is shown to be dependent upon expression of cytosolic tau rather than the accumulation of tangles.[34]

The progressive accumulation of hyperphosphorylated tau leads to progressive cortical neuronal cell loss, manifest as dementia that is the characteristic clinical feature of tau-related disorders or 'tauopathies'.[35,36] The most prevalent tauopathy is Alzheimer's disease. Patients with Alzheimer's disease have brain levels of hyperphopshorylated tau 3–4 times that of normal brains, and the number of neurofibrillary tangles correlates well with the presence and degree of dementia.[30,37] Similarly, missense mutations in the tau gene that promote hyperphosphorylation cosegregate with at least half of the cases of hereditary frontotemporal dementia.[38,39] Further evidence of the clinical importance of tau phosphorylation is from Downs Syndrome, where Dryk1A is overexpressed leading to tau hyperphosphorylation and early onset dementia.[40–42]

10.2.1 Pathological Tau Hyperphosphorylation: The Key Role of PP2A

Whilst a number of phosphatases have been implicated *in vitro* in inducing tau dephosphorylation, PP2A and in particular, the specific Bα regulatory subunit form of the PP2A heterotrimeric complex has now emerged as a key player in tau dephosphorylation and by implication, abnormal tau phosphorylation. There is a clear decline in the activity of PP2A in the brains of Alzheimer's disease patients[43] and impaired Bα expression correlates with abnormal phosphotau deposition in Alzheimer's disease brains.[44–46]

The key role of the Bα regulatory subunit in tau phosphoregulation is reinforced by the observation that the Bα subunit, alone amongst the regulatory subunit family members can bind to tau directly, where it directs PP2A localisation to microtubules, facilitating tau dephosphorylation and orchestrating the activity of multiple tau kinases.[44,47–49]

The recently solved crystal structure of the Bα form of the PP2A heterotrimeric complex,[49] which has been implicated in tau dephosphorylation, when compared with the structure of the B56 gamma form of the PP2A heterotrimer,[50] has permitted a clearer understanding of the crucial role played by the B regulatory subunit, in forming an extended interface to position specific substrates adjacent to the catalytic groove of the common PP2A$_C$ subunit.

The Bα subunit has an acidic central groove on its surface surrounded by seven β-sheets in a propeller like structure. The acidic central groove is positioned in the PP2A heterotrimeric complex adjacent to the catalytic pocket of the PP2A$_C$ subunit. Changing the charge composition of the Bα acidic surface groove severely impeded tau dephosphorylation confirming that the juxtaposition of these two subunits creates a unique protein interaction region that would explain the substrate specificity of different PP2A regulatory domain containing complexes.[49]

10.3 Activating PP2A with Chemical Agents. An Alternative Strategy to Limit Toxic Phosphotau

A growing body of work then, places tau at a signalling crux, possibly receiving cues from neuronal β-amyloid species, orchestrating the eventual demise of selected neuronal populations. Whilst many efforts by the pharmaceutical industry have been directed to discover and develop agents that would interfere with β-amyloid deposition, comparatively little has been done to identify druggable candidates that would act to reduce tau protein phosphorylation and/or levels.

Tau protein phosphorylation could theoretically either be reduced by inhibiting the key kinase/s responsible for the Alzheimer's signature phosphotau epitopes or by boosting the activity of the key phosphatase/s that mediate dephosphorylation of those same phosphoepitopes. Whilst the concept of activating a key phosphatase such as PP2A is attractive in principal, it is much harder to instigate this process in practice. The major dilemma concerns the common, catalytic C subunit that, as we have discussed, is shared between many different PP2A species. Any drug targeting this domain and potentially increasing its activity, is highly likely to induce many different PP2A heterotrimers with consequent undesirable side effects. However, some compounds have already been found that can influence specific post-translational modifications that bias the formation of tau-relevant, PP2A species. It is these compounds that we will now focus our attention on.

10.3.1 Folate and Xylulose-5-Phosphate

Numerous studies have now extensively documented that the methylation status of the $PP2A_C$ catalytic subunit, directly affects the recruitment of specific B regulatory subunits to the PP2A heterotrimeric complex and thereby would be expected to promote the dephosphorylation of certain PP2A substrates to the exclusion of others.[17,51,52]

Specifically, methylation of the C-terminal leucine309 promotes assembly of the Bα, PP2ABαC complex, and hence promotes dephosphorylation of tau. The regulation of the $PP2A_C$ catalytic subunit is dependent on a S-adenosyl-L-methionine (SAM) dependent methyltransferase LCMT-1, and hence the supply of SAM itself available for methylation is a rate-limiting step in this process. This has given rise to the concept of boosting the levels of SAM precursors such as folate and especially methyl-folate to promote methylation of $PP2A_C$ and subsequent formation of an active Bα containing PP2A complex that theoretically should lead to increased dephosphorylation of tau.[52,53]

The attractiveness of this approach to other chemical-based agonistic methods of boosting PP2A activity, is that it would apparently permit selectivity in the activation of specific PP2A complexes to the exclusion of others, and therefore would not be likely to induce deleterious consequences that might be expected to ensue from indiscriminate pleiotropic activation of PP2A.

Another potential agent based on the same principle of boosting PP2A activation indirectly, by enhancing methylation of the $PP2A_C$ catalytic subunit, is that of xylulose-5-phosphate. Xylulose-5-phosphate is a glucose metabolite in the glycolytic and lipogenic pathway acting as a glucose-signalling compound that recruits and activates PP2A.[54,55]

In the original report describing this phenomenon, it was determined that the PP2A activity in the liver that was enhanced by xylulose-5-phosphate, specifically contained a PP2A Bα subunit form.[55] This is a finding with obvious ramifications for the potential use of this agent as an inducer of tau dephosphorylation in the CNS, particularly as xylulose-5-phosphate is a low molecular weight substance that should freely cross the blood/brain barrier. However, subsequent studies,[56] in fact strongly suggested that it was the Bγ regulatory domain form of PP2A that was boosted by xylulose-5-phosphate, which would limit its use as an agent to induce tau dephosphorylation in the CNS. Further work, however, exploring the mechanism of action of xylulose-5-phosphate in inducing PP2A, has indicated that this compound can induce the methylation of the PP2A$_C$ catalytic subunit, which should be expected to promote the formation of the tau-relevant, Bα subunit containing form of PP2A.[57] It therefore remains to be determined whether xylulose-5-phosphate could be used as an agent to reduce tau dephosphorylation in Alzheimer's disease. Further studies in *in vitro* and preferably *in vivo* tau transgenic models are needed before the potential utility of this compound for the treatment of neurodegenerative conditions can be properly assessed.

10.3.2   Ceramide-Palmitate

Cellular membranes are composed of a lipid milieu, containing phospholipids, cholesterol and sphingolipids. Sphingolipids, similar to other lipid molecules can serve as precursors for downstream secondary messengers metabolised from the precursor molecules. One of the products of glycosphingolipid catabolism is ceramide. Ceramide is a cellular signalling moiety implicated in a plethora of cellular functions, however, it has also been reported to stimulate PP2A activity.[58] Some studies have suggested that ceramide shows some specificity in activating certain PP2A complexes, in particular by inducing the recruitment of the Bα regulatory subunit of PP2A into a PP2A complex in mitochondria, where it is thought to have a role in regulating mitochondrial apoptosis.[59,60]

As one might predict, the propensity to induce recruitment of the Bα subunit of PP2A into the PP2A heterocomplex suggests a possible role in PP2A$_C$ methylation, and indeed a recent report confirmed that ceramide can induce the methylation of the catalytic subunit of PP2A$_C$ in human T cells.[61] However, in an earlier report, the authors were unable to demonstrate any induction in the methylation of the PP2A$_C$ subunit using an *in vitro* insulin secreting INS-1 cell cytosolic fraction that was treated with ceramide.[62] This may indicate that there is a cell-type-specific cofactor or cofactors required for ceramide-induced methylation of PP2A$_C$ subunit. This would suggest that it is unlikely that ceramide acts directly on the PP2A$_C$ subunit, but rather indirectly to modulate methylation activity. This supposition has been supported by a very recent report that describes ceramide binding to and inactivating the inhibitor 2 protein of PP2A, I$_2$PP2A or SET by binding to a specific motif on I$_2$PP2A and subsequently preventing its association with PP2A.[63]

Clearly, the molecular mechanisms by which ceramide can induce PP2A activity is still controversial, and further developments are anticipated that will hopefully shed further light on this issue.

A number of lines of evidence then, suggest that ceramide may activate PP2A and quite possibly the relevant tau phosphatase form of PP2A, namely the PP2A Bα subunit form. At this point in time, we have not been able to discern any reports as to whether ceramide has in fact been tested in any neurodegenerative animal models to determine whether it might, in principle, be useful as a potential tau therapeutic. However, even if this were the case, any enthusiasm in the use of ceramide-based strategies for tau-based neurodegenerative disorders, must be tempered by the lack of specificity, and therefore unwanted side effects, such an approach is likely to have.

Physiologically, ceramide is an important regulator of a variety of stress responses and cellular growth pathways. Numerous cytokines and proinflammatory compounds, chemotherapeutics, growth factors, and toxic insults have all been shown to induce the formation of ceramide from the hydrolysis of sphingomyelin or from *de novo* synthesis pathways.[64–66,67–72]

It has also been demonstrated unequivocally that addition of exogenous ceramide or raising intracellular levels of this pleiotropic second messenger, induces cell-cycle arrest, apoptosis and even cellular senescence.[64,73,74] To accomplish this diversity of effects, it is unlikely that ceramide is acting solely through the mechanism of activating PP2A or even less as a specific inducer of the Bα subunit form of PP2A and indeed ceramide has been shown to activate protein phosphatase 1, PP1, as well as a variety of kinases.[55,75,76] *In vitro* evidence also demonstrates the ceramide can activate different subunit forms of PP2A, such as that of the PP2A Bα heterotrimeric complex.[77] This broad action in boosting different PP2A heterotrimers, as well as other phosphatases such as PP1 can be reconciled with the report that ceramide can bind directly to the catalytic domains of both PP1 and PP2A[77] A similar concern is likely to be found with other cell-permeable precursors of ceramide such as palmitate, which has similarly been shown *in vitro* to increase PP2A activity in mouse aorta tissue, most likely by acting to boost intracellular ceramide levels.[78]

In addition, another study has shown that the toxic amyloid Aβ peptide, which can induce apoptotic cell death in cerebral endothelial cells and is thought to be the causative agent of cerebral amyloid angiopathy,[79,80] does this by increasing intracellular levels of ceramide in cerebral endothelial cells. This leads to activation of the B″ family of PP2A heterotrimers, which have been reported to control dephosphorylation of proapoptotic cascade regulators in these cells and thereby to induce cerebral endothelial cell death.[81]

Clearly, the use of pleiotropic, indiscriminate activators of PP2A such as ceramide, as a therapeutic strategy to treat neurodegenerative conditions, is likely to have many unwanted side effects that will limit the clinical utility of such compounds.

10.3.3 Memantine

Memantine was developed as an NMDA receptor antagonist, which has, however, only low to moderate NMDA receptor antagonist properties but that somewhat surprisingly, clearly demonstrates a real if modest efficacy profile in moderate to severe Alzheimer's disease patients, improving mental function and quality of daily living in this patient cohort.[82,83]

In some elegant studies it is now been demonstrated that memantine, in addition to its NMDA receptor antagonistic properties, also possesses the ability to enhance PP2A in an indirect manner, by interfering with the ability of I$_2$PP2A or SET to block PP2A function and hence restore PP2A activity in cell-culture models of tau hyperphosphorylation.[84,85] It is certainly tempting to speculate that this extra PP2A-related property of memantine, unique amongst other NMDA receptor antagonist molecules, may explain its beneficial clinical effects in Alzheimer's disease patients. This assumption, if correct, would lend clinical credence to the concept of tau hyperphosphorylation as a pathological lynchpin of Alzheimer's disease formation and give impetus to the clinical development of specific PP2A activators for treating this condition and neurodegenerative disorders in general.

In this regard, it has been noted that memantines physicochemical properties most likely restrict its penetration of affected neurons in the CNS, and the development of memantine-like derivatives with improved CNS penetrating properties might yield derivatives with improved clinical benefits superior to that observed with the parent molecule.[86]

10.3.4 B$_2$ Adrenergic Receptor Agonists

The G-protein-coupled receptor B$_2$-adrenergic receptor agonist, isoproterenol, in some cell types is capable of activating the important cellular signalling kinase ERK.[87,88] However, in keratinocytes the same agonist has been reported to do the direct opposite, that is to deactivate ERK, by inducing association of the B$_2$-adrenergic receptor with PP2A and thereby stimulating PP2A activity two-fold.[87] Importantly, from a tau-dephosphorylation perspective, the same property has also been observed in hippocampal neurons,[88] raising the possibility that B$_2$-adrenergic receptor agonists might have some efficacy in tau-based pathological disorders. However, similar to the ceramide story as discussed above, the potential use of these agents is likely to be compromised by the potential for diverse effects upon different physiological processes, and clearly further research into whether the activity or recruitment of the Bα form of PP2A could be induced by more selective B$_2$-adrenergic receptor agonists is required.

10.4 Conclusions

The unbridled phosphorylation of tau is now recognised as a key step in the development of tauopathies and neurodegenerative conditions epitomised by Alzheimer's disease. Whilst it is theoretically simpler to inhibit the kinases that phosphorylated tau, the attractiveness of activating the PP2A phosphatase is that only a single protein complex needs to be targeted, compared with multiple kinases. There are, however, multiple challenges that need to be addressed before this strategy becomes a therapeutic reality. Insights into the complex post-translational regulation of the PP2A heterotrimer have indicated specific processes that can be exploited to promote the formation and consequent activity of specific, tau-relevant PP2A species. Some compounds that appear to be able to induce this process have already been characterised and it is to be hoped that more potent and specific derivatives with enhanced properties will be developed in the near future. This should eventually increase the armentarium of useful therapeutics to treat one of the underlying causes of Alzheimer's disease.

References

1. J. Gotz, A. Probst, E. Ehler, B. Hemmings and W. Kues, *Proc. Nature Acad. Sci. USA*, 1998, **95**, 12370.
2. J. Gotz, A. Probst, C. Mistl, R. M. Nitsch and E. Ehler, *Mechanisms Dev.*, 2000, **93**, 83.
3. J. Gotz and A. Schild, *Methods Enzymol.*, 2003, **366**, 390.
4. M. Goudreault, L. M. D'Ambrosio, M. J. Kean, M. J. Mullin, B. G. Larsen, A. Sanchez, S. Chaudhry, G. I. Chen, F. Sicheri, A. I. Nesvizhskii, R. Aebersold, B. Raught and A. C. Gingras, *Mol. Cell Proteomics*, 2009, **8**, 157.
5. B. McCright, A. M. Rivers, S. Audlin and D. M. Virshup, *J. Biol. Chem.*, 1996, **271**, 22081.
6. C. S. Moreno, S. Park, K. Nelson, D. Ashby, F. Hubalek, W. S. Lane and D. C. Pallas, *J. Biol. Chem.*, 2000, **275**, 5257.
7. S. Strack, J. A. Zaucha, F. F. Ebner, R. J. Colbran and B. E. Wadzinski, *J. Comp. Neurol.*, 1998, **392**, 515.
8. A. H. Schonthal, *Cancer Lett.*, 2001, **170**, 1.
9. K. Lechward, O. S. Awotunde, W. Swiatek and G. Muszynska, *Acta Biochim. Polon.*, 2001, **48**, 921.
10. A. Cegielska, S. Shaffer, R. Derua, J. Goris and D. M. Virshup, *Mol. Cell Biol.*, 1994, **14**, 4616.
11. V. Janssens and J. Goris, *Biochem. J.*, 2001, **353**, 417.
12. A. M. Healy, S. Zolnierowicz, A. E. Stapleton, M. Goebl, A. A. DePaoli-Roach and J. R. Pringle, *Mol. Cell Biol.*, 1991, **11**, 5767.
13. R. E. Mayer-Jaekel, H. Ohkura, R. Gomes, C. E. Sunkel, S. Baumgartner, B. A. Hemmings and D. M. Glover, *Cell*, 1993, **72**, 621.
14. A. Schild, K. Schmidt, Y. A. Lim, Y. Ke, L. M. Ittner, B. A. Hemmings and J. Gotz, *Int. J. Dev. Neurosci.*, 2006, **24**, 437.
15. K. Okamoto, H. Li, M. R. Jensen, T. Zhang, Y. Taya, S. S. Thorgeirsson and C. Prives, *Molec. Cell*, 2002, **9**, 761.
16. Y. Xu, Y. Xing, Y. Chen, Y. Chao, Z. Lin, E. Fan, J. W. Yu, S. Strack, P. D. Jeffrey and Y. Shi, *Cell*, 2006, **127**, 1239.
17. S. Longin, K. Zwaenepoel, J. V. Louis, S. Dilworth, J. Goris and V. Janssens, *J. Biol. Chem.*, 2007, **282**, 26971.

18. M. Li and Z. Damuni, *Methods Mol. Biol.*, 1998, **93**, 59–66.
19. B. Kovacech, E. Kontsekova, N. Zilka, P. Novak, R. Skrabana, P. Filipcik, K. Iqbal and M. Novak, *FEBS Lett.*, 2007, **581**, 617.
20. Y. Adachi, G. N. Pavlakis and T. D. Copeland, *J. Biol. Chem.*, 1994, **269**, 2258.
21. H. T. Adler, F. S. Nallaseth, G. Walter and D. C. Tkachuk, *J. Biol. Chem.*, 1997, **272**, 28407.
22. P. Neviani, R. Santhanam, R. Trotta, M. Notari, B. W. Blaser, S. Liu, H. Mao, J. S. Chang, A. Galietta, A. Uttam, D. C. Roy, M. Valtieri, R. Bruner-Klisovic, M. A. Caligiuri, C. D. Bloomfield, G. Marcucci and D. Perrotti, *Cancer Cell*, 2005, **8**, 355.
23. P. Neviani, R. Santhanam, J. J. Oaks, A. M. Eiring, M. Notari, B. W. Blaser, S. Liu, R. Trotta, N. Muthusamy, C. Gambacorti-Passerini, B. J. Druker, J. Cortes, G. Marcucci, C. S. Chen, N. M. Verrills, D. C. Roy, M. A. Caligiuri, C. D. Bloomfield, J. C. Byrd and D. Perrotti, *J. Clin. Investig.*, 2007, **117**, 2408.
24. I. Tsujio, T. Zaidi, J. Xu, L. Kotula, I. Grundke-Iqbal and K. Iqbal, *FEBS Lett.*, 2005, **579**, 363.
25. H. Tanimukai, I. Grundke-Iqbal and K. Iqbal, *Am. J. Pathol.*, 2005, **166**, 1761.
26. M. L. Billingsley and R. L. Kincaid, *Bioch. J.*, 1997, **323**(Pt 3), 577.
27. L. Dehmelt and S. Halpain, *Genome Biol.*, 2005, **6**, 204.
28. A. Harada, K. Oguchi, S. Okabe, J. Kuno, S. Terada, T. Ohshima, R. Sato-Yoshitake, Y. Takei, T. Noda and N. Hirokawa, *Nature*, 1994, **369**, 488.
29. C. X. Gong, F. Liu, I. Grundke-Iqbal and K. Iqbal, *J. Neural. Transm.*, 2005, **112**, 813.
30. H. Ksiezak-Reding, W. K. Liu and S. H. Yen, *Brain Res.*, 1992, **597**, 209.
31. K. Iqbal, C. Alonso Adel and I. Grundke-Iqbal, *J. Alzheimers Dis.*, 2008, **14**, 365.
32. A. C. Alonso, I. Grundke-Iqbal and K. Iqbal, *Nature Med.*, 1996, **2**, 783–787.
33. J. Z. Wang, C. X. Gong, T. Zaidi, I. Grundke-Iqbal and K. Iqbal, *J. Biol. Chem.*, 1995, **270**, 4854.
34. K. Santacruz, J. Lewis, T. Spires, J. Paulson, L. Kotilinek, M. Ingelsson, A. Guimaraes, M. DeTure, M. Ramsden, E. McGowan, C. Forster, M. Yue, J. Orne, C. Janus, A. Mariash, M. Kuskowski, B. Hyman, M. Hutton and K. H. Ashe, *Science*, 2005, **309**, 476.
35. K. Iqbal, F. Liu, C.X. Gong, A.D. Alonso, I. Grundke-Iqbal, *Acta Neuropathol.*, 2009.
36. S. Khatoon, I. Grundke-Iqbal and K. Iqbal, *J. Neurochem.*, 1992, **59**, 750.
37. P. V. Arriagada, J. H. Growdon, E. T. Hedley-Whyte and B. T. Hyman, *Neurology*, 1992, **42**, 631.
38. M. Hutton, C. L. Lendon, P. Rizzu, M. Baker, S. Froelich, H. Houlden, S. Pickering-Brown, S. Chakraverty, A. Isaacs, A. Grover, J. Hackett, J. Adamson, S. Lincoln, D. Dickson, P. Davies, R. C. Petersen, M. Stevens, E. de Graaff, E. Wauters, J. van Baren, M. Hillebrand, M. Joosse, J. M.

Kwon, P. Nowotny, L. K. Che, J. Norton, J. C. Morris, L. A. Reed, J. Trojanowski, H. Basun, L. Lannfelt, M. Neystat, S. Fahn, F. Dark, T. Tannenberg, P. R. Dodd, N. Hayward, J. B. Kwok, P. R. Schofield, A. Andreadis, J. Snowden, D. Craufurd, D. Neary, F. Owen, B. A. Oostra, J. Hardy, A. Goate, J. van Swieten, D. Mann, T. Lynch and P. Heutink, *Nature*, 1998, **393**, 702.

39. M. G. Spillantini, J. R. Murrell, M. Goedert, M. R. Farlow, A. Klug and B. Ghetti, *Proc. Nature Acad. Sci. USA*, 1998, **95**, 7737.

40. R. Kimura, K. Kamino, M. Yamamoto, A. Nuripa, T. Kida, H. Kazui, R. Hashimoto, T. Tanaka, T. Kudo, H. Yamagata, Y. Tabara, T. Miki, H. Akatsu, K. Kosaka, E. Funakoshi, K. Nishitomi, G. Sakaguchi, A. Kato, H. Hattori, T. Uema and M. Takeda, *Human Mol. Gen.*, 2007, **16**, 15.

41. F. Liu, Z. Liang, J. Wegiel, Y. W. Hwang, K. Iqbal, I. Grundke-Iqbal, N. Ramakrishna and C. X. Gong, *Faseb. J.*, 2008, **22**, 3224.

42. J. Shi, T. Zhang, C. Zhou, M. O. Chohan, X. Gu, J. Wegiel, J. Zhou, Y. W. Hwang, K. Iqbal, I. Grundke-Iqbal, C. X. Gong and F. Liu, *J. Biol. Chem.*, 2008, **283**, 28660.

43. F. Liu, I. Grundke-Iqbal, K. Iqbal and C. X. Gong, *Eur. J. Neurosci.*, 2005, **22**, 1942.

44. E. Sontag, V. Nunbhakdi-Craig, G. Lee, R. Brandt, C. Kamibayashi, J. Kuret, C. L. White 3rd, M. C. Mumby and G. S. Bloom, *J. Biol. Chem.*, 1999, **274**, 25490.

45. E. Sontag, A. Luangpirom, C. Hladik, I. Mudrak, E. Ogris, S. Speciale and C. L. White 3rd, *J. Neuropathol. Exp. Neurol.*, 2004, **63**, 287.

46. C. X. Gong, T. J. Singh, I. Grundke-Iqbal and K. Iqbal, *J. Neurochem.*, 1993, **61**, 921.

47. A. Hiraga and S. Tamura, *Biochem. J.*, 2000, **346**(Pt 2), 433.

48. E. Sontag, V. Nunbhakdi-Craig, G. Lee, G. S. Bloom and M. C. Mumby, *Neuron.*, 1996, **17**, 1201.

49. Y. Xu, Y. Chen, P. Zhang, P. D. Jeffrey and Y. Shi, *Molecular Cell*, 2008, **31**, 873.

50. U. S. Cho and W. Xu, *Nature*, 2007, **445**, 53.

51. C.E. Zhang, Q. Tian, W. Wei, J.H. Peng, G.P. Liu, X.W. Zhou, Q. Wang, D.W. Wang, J.Z. Wang, *Neurobiol. Aging*, 2007.

52. E. Sontag, V. Nunbhakdi-Craig, J. M. Sontag, R. Diaz-Arrastia, E. Ogris, S. Dayal, S. R. Lentz, E. Arning and T. Bottiglieri, *J. Neurosci.*, 2007, **27**, 2751.

53. J. M. Sontag, V. Nunbhakdi-Craig, L. Montgomery, E. Arning, T. Bottiglieri and E. Sontag, *J. Neurosci.*, 2008, **28**, 11477.

54. M. A. Davare, V. Avdonin, D. D. Hall, E. M. Peden, A. Burette, R. J. Weinberg, M. C. Horne, T. Hoshi and J. W. Hell, *Science*, 2001, **293**, 98.

55. R. T. Dobrowsky, C. Kamibayashi, M. C. Mumby and Y. A. Hannun, *J. Biol. Chem.*, 1993, **268**, 15523.

56. T. Kabashima, T. Kawaguchi, B. E. Wadzinski and K. Uyeda, *Proc. Nature Acad. Sci. USA*, 2003, **100**, 5107.

57. S. Guenin, L. Schwartz, D. Morvan, J. M. Steyaert, A. Poignet, J. C. Madelmont and A. Demidem, *Inter. J. Oncol.*, 2008, **32**, 49.
58. P. P. Ruvolo, *Pharmacol. Res.*, 2003, **47**, 383.
59. P. P. Ruvolo, X. Deng, T. Ito, B. K. Carr and W. S. May, *J. Biol. Chem.*, 1999, **274**, 20296.
60. P. P. Ruvolo, W. Clark, M. Mumby, F. Gao and W. S. May, *J. Biol. Chem.*, 2002, **277**, 22847.
61. C. L. Chen, C. F. Lin, C. W. Chiang, M. S. Jan and Y. S. Lin, *Molec. Pharmacol.*, 2006, **70**, 510.
62. A. Kowluru and S. A. Metz, *FEBS Lett.*, 1997, **418**, 179.
63. A. Mukhopadhyay, S. A. Saddoughi, P. Song, I. Sultan, S. Ponnusamy, C. E. Senkal, C. F. Snook, H. K. Arnold, R. C. Sears, Y. A. Hannun and B. Ogretmen, *Faseb. J.*, 2009, **23**, 751.
64. T. Okazaki, R. M. Bell and Y. A. Hannun, *J. Biol. Chem.*, 1989, **264**, 19076.
65. M. Y. Kim, C. Linardic, L. Obeid and Y. Hannun, *J. Biol. Chem.*, 1991, **266**, 484.
66. L. R. Ballou, *Immunol. Today*, 1992, **13**, 339.
67. M. Liscovitch, *Trends Biochem. Sci.*, 1992, **17**, 393.
68. Y. A. Hannun, *J. Biol. Chem.*, 1994, **269**, 3125.
69. J. Quintans, J. Kilkus, C. L. McShan, A. R. Gottschalk and G. Dawson, *Biochem. Biophys. Res. Commun.*, 1994, **202**, 710.
70. M. G. Cifone, R. De Maria, P. Roncaioli, M. R. Rippo, M. Azuma, L. L. Lanier, A. Santoni and R. Testi, *J. Exp. Med.*, 1994, **180**, 1547.
71. J. C. Strum, G. W. Small, S. B. Pauig and L. W. Daniel, *J. Biol. Chem.*, 1994, **269**, 15493.
72. A. Haimovitz-Friedman, C. C. Kan, D. Ehleiter, R. S. Persaud, M. McLoughlin, Z. Fuks and R. N. Kolesnick, *J. Exp. Med.*, 1994, **180**, 525.
73. T. Okazaki, A. Bielawska, R. M. Bell and Y. A. Hannun, *J. Biol. Chem.*, 1990, **265**, 15823.
74. L. M. Obeid, C. M. Linardic, L. A. Karolak and Y. A. Hannun, *Science*, 1993, **259**, 1769.
75. R. A. Wolff, R. T. Dobrowsky, A. Bielawska, L. M. Obeid and Y. A. Hannun, *J. Biol. Chem.*, 1994, **269**, 19605.
76. A. H. Merrill Jr., E. M. Schmelz, D. L. Dillehay, S. Spiegel, J. A. Shayman, J. J. Schroeder, R. T. Riley, K. A. Voss and E. Wang, *Toxicol. Applied Pharmacol.*, 1997, **142**, 208.
77. C. E. Chalfant, K. Kishikawa, M. C. Mumby, C. Kamibayashi, A. Bielawska and Y. A. Hannun, *J. Biol. Chem.*, 1999, **274**, 20313.
78. Y. Wu, P. Song, J. Xu, M. Zhang and M. H. Zou, *J. Biol. Chem.*, 2007, **282**, 9777.
79. L. S. Perlmutter, M. A. Myers and E. Barron, *Microsc. Res. Tech.*, 1994, **28**, 204.
80. H. M. Wisniewski, J. Wegiel, A. W. Vorbrodt, B. Mazur-Kolecka and J. Frackowiak, *Annals NY Acad. Sci.*, 2000, **903**, 6.
81. K. J. Yin, C. Y. Hsu, X. Y. Hu, H. Chen, S. W. Chen, J. Xu and J. M. Lee, *J. Neurosci.*, 2006, **26**, 2290.

82. B. Winblad and N. Poritis, *Internat. J. Geriatric Psychiatry*, 1999, **14**, 135.
83. B. Reisberg, R. Doody, A. Stoffler, F. Schmitt, S. Ferris and H. J. Mobius, *New England J. Med.*, 2003, **348**, 1333.
84. M. O. Chohan, S. Khatoon, I. G. Iqbal and K. Iqbal, *FEBS Lett.*, 2006, **580**, 3973.
85. L. Li, A. Sengupta, N. Haque, I. Grundke-Iqbal and K. Iqbal, *FEBS Lett.*, 2004, **566**, 261.
86. K. Iqbal and I. Grundke-Iqbal, *Cell Mol. Life Sci.*, 2007, **64**, 2234.
87. C. E. Pullar, J. Chen and R. R. Isseroff, *J. Biol. Chem.*, 2003, **278**, 22555.
88. M. A. Davare, M. C. Horne and J. W. Hell, *J. Biol. Chem.*, 2000, **275**, 39710.

Rationale for Tau-Aggregation Inhibitor Therapy in Alzheimer's Disease and Other Tauopathies

CLAUDE M. WISCHIK,[1] DAMON J. WISCHIK,[2] JOHN M.D. STOREY[3] AND CHARLES R. HARRINGTON[1]

[1] TauRx Therapeutics Ltd. and School of Medicine and Dentistry, University of Aberdeen, Foresterhill, Aberdeen, AB25 2ZD, Scotland, UK; [2] Department of Computer Science, University College, Gower Street, London, WC1E 6BT, UK; [3] TauRx Therapeutics Ltd. and Department of Chemistry, University of Aberdeen, Meston Walk, Aberdeen, AB24 3UE, Scotland, UK

11.1 The Tau-Aggregation Pathology of Alzheimer's Disease

Alzheimer's disease (AD) is an irreversible, neurodegenerative disorder characterised by the progressive loss of memory and thinking skills. It was first presented at a meeting in 1906 by Dr. Alois Alzheimer, a German psychiatrist, who discovered "neurofibrillary tangles" in the brain tissue of a woman who died with dementia at the age of 55.[1] It was not until the tangle could be isolated and purified that its structure[2–4] and composition could be determined.[5,6] The neurofibrillary tangles are composed predominantly of tau protein, a protein essential for neuronal shape and axonal transport and the aggregation of tau is closely linked both to clinical dementia and cell death.[7–9] These findings have also been affirmed by others in the field.[10–14]

RSC Drug Discovery Series No. 2

Emerging Drugs and Targets for Alzheimer's Disease

Volume 1: Beta-Amyloid, Tau Protein and Glucose Metabolism

Edited by Ana Martinez

© Royal Society of Chemistry 2010

Published by the Royal Society of Chemistry, www.rsc.org

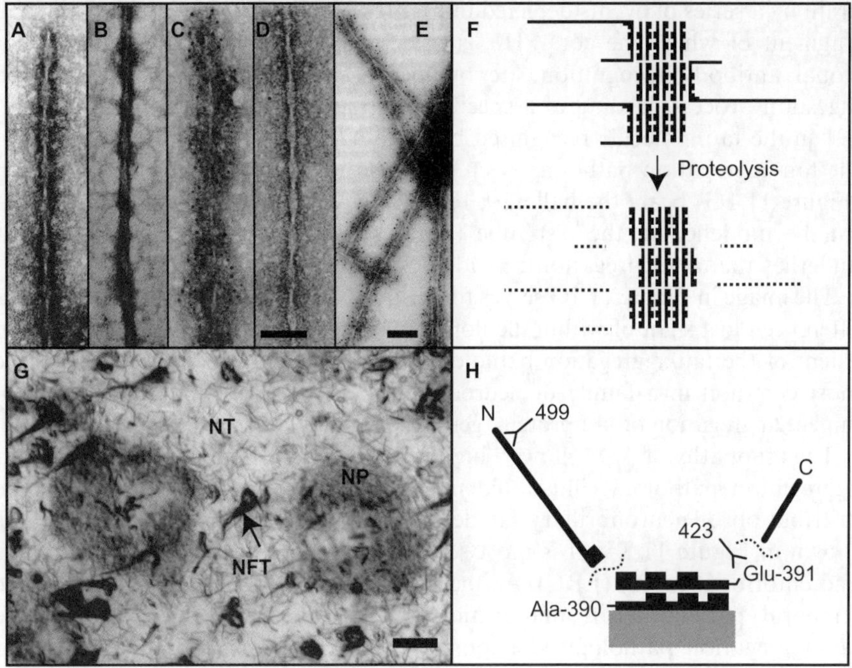

Figure 11.1 Neurofibrillary pathology is composed of aggregated tau in filaments. The filaments have a fuzzy coat (A) that can be removed by proteolysis (B). Immunolabelling shows that the fuzzy coat contains N-terminal parts of tau (C) that are removed by proteolysis (D), while the remaining PHF core contains fragments of tau truncated at Glu-391 and recognised by mAb 423 (E). mAb 423 immunoreactivity can be demonstrated in PHFs and in intracellular pretangle tau oligomers without prior treatment with proteases and this is represented schematically (F). All neurofibrillary pathology is visualised in AD brain tissue using mAb 423 (NFT, neurofibrillary tangle; NT, neuropil threads; NP, neuritic plaque) (G). The process of aggregation has been simulated in an *in vitro* assay, using antibodies specific to the N-terminus (499) and C-terminal truncation of tau (423); dotted lines indicate sites of proteolytic cleavage (H). Scale bars, 100 nm for A-D and E and 50 μm for G.

Neurofibrillary tangles are intraneuronal clusters of aberrant tau protein polymers that consist of "paired helical filaments" (PHFs), thus termed from their characteristic double-twisted ribbon shape. The neurofibrillary tangle and its constituent PHFs are shown in Figure 11.1. PHFs have a fuzzy outer coat accounting for about 20% of the mass of the filament that can be removed by digestion with proteases leaving behind a stable inner core (Figures 11.1A–D) which retains the characteristic twisted ribbon structure. A monoclonal antibody (mAb 423), raised against core-PHFs, labels both isolated core filaments and intact PHFs (Figure 11.1E). Further research led to the discovery that the molecular configuration recognised by this monoclonal had been created in the brain by a series of events depicted in Figures 11.1F and 11.7A.[15] The short tau fragment of which the core-PHF is principally composed creates the mono-clonal antibody recognition site by means of a high-affinity binding and digestion process, leading to a specific C-terminal truncation at position Glu-391 in the tau molecule recognised by mAb 423. Thus, the entire spectrum of the tau-aggregation pathology seen in a histological section of the AD brain (Figure 11.1G) bears the hallmark of this aggregation/truncation process that can be modelled in the test tube (Figures 11.1F and 11.7A). This process underlies the tau-aggregation cascade described further in Section 11.2.

The image in Figure 11.1G serves to illustrate very clearly a simple fact that has often been lost sight of within the dominant β-amyloid theory, namely the sheer extent of the tau-aggregation pathology that is characteristic of AD. AD is the most common in a family of neurodegenerative disease characterised by pro-minent aggregation of tau protein, generally termed "tauopathies" (Figure 11.2).

The tauopathy of AD follows a highly characteristic pattern of spread of tau-aggregation pathology, illustrated in Figure 11.3. The characteristic overall distribution of neurofibrillary tangle pathology in the human brain in AD is shown in Figure 11.3A. It is most severe where it begins in the hippocampus and entorhinal cortex (ERC). As the disease progresses, tangles spread to the temporal, parietal and frontal cortices. This characteristic distribution of the tau-aggregation pathology seen postmortem closely matches the pattern of defects that can be demonstrated by functional brain scans during life. These scans reveal patterns of reduced neuronal function, either by way of reduced

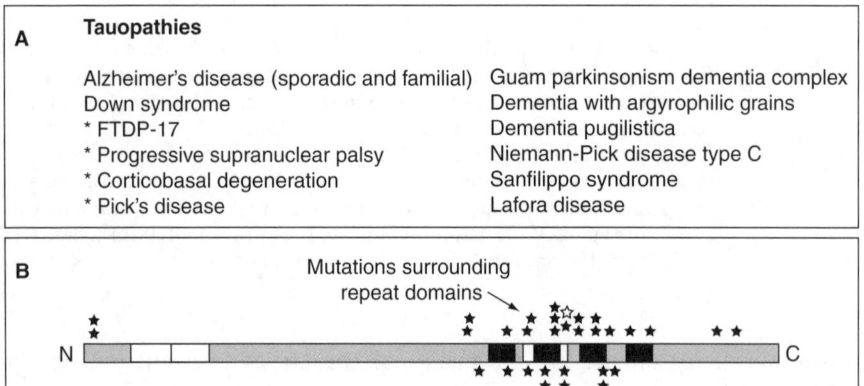

Figure 11.2   The tauopathies are neurodegenerative diseases characterised by tau-aggregation pathology (A). Those denoted with an asterisk have cases directly linked to mutations in the tau-protein gene. These mutations are typically clustered around the same repeat domains (B) found in the core-PHF structure (see Figures 11.1 and 11.7). Nearly 40 mutations have been identified, including 7 intronic mutations (open asterisk) that affect alternative splicing of exon 10 within the repeat domains.

blood flow (HMPAO-SPECT) or reduced glucose utilisation (FDG-PET). As shown in Figure 11.3B, there is progressive deterioration in neuronal function in the regions of progressive accumulation of aggregated tau. These scan deficits match the clinical progression of AD as measured by the mini-mental state examination (MMSE).[16,17]

This stereotyped pattern of spread of pathology has been formalised into the Braak staging system,[18] illustrated in Figure 11.3C. By taking a single brain slice through the brain region shown by the vertical line in Figure 11.3A, the spread of tau aggregation can be staged on the basis of microscopic examination of the brain into the six Braak stages illustrated in Figure 11.3C. It should be noted that Braak attempted to devise a staging system based on the β-amyloid pathology of AD, but showed that this was not possible because it appears to be a general feature of the aging brain, and does not appear to follow any clear pattern of progression or spread. It has since been confirmed by numerous studies that there is minimal if any systematic relationship between β-amyloid pathology and cognitive decline.[19–22]

Braak staging therefore provides a useful unifying schema for bringing together all of the key processes that characterise the evolution of AD from its earliest detectable stages through to end-stage dementia. This is shown in Figure 11.4. Figure 11.3B shows the approximate relationship between Braak stages and the defects shown by HMPAO-SPECT scan.[16] In Figure 11.4A, the relationship between Braak stage and decline in MMSE score is shown.[8] This data comes from a prospective clinicopathological study conducted in an epidemiologically defined cohort in the Cambridge area in which patients were followed in the community by means of repeat assessment every 12–24 months. Because cases were followed irrespective of disease severity or diagnosis, it became possible to establish a calibration between level of cognitive function and Braak stage, once cases with vascular pathology were excluded. What is striking in Figure 11.4B is that already at preclinical and early clinical stages of AD, brain pathology has already advanced to Braak stages 2 and 3. The transition to Braak stage 4 corresponds to disease severity generally meeting agreed clinical criteria for a diagnosis of AD.

The process of tau aggregation begins in the neocortical regions well before neurofibrillary tangles appear (Figure 11.4). Tau aggregation, in the form of proteolytically stable PHFs, can be detected from Braak stage 2 onwards. Tangles that can be visualised by conventional microscopy do not appear until Braak stage 4 in the neocortex. As will be shown below, the time interval between Braak stages 2 and 4 is about 20 years. There is therefore a long pretangle/preclinical and early clinical period during which therapeutic intervention in the tau-aggregation cascade can be achieved.

Tau aggregation begins in the form of submicroscopic formations referred to as tau oligomers (Figure 11.5). This progresses to filaments (*i.e.* PHFs) that, when they occupy the whole of the intracellular space, are referred to as a neurofibrillary tangle ("NFT"). This eventually chokes the neurone by eliminating the possibility of normal neuronal metabolism and in the process eventually leaves behind an extracellular "ghost" tangle as the only marker of a previously existing neurone.

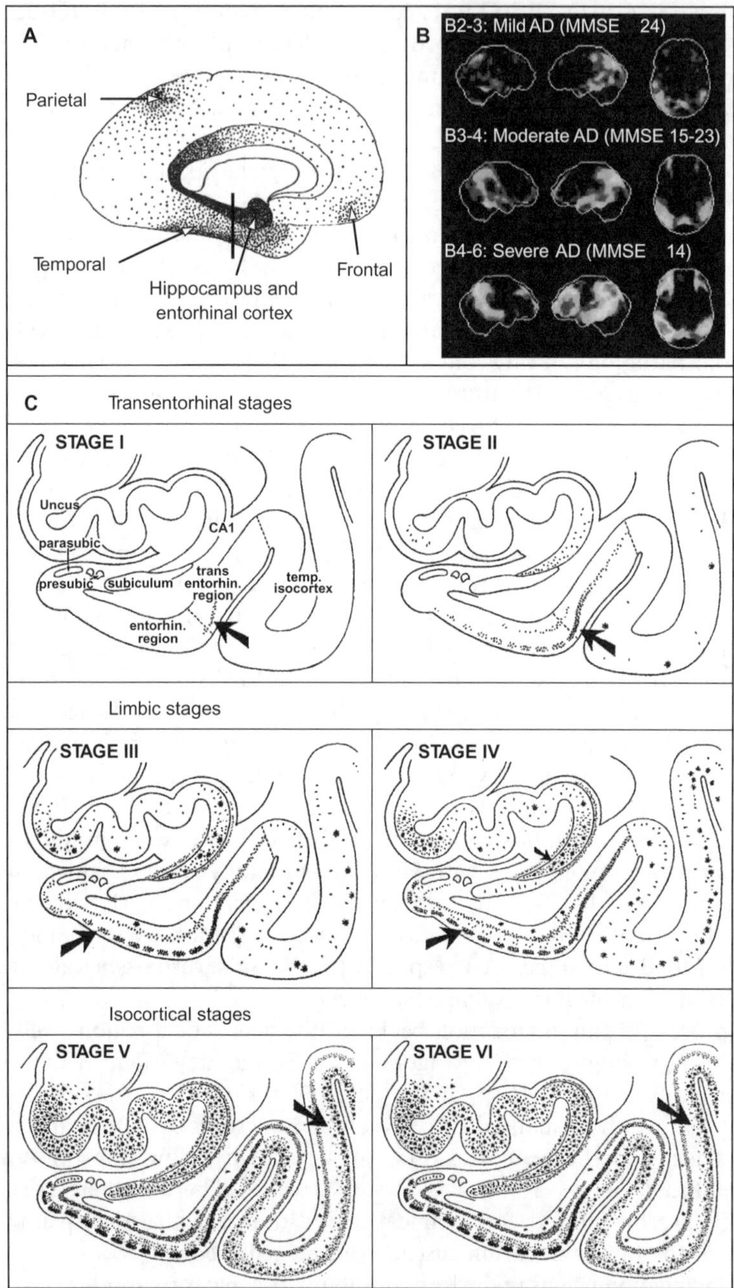

Figure 11.3 Braak staging of Alzheimer's disease. Tau aggregation spreads in a characteristic neuroanatomical pattern (A) and the distribution of tau pathology matches the pattern of SPECT scan defects seen in early stages of AD (B). The Braak staging of the spread of tangles from medial temporal lobe to neocortex can be assessed in a single brain slice (C). [B from Nishimura *et al.* 2007[16] and C from Braak and Braak, 1991;[18] with permissions].

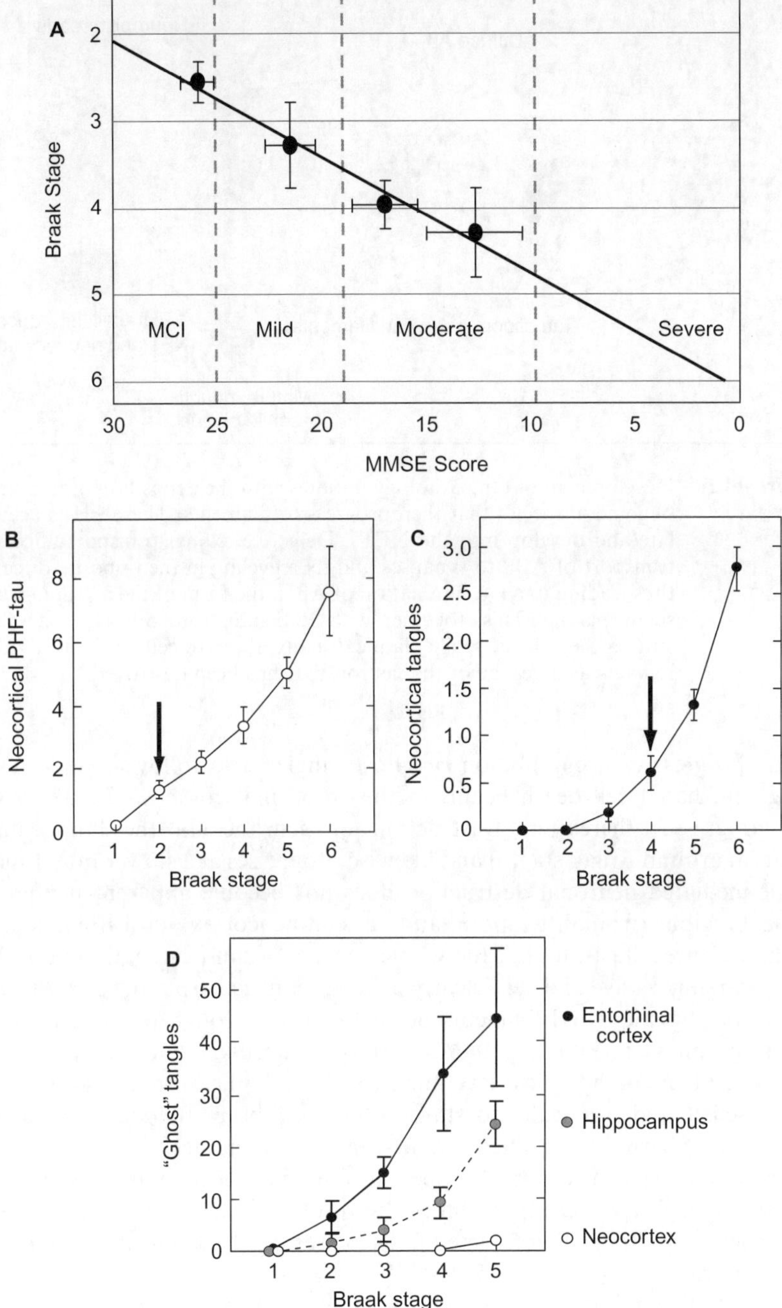

Figure 11.4 Braak staging measures cognitive decline (A), tau aggregation (B), neurofibrillary tangle pathology (C) and tangle-mediated neuronal destruction (D). Results from a prospective clinicopathological study.[7,8]

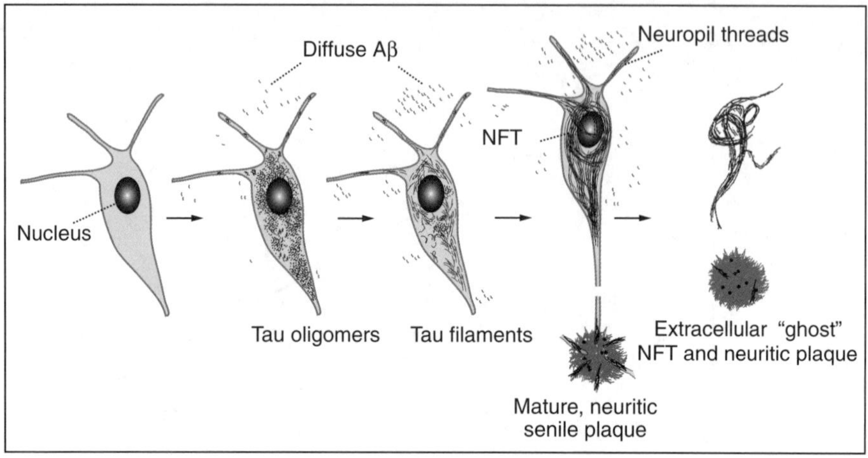

Figure 11.5 Development of tau pathology initiates with the deposition of tau, first as oligomeric species that then progress to filaments. These PHFs accumulate and develop into the NFTs. Defective axonal transport affects the transport of APP to synapses and its recycling in the opposite direction, thus leading to an accumulation of Aβ in the form of the mature, neuritic senile plaque. This, together with PHF-tau from other neurites, constitutes the neuritic senile plaque. Finally, the extracellular "ghost" tangle remains as a vestige of the neuron that has been destroyed.

The progressive accumulation of ghost tangles also follows Braak stages. Tangle-mediated cell death begins in the entorhinal cortex at Braak stage 2 and progresses thereafter. Destruction of neurones in the hippocampus begins at around Braak stage 3 and likewise progresses at a slower rate. Finally, tangle-mediated neuronal destruction does not become apparent in the neocortical regions (frontal, temporal and parietal neocortex) until Braak stage 5.

These figures illustrate that the whole process of tau-aggregation pathology follows an impressive and very clearly defined sequence, from the earliest stages when tau oligomers and filaments begin to form, through to the final stages when neurones are killed by the mature neurofibrillary tangles. All of these processes at the pathological level map to clinically measurable stages of cognitive deficit and physically to stages of loss of brain function that can be measured by currently available brain-scanning techniques.

Braak staging can also be mapped throughout the human lifespan. In a major study, Braak's group reported the results of one of the largest human postmortem studies ever conducted in 847 cases in whom age and Braak stage were reported between ages 45–95.[23] There were approximately 17 cases per year of life throughout this age span. A Kaplan–Meier survival analysis using this data is shown in Figure 11.6. The probability of transition from one stage to a later stage demonstrates that there is a simple age-dependent transition to higher Braak stages with advancing age (Figure 11.6A). This indicates that, irrespective of the claims of the β-amyloid theory regarding specific genetic

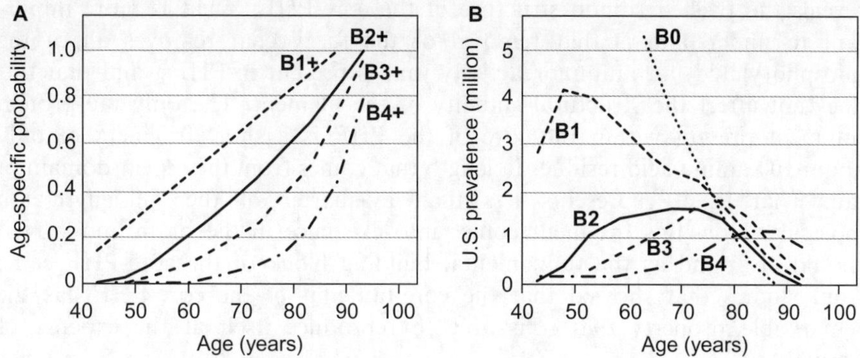

Figure 11.6 Epidemiological implications of Braak staging with respect to age. (A) Age-specific probability of Braak stage transitions calculated from the data derived from 847 postmortems.[3] (B) Estimated prevalence of different Braak stages by age for the US population.

causation of AD, the entire process is clearly embedded within the blueprint of the general aging human population and regardless of specific genetic factors that may accelerate the process in some individuals.

The survival plot can be transformed into a prevalence plot at specific Braak stages. This gives a cross-sectional view of population progression by age through the Braak stages through the ages 40–95. The peak for Braak stage 1 is around 50 years. Considering the population aged 65 and over, only 24% will remain at Braak stage 0 for the remainder of their lives. Of the 41% of the population who progress to Braak stage 2, which peaks at around age 65, 92% will progress to Braak stage 3, which peaks at around age 75. Progression to Braak stage 4 is limited by the reduction in survival population beyond age 75. The most surprising conclusion to be drawn from Figure 11.6B is that Braak stage 2, which appears clinically innocuous from Figure 11.4A (*i.e.* corresponds to the maximum MMSE score of 30), is actually already on what appears to be a largely irreversible pathway of future inexorable disease progression. This analysis is fundamentally at odds with the claims made by the β-amyloid school of thought, to the effect that tau-aggregation pathology is a late-stage event and therefore not amenable to therapeutic intervention. On the contrary, the real challenge facing the field is to identify those individuals at Braak stage 2 in whom prevention of the further progression of tau pathology could essentially eliminate AD from the aging human blueprint.

11.2 The Molecular Mechanics of Tau Aggregation in Alzheimer's Disease

As indicated above, PHFs which come from intracellular tangles are associated with a fuzzy outer coat that can be removed by proteolysis, which reveals the twisted ribbon structure of the core-PHF. What is more important to understand is that removal of the fuzzy coat removes all of the phosphorylated tau immunoreactivity that is present in PHFs, and that this does not affect the structural stability of the filament. The only tau-protein fragment present within the core of the PHF is a short fragment of only about 100 amino acid residues in length that comes from the repeat-domain of tau. That is, it represents less than a quarter of the full-length tau molecule. How this fragment comes into existence in the brain and why it should be found as the fundamental building block of the core-PHF came from studies that showed that the core-tau unit of the core-PHF has the remarkable property that it is able to reproduce itself at the expense of normal tau.[8]

The core fragment of tau attached to a solid-phase permits a high-affinity binding interaction with full-length tau (Figure 11.7A). This locks the repeat domain of the bound full-length molecule into a proteolytically stable

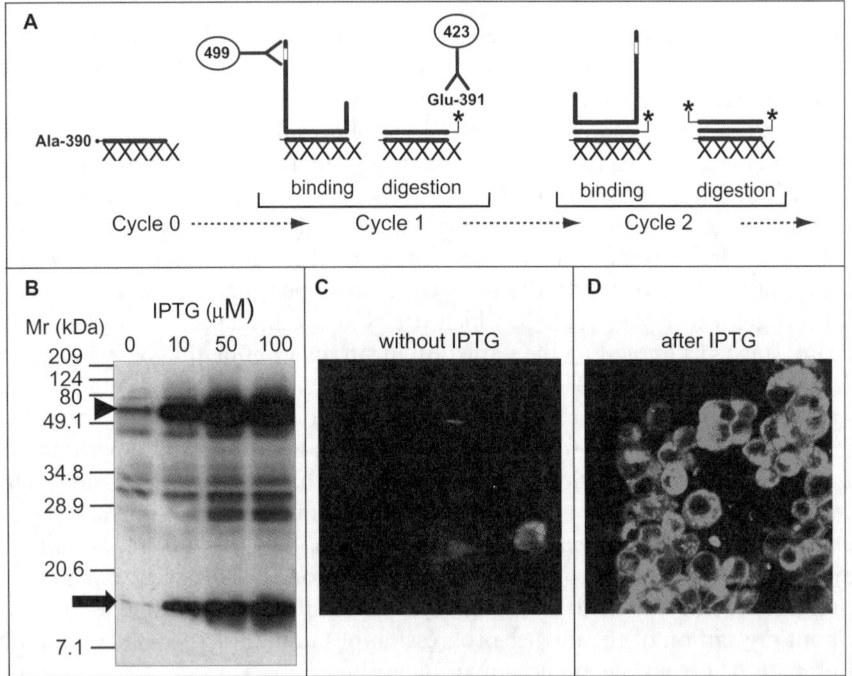

Figure 11.7 The core-PHF is composed of truncated tau protein that aggregates via an autocatalytic process of binding and proteolysis, depicted schematically (A). In a cellular model (B), induction of expression of full-length tau (arrowhead) leads to the accumulation of 12-kDa truncated tau (arrow) in cells constitutively expressing low levels of truncated tau that serves as a seed for tau capture and propagation. Cells are labelled by a fluorescent PHF-ligand only after induction (C) and not in the absence of IPTG (D).

configuration that is essentially identical to the starting fragment. That is, a digestion step with a broad spectrum exoprotease recreates the core-tau fragment. Digestion removes the other parts of the molecule, but fails to remove the bound segment that constitutes the new replicate of the original core-tau template. The fundamental problem is that the proteolytically stable aggregate retains the ability to bind yet more normal tau protein through subsequent binding/digestion cycles. This process creates the mAb 423 antibody recognition site already alluded to in Figure 11.1, and shown to be such a prominent feature of the AD tauopathy.

The process shown schematically in a cell-free environment (Figure 11.7A) was also confirmed in a cellular environment in the form of a stable cell line suitable for screening of tau-aggregation inhibitors (Figure 11.7B).[24] In this cell line, expression of normal full-length tau is under the control of an inducer that can be turned on in the presence of IPTG (arrow-head). The cell line also has very low level constitutive expression of the core-tau PHF unit (arrow). The levels of the latter have to be extremely low, otherwise toxic aggregates are readily formed. When tau expression is turned on, by the addition of IPTG, there is a conversion of the full-length tau into the truncated core-tau PHF unit. This induction is associated with the appearance of tau protein aggregates that are recognised by a proprietary, fluorescent aggregation-dependent ligand (Figures 11.7C and D). Simple overexpression of full-length tau does not lead to aggregation in this model.

The capacity of tau-protein pathology to propagate itself at the expense of normal tau was confirmed in recent independent studies. Aggregated fibrils of tau243-375 can be taken up into cells *in vitro* containing full-length tau and passed on to neighbouring cells to seed aggregation.[25] Similarly, the transmission and spread of tau pathology from tau-mutant mice into normal mice was achieved, provided human full-length tau protein was already expressed in the latter.[26]

The place of tau-aggregation pathology and of tau aggregation inhibitor therapy is illustrated schematically in Figure 11.8. In general, the critical event in triggering the tau-aggregation cascade is the nucleation or seeding event (Figure 11.8A). Once the seeding event has occurred, the tau-aggregation cascade is self-propagating, and leads to two deleterious outcomes. It converts normal functional tau protein into the truncated aggregated form found in PHFs. This leads to a loss of normal tau protein,[27] needed to stabilise axonal microtubules for linking different parts of the brain. More importantly, as alluded to above, the tau aggregates are directly neurotoxic and eventually lead to neuronal death.

Although the β-amyloid school of thought sets great store by the abnormal processing of amyloid precursor protein (APP) and/or presenilin as a critical upstream trigger for the tau-aggregation cascade, a more general age-related abnormality in the endosomal-lysosomal processing pathway may be much more important, particularly the processing of mitochondria.[28] There are now two examples of tauopathies caused by genetically inherited abnormalities in the endosomal-lysosomal processing pathway: Niemann–Pick Type C[29] and

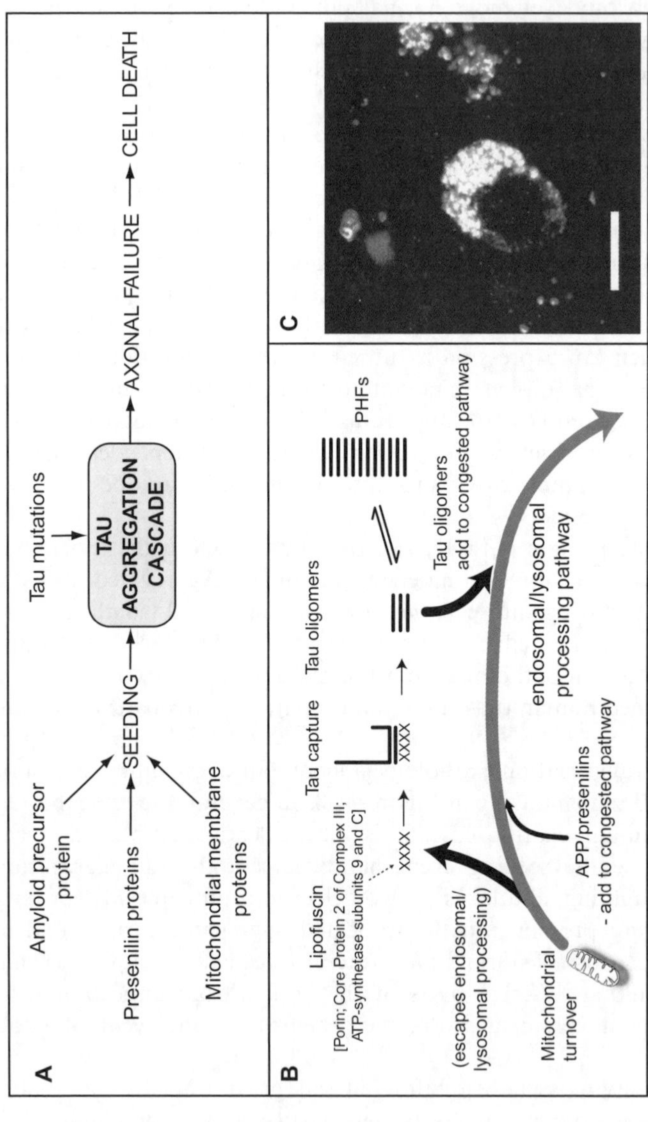

Figure 11.8 Tau aggregation is in a pivotal part of the degenerative process in AD and other tauopathies. (A) The initial seeding of tau protein may be initiated by any one or more events, *e.g.* mutations in APP, presenilin proteins, altered processing of mitochondrial membrane proteins or, more directly, through mutations in tau itself. (B) The aggregation of tau protein adds to the age-related congestion of the endosomal-lysosomal pathway. (C) The capture of tau by mitochondrial membrane proteins is illustrated in neurons in AD where there is colocalisation of tau and porin (white) in lysosomes (scale bar, 25 µm).

Sanfilippo Syndrome Type B,[30] neither related to any abnormality in APP processing.

In the case of sporadic AD, we consider that an age-related defect in the turnover of mitochondria is more important than APP.[28] This is illustrated in Figure 11.8B. Mitochondria are the obligate energy source for neuronal metabolism and their turnover period is approximately 6 months. Long-lived, nondividing cells such as neurones show signs of failure in the turnover of mitochondria-derived proteins as they age. These take the form of so-called aging pigments, or lipofuscin, which are in fact composed predominantly of unprocessed products of mitochondrial turnover.[31] These same proteins are found in close association with early-stage tau aggregates in the AD brain, and biochemical analyses show they have the ability to form proteolytically resistant stable complexes with tau. We have observed colocalisation of the mitochondrially derived proteins (porin, core protein 2 of complex III and ATP synthase subunit 9) with early tau aggregates. An example is shown for porin and tau (Figure 11.8C). Essentially identical images have been produced in a mouse model of Sanfilippo syndrome showing colocalisation of tau with lysozyme and a lysosomal form of ATP-synthase subunit C.[30]

The system whose failure is responsible for clearance of mitochondrially derived proteins is the endosomal-lysosomal processing pathway, and is exactly the same pathway required for processing mutant membrane-bound proteins, such as APP and the presenilin proteins.[32] Therefore, mutations in these proteins may add to the congestion and dysfunction in this pathway without actually being directly causative of AD.

Once tau aggregation has been initiated, the only pathway available for clearance of the tau oligomers is the same congested endosomal-lysosomal pathway. The capacity of the neuron to clear proteolytically stable tau oligomers is therefore critically compromised, leading to uncontrolled progression of the tau-aggregation process. The progressive rate of accumulation of aggregated tau in the form of PHFs is, in fact, exponential over time (Figure 11.9). This can be calculated on the basis of levels of aggregated tau protein measured in neocortex at different Braak stages (as shown for example in Figure 11.4B) and real time between Braak stages, calculated for example from the data shown in Figure 11.6A.[23] From this it is possible to deduce the level of aggregated tau in neocortex as a function of time in years (Figure 11.9).

The critical transition points such as conversion to Braak stages 2–4 and the appearance of clinically visible dementia occur about 7 years after the transition to the steep exponential phase of tau aggregation in neocortex (Figure 11.9, arrow). This occurs about 25 years after the transition to Braak stage 1.

11.3   The Basis of Tau-Aggregation Inhibitor Therapy

Given the autocatalytic/self-propagating nature of the tau-aggregation cascade, it is possible to construct a mathematical model of a system that contains a self-

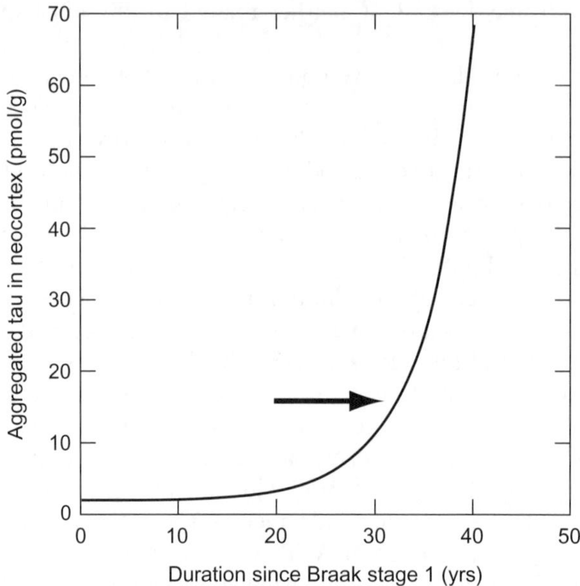

Figure 11.9 The exponential accumulation of aggregated tau in neocortex in humans calculated from a prospective clinicopathological study based on Braak staging. The arrow indicates stage at which clinically visible dementia is observed.

propagating component (Figure 11.10A). In this model, normal tau protein is diverted into an aggregation cascade with a positive feedback component. There are two different timeframes operating in the data incorporated into the model. The times required for aggregation and clearance phenomena *in vitro* are relatively short (hours/days) whereas the time scale over which aggregates build up in the brain is measured in years, *i.e.* the time scale of Braak staging. Since production and clearance can be shown *in vitro* and in cell models to reach equilibrium over a short time (hours/days), the progressive slow build-up of aggregated tau over years cannot be explained by short-acting internal equilibrium processes. Rather, the progressive accumulation over time must be due to progressive failure of clearance that undergoes very gradual degradation. This is envisaged as a slow deterioration in the capacity of the endosomal-lysosomal system to deal with clearance of oligomers and larger tau aggregates. We have modelled this by incorporating the time required for the aggregation and dissolution processes observed *in vitro*, and an underlying long-term component that leads to decreased clearance capacity over time.

It then becomes possible to calculate the predicted consequences of intervention on the input side or enhancement on the clearance side (Figure 11.10A). Although it seems intuitively obvious that inhibition on the input side should be beneficial, the model output indicates that the benefit of input-side intervention would be short lived (Figure 11.10B). The two axes represent time (in years) and corresponding PHF levels (*e.g.* in neocortex based on data such as that shown in

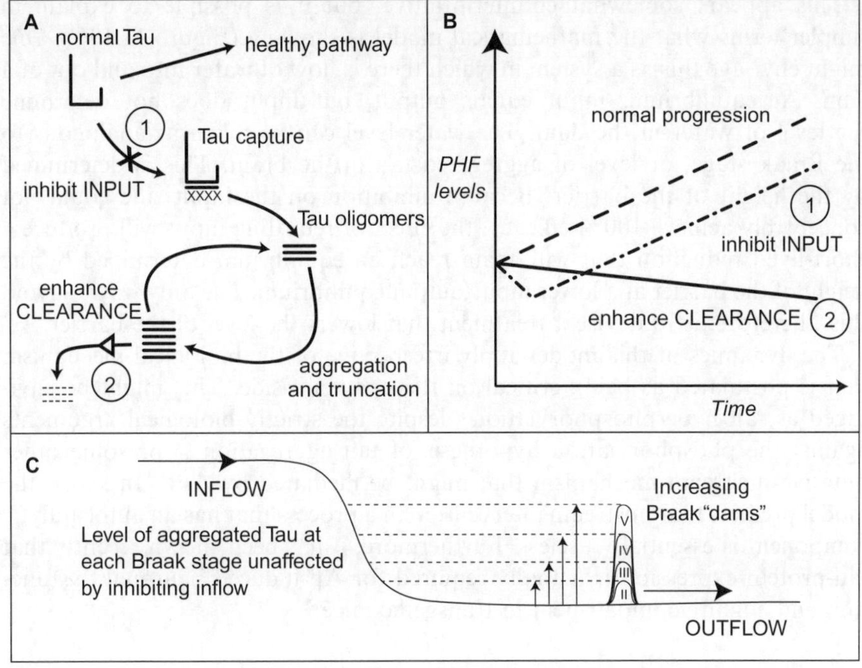

Figure 11.10 Opening up a new clearance pathway via tau disaggregation of oligomers via a tau-aggregation inhibitor (TAI) is the only way to affect the rate of disease progression. The tau-aggregation cascade proceeds by an autocatalytic process of binding and proteolysis of tau. (A) Braak progression is likely driven by age-related impairment in clearance of aggregated tau (*i.e.* progressive degradation of the endosomal-lysosomal pathway (see Figure 11.8B)). (B) Only by enhancing clearance of oligomers by means of TAI treatment is the rate of progression of AD likely to be altered. (C) Inhibiting the inflow to the cycle will not affect the levels of aggregated tau accumulating at each Braak stage. Thus, upstream inhibition of factors that initiate tau capture are unlikely to alter the rate of progression of AD.

Figure 11.9). The surprising conclusion is that a treatment that reduces the influx of tau into the aggregating cascade would be expected to have clinical characteristics of a purely symptomatic treatment. Despite input-side intervention appearing to act mechanistically on what might be postulated to be a rate-limiting aspect of the disease process, the net effect would be to produce only a transient reduction in levels of aggregated tau, but would have no effect on the long-term rate of accumulation of aggregated tau in the brain over time (Figure 11.10B). In other words, the intervention would be expected to produce only transient benefit, and not impact on the subsequent rate of disease progression. On the other hand, an intervention on the output side of the process (*i.e.* enhanced clearance) would have the effect of altering the disease progression trajectory.

This appears somewhat counterintuitive, but it is possible to explain in simpler terms what the mathematical model is showing (Figure 11.10C). One might envisage this as a system in which there is flow of water into and out of a dam. At equilibrium, input equals output, but input does not determine the level of water in the dam. The water level can be taken as analogous to the Braak stage, or level of aggregated tau in the brain. This is determined by the height of the barrier. Because inhibition on the input side could not conceivably achieve 100% efficacy, the effect of retarding input will produce a short-lived reduction that will again reach an equilibrium determined by the height of the barrier at a lower input/output equilibrium. The only way to avoid this, therefore, is to devise a treatment that lowers the level of the barrier.

The dynamics of this model apply irrespective of the biological mechanism that is postulated as being critical on the upstream side. This might be envisaged as tau hyperphosphorylation (despite the strictly biological arguments against the phosphorylation hypothesis of tau aggregation[28]) or some other putative upstream mechanism that might be mediated by APP. In short, the model predicts that upstream intervention in a process that has an autocatalytic component is essentially useless. Furthermore, it has been shown recently that tau-protein expression is actually required for Aβ-induced neuronal dysfunction and cognitive impairment in transgenic mice.[33]

11.4   Methylthioninium Chloride as a Tau-Aggregation Inhibitor

Methylthioninium chloride (MTC; Structure 11.1) was the first tau-aggregation inhibitor (TAI) reported[15] and the only TAI yet to be tested in clinical trials for AD.

11.1

TAI therapy, such as that provided by MTC, acts essentially on the output side by enhancing clearance. The fundamental kinetic block in the clearance of aggregated tau is envisaged as being due to the state of aggregation itself. MTC was shown to reverse the proteolytic stability of the tau-protein fragment of the core-PHF by disaggregating the polymer, permitting the release of the core-tau monomer.[15] In its monomeric state, the core-tau unit of the PHF is extremely sensitive to proteases. PHFs disaggregate in the presence of MTC and MTC is effective at submicromolar concentrations in cell-free and cellular assays used to screen TAI activity (Figure 11.11).

The activity of MTC has been tested *in vivo* in two proprietary transgenic mouse models. In the first model, "Line 1" mice express the core-tau unit of the

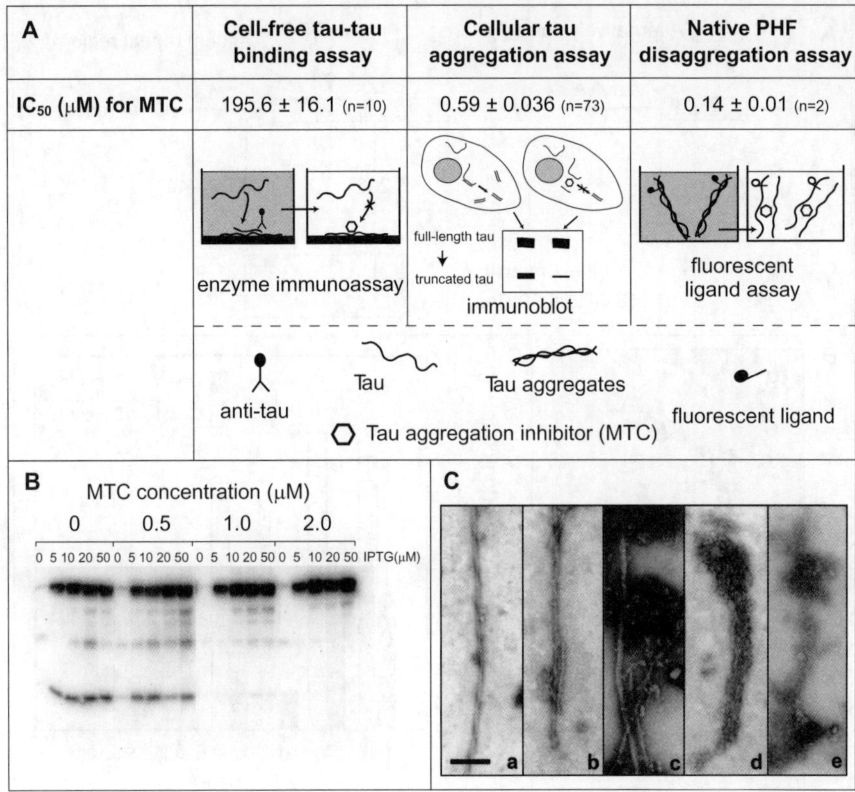

Figure 11.11 MTC is a tau-aggregation inhibitor that dissolves both oligomers and filaments. (A) Inhibitory activity of MTC in three assays shown diagrammatically. (B) Oligomeric tau produced in cells, as indicated by a 12-kDa fragment, is removed by MTC. (C) PHFs isolated from AD brain tissue are dissolved by increasing concentrations of MTC (a, 0.01%; b, 0.1% and c–e, 1%; scale bar, 100 nm).

PHF linked with a short N-terminal signal-sequence and under the control of a neuron-specific Thy-1 promoter (Figure 11.12A).[34] The purpose of the signal sequence is to force tau aggregation by targeting expression to a system capable or providing an initiation binding substrate, *i.e.* the endoplasmic reticulum, and simultaneously forcing tau clearance into the endosomal-lysosomal pathway reproducing the conditions for a failure of this clearance pathway in respect of tau. The net effect is to produce a pathology that reaches the stage of oligomers but not of fully formed filaments or neurofibrillary tangles (Figure 11.12B). The mice reproduce the essential phenomena of Braak staging (Figure 11.12C). In mice less than 12 months of age, the aggregated-tau pathology is located predominantly in the entorhinal cortex and hippocampus, but little in other brain regions. As mice age, the pathology spreads into isocortical brain regions (retrosplenial, visual and auditory cortices and

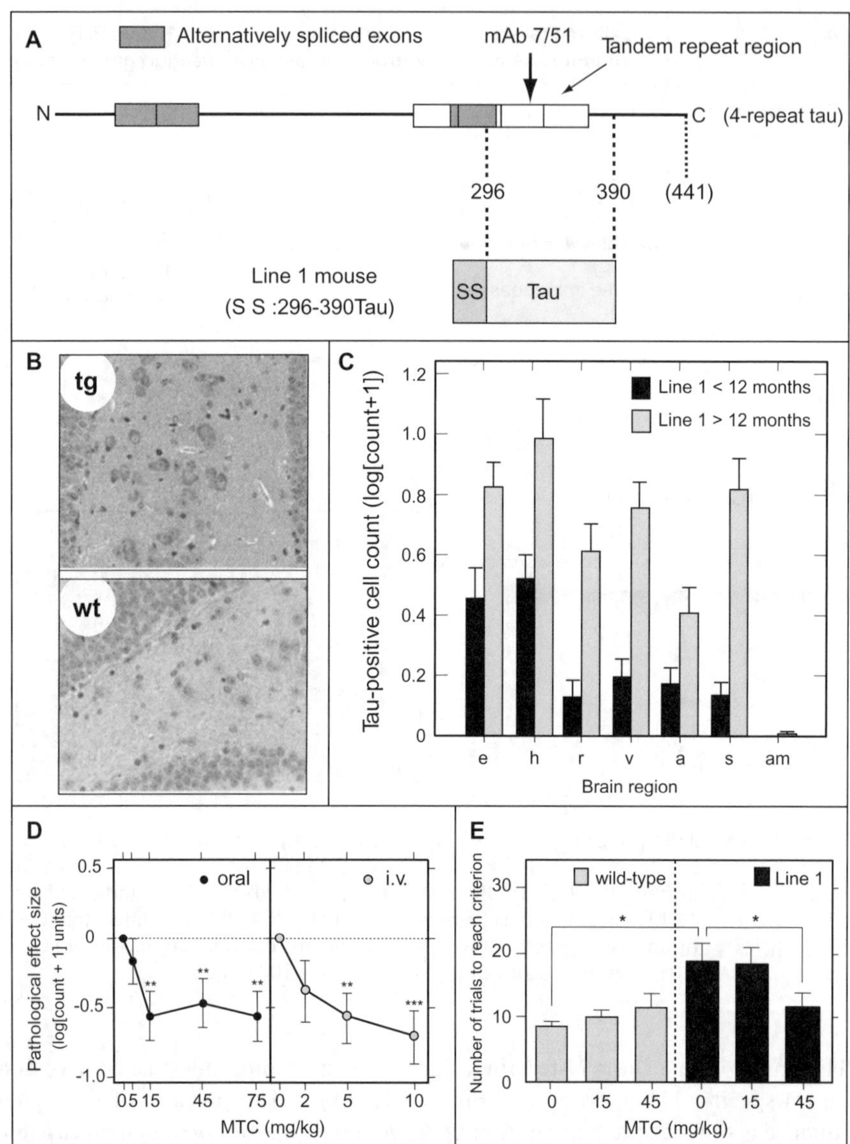

Figure 11.12   Line 1 mouse expresses truncated tau targeted to membrane via a signal sequence peptide (SS:296-390Tau; A). Pathology develops first in hippocampus and entorhinal cortex before progressing to other cortical areas (B, C). Pathology (D) and behavioural deficits in water maze learning (E) respond to treatment with MTC. (tg, transgenic; wt, wild type; e, entorhinal cortex; h, hippocampus; r, retrosplenial cortex; v, visual cortex; a, auditory cortex; s, subiculum; am, amygdala).

subiculum). There is a reduction in the counts of tau-positive neurones containing oligomeric tau after mice have been treated with MTC (administered either orally or intravenously) (Figure 11.12D).

Line 1 mice develop a learning impairment after about 7 months of age, as shown for example in a modified Morris water maze,[35] in which the task is to find a hidden platform with a minimum number of trials. Line 1 mice require about twice as many trials to reach a given learning criterion while oral treatment with MTC reverses this learning defect (Figure 11.12E).

Truncated tau was directed to the endoplasmic reticulum in Line 1 mice to enable the initiation of tau nucleation and avoid the toxic effects observed by overexpressing truncated tau within cells. Mice expressing mutant tau exhibit increased tau associated with rough ER in motor neurons and a greater number of contacts between rough ER and mitochondria was observed by others.[36] Increased tau was also associated with a rough ER fraction extracted from AD brains compared with controls in the latter study.

MTC also exhibits efficacy in a second proprietary mouse model ("Line 66"; Figure 11.13). In this case, a full-length tau construct with two mutations, at P301S (associated with frontotemporal dementia with Parkinsonism linked with chromosome 17, FTDP-17)[37] and a further mutation at G335D, shown to further enhance tau aggregation *in vitro*. Whereas the Line 1 mouse exhibits diffuse neuronal tau pathology, the Line 66 mouse, has severe tau pathology, to an extent similar to that noted by other researchers expressing tau with FTDP-17 mutations in transgenic mice.[38–40] Tau tangles that stain positively with Bielschowsky silver and with thioflavin S indicate their filamentous nature. Tau pathology is observed in CA1 and CA3 of the hippocampus, entorhinal and other cortical areas (Figure 11.13B).

Line 66 mice show a severe abnormality of motor learning whereby they are unable to learn how to remain on a rotating rod (Figure 11.13C). Following oral treatment with MTC (1 mg/kg), however, this learning deficit is reversed. Somewhat higher doses (10 mg/kg) were able to reverse tau pathology in hippocampus and entorhinal cortex (Figure 11.13D). It is possible to extract protease-resistant 12-kDa tau from brains of Line 66 mice, indicating the presence of AD-like filaments of tau. This protease-resistant tau can be dissolved in the presence of MTC (Figure 11.13E).

Thus, MTC is able to reverse both the behavioural and pathological effects that arise *in vivo* in two transgenic mouse models of tau aggregation: a cognitive phenotype model in which tau oligomers predominate (Line 1) and a frontotemporal-like motor phenotype in which abundant filamentous tau accumulates (Line 66).

11.5 Clinical Application of Tau-Aggregation Inhibitor Therapies

MTC was the first tau-aggregation inhibitor described,[15] and several other polymerisation inhibitors have since been reported. These include benzothiazole

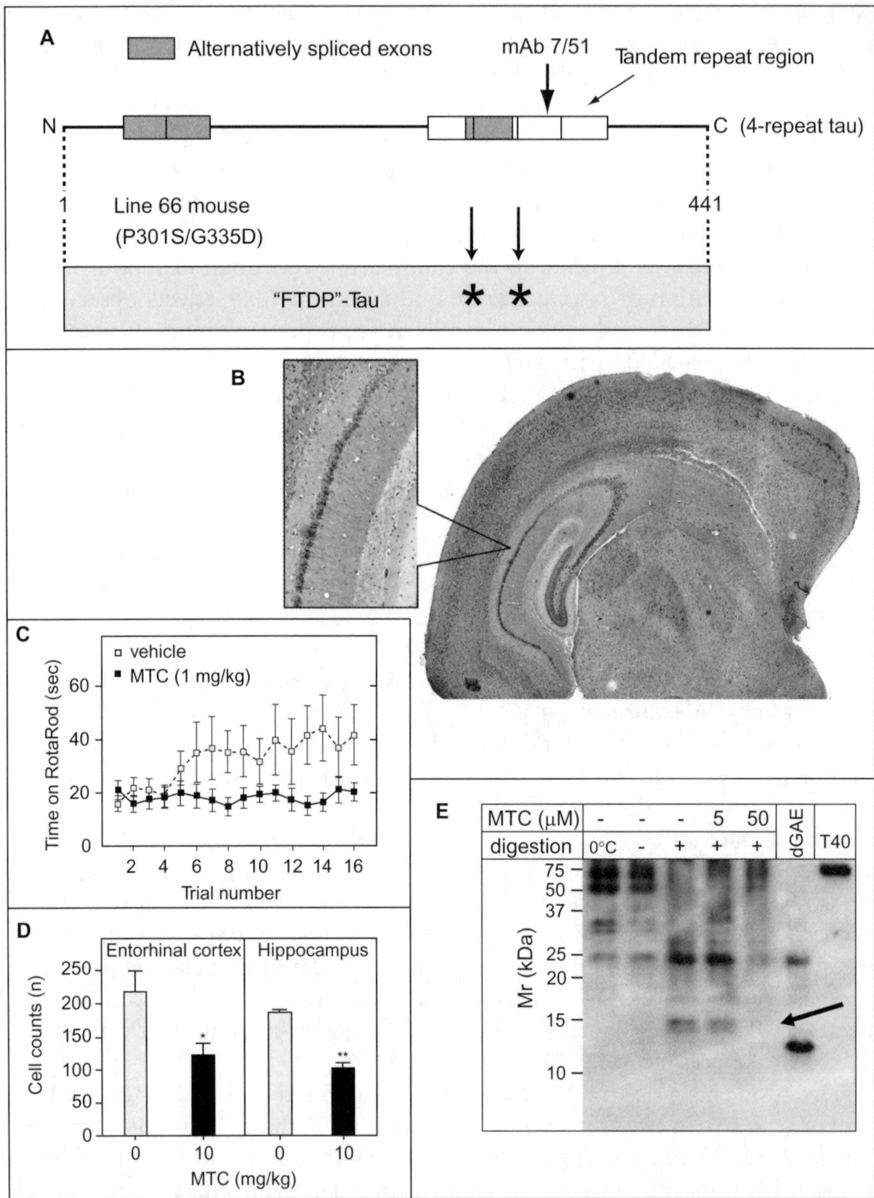

Figure 11.13 Transgenic "FTDP-type" tau mice (A) exhibit tau pathology throughout brain including hippocampus (B). Motor learning on a rotating rod (C) and pathology in extorhinal cortex and hippocampus (D) are both improved following oral treatment of Line 66 mice with MTC. Protease-resistant 12-kDa tau extracted from mouse brains is also sensitive to MTC (E, arrow).

derivatives,[41,42] Congo red derivatives and anthraquinones,[43] 2,3-di(furan-2-yl)-quinoxalines,[44] phenylthiazolyl-hydrazide,[45] polyphenols and porphyrins,[46] cyanin dyes[47] and aminothienopyridazines.[48]

Importantly, MTC selectively avoids disruption of the normal tau–tubulin interaction.[15] MTC has been used clinically since Paul Ehrlich first reported its use as an analgesic and in the treatment of malaria over a century ago.[49] Its main therapeutic use has been in the intravenous treatment of methaemoglobinaemia[50] and as an oral urinary antiseptic. It has also had clinical use in CNS, where it has been used to treat ifosfamide encephalopathy[51] and manic depression and psychosis.[52] MTC is not only a tau-aggregation inhibitor *in vitro*, but a compound that crosses the blood/brain barrier and demonstrates efficacy in reducing tau pathology and in relieving cognitive and motor learning symptoms in tau transgenic mice.

Furthermore, the rationale for using the TAI therapeutic approach has been confirmed by the results from a phase 2 clinical trial in 321 patients diagnosed with mild or moderate AD.[53] MTC reduced the rate of disease progression by 84% over 50 weeks when measured by the ADAS-cog scale and a 100% reduction on the MMSE scale, the two most common means to evaluate the usefulness of AD therapeutics. This result is a marked increase over the 30–50% reduction defined as desirable for an AD disease modifying agent by a European Task Force Consensus Statement.[54] These results were further supported by brain scans that allowed the visualisation of MTC-related prevention of loss of neuronal function as shown by a HMPAO-SPECT scan in a 55% subpopulation of the primary clinical study.[55] The accumulated data suggests that MTC has the potential not only to slow the rate of AD progression, but may even halt it and restore neuronal function, particularly at early stages of the disease.

References

1. A. Alzheimer, *Allg. Z. Psych. Psych.-gerich. Med.*, 1907, **64**, 146–148.
2. R. A. Crowther and C. M. Wischik, *EMBO J.*, 1985, **4**, 3661–3665.
3. C. M. Wischik, R. A. Crowther, M. Stewart and M. Roth, *J. Cell Biol.*, 1985, **100**, 1905–1912.
4. C. M. Wischik and R. A. Crowther, *Br. Med. Bull.*, 1986, **42**, 51–56.
5. C. M. Wischik, M. Novak, H. C. Thøgersen, P. C. Edwards, M. J. Runswick, R. Jakes, J. E. Walker, C. Milstein, R. M. and A. Klug, *Proc. Natl. Acad. Sci. USA*, 1988, **85**, 4506–4510.
6. C. M. Wischik, M. Novak, P. C. Edwards, A. Klug, W. Tichelaar and R. A. Crowther, *Proc. Natl. Acad. Sci. USA*, 1988, **85**, 4884–4888.
7. F. García-Sierra, C. M. Wischik, C. R. Harrington, J. Luna-Muñoz and R. Mena, *J. Chem. Neuroanat.*, 2001, **22**, 65–77.
8. E. B. Mukaetova-Ladinska, F. Garcia-Sierra, J. Hurt, H. J. Gertz, J. H. Xuereb, R. Hills, C. Brayne, F. A. Huppert, E. S. Paykel, M. McGee,

R. Jakes, W. G. Honer, C. R. Harrington and C. M. Wischik, *Am. J. Pathol.*, 2000, **157**, 623–636.

9. C. R. Harrington, E. B. Mukaetova-Ladinska, R. Hills, P. C. Edwards, E. Montejo de Garcini, M. Novak and C. M. Wischik, *Proc. Natl. Acad. Sci. USA*, 1991, **88**, 5842–5846.

10. N. Zilka, P. Filipcik, P. Koson, L. Fialova, R. Skrabana, M. Zilkova, G. Rolkova, E. Kontsekova and M. Novak, *FEBS Lett.*, 2006, **580**, 3582–3588.

11. M. von Bergen, S. Barghorn, L. Li, A. Marx, J. Biernat, E. M. Mandelkow and E. Mandelkow, *J. Biol. Chem.*, 2001, **276**, 48165–48174.

12. T. C. Gamblin, M. E. King, H. Dawson, M. P. Vitek, J. Kuret, R. W. Berry and L. I. Binder, *Biochemistry*, 2000, **39**, 6136–6144.

13. E. Grober, D. Dickson, M. J. Sliwinski, H. Buschke, M. Katz, H. Crystal and R. B. Lipton, *Neurobiol. Aging*, 1999, **20**, 573–579.

14. D. R. Thal, T. Arendt, G. Waldmann, M. Holzer, D. Zedlick, U. Rüb and R. Schober, *Neurobiol. Aging*, 1998, **19**, 517–525.

15. C. M. Wischik, P. C. Edwards, R. Y. K. Lai, M. Roth and C. R. Harrington, *Proc. Natl. Acad. Sci. USA*, 1996, **93**, 11213–11218.

16. T. Nishimura, K. Hashikawa, H. Fukuyama, T. Kubota, S. Kitamura, H. Matsuda, H. Hanyu, H. Nabatame, N. Oku, H. Tanabe, Y. Kuwabara, S. Jinnouchi and A. Kubo, *Ann. Nuc. Med.*, 2007, **21**, 15–23.

17. M. F. Folstein, S. E. Folstein and P. R. McHugh, *J. Psychiatr. Res.*, 1975, **12**, 189–198.

18. H. Braak and E. Braak, *Acta Neuropathol.*, 1991, **82**, 239–259.

19. G. K. Wilcock and M. M. Esiri, *J. Neurol. Sci.*, 1982, **56**, 407–417.

20. C. R. Harrington, J. Louwagie, R. Rossau, E. Vanmechelen, R. H. Perry, E. K. Perry, J. H. Xuereb, M. Roth and C. M. Wischik, *Am. J. Pathol.*, 1994, **145**, 1472–1484.

21. H. Crystal, D. Dickson, P. Fuld, D. Masur, R. Scott, M. Mehler, J. Masdeu, C. Kawas, M. Aronson and L. Wolfson, *Neurology*, 1988, **38**, 1682–1687.

22. P. W. Arriagada, J. H. Growdon, E. T. Hedley-White and B. T. Hyman, *Neurology*, 1992, **42**, 631–639.

23. T. G. Ohm, H. Müller, H. Braak and J. Bohl, *Neuroscience*, 1995, **64**, 209–217.

24. C. M. Wischik, D. Horsley, J. E. Rickard and C. R. Harrington, *PCT International Application*, 2002, WO02/055720.

25. B. Frost, R. L. Jacks and M. I. Diamond, *J. Biol. Chem.*, 2009, **284**, 12845–12852.

26. F. Clavaguera, T. Bolmont, R. A. Crowther, D. Abramowski, S. Frank, A. Probst, G. Fraser, A. K. Stalder, M. Beibel, M. Staufenbiel, M. Jucker, M. Goedert and M. Tolnay, *Nature Cell Biol.*, 2009, **11**, 909–914.

27. R. Y. K. Lai, H.-J. Gertz, D. J. Wischik, J. H. Xuereb, E. B. Mukaetova-Ladinska, C. R. Harrington, P. C. Edwards, R. Mena, E. S. Paykel, C. Brayne, F. A. Huppert, M. Roth and C. M. Wischik, *Neurobiol. Aging*, 1995, **16**, 433–445.

28. C. M. Wischik, R. Y. K. Lai and C. R. Harrington, in *Microtubule-Associated Proteins: Modifications in Disease.*, ed. J. Avila, R. Brandt and K. S. Kosik, Harwood Academic Publishers, Amsterdam, 1997, pp. 185–241.
29. S. Love, L. R. Bridges and C. P. Case, *Brain*, 1995, **118**, 119–129.
30. K. Ohmi, L. C. Kudo, S. Ryazantsev, H.-Z. Zhao, S. L. Karsten and E. F. Neufeld, *Proceedings of the National Academy of Sciences*, 2009, **106**, 8332–8337.
31. D. N. Palmer, R. D. Martinus, S. M. Cooper, G. G. Midwinter, J. C. Reid and R. D. Jolly, *J. Biol. Chem.*, 1989, **264**, 5736–5740.
32. R. A. Nixon and A. M. Cataldo, *Trends Neurosci.*, 1995, **18**, 489–496.
33. E. D. Roberson, K. Scearce-Levie, J. J. Palop, F. Yan, I. H. Cheng, T. Wu, H. Gerstein, G.-Q. Yu and L. Mucke, *Science*, 2007, **316**, 750–754.
34. C. M. Wischik, J. E. Rickard, D. Horsley, C. R. Harrington, F. Theuring, K. Stamer and C. Zabke, *PCT International Application*, 2002, WO02/059150.
35. G. Chen, K. S. Chen, J. Knox, J. Inglis, A. Bernard, S. J. Martin, A. Justice, L. McConlogue, D. Games, S. B. Freedman and R. G. M. Morris, *Nature*, 2000, **408**, 975–979.
36. S. Perreault, O. Bousquet, M. Lauzon, J. Paiement and N. Leclerc, *J. Neuropathol. Exptl. Neurol.*, 2009, **68**, 503–514.
37. O. Bugiani, J. R. Murrell, G. Giaccone, M. Hasegawa, G. Ghigo, M. Tabaton, M. Morbin, A. Primavera, F. Carella, C. Solaro, M. Grisoli, M. Savoiardo, M. G. Spillantini, F. Tagliavini, M. Goedert and B. Ghetti, *J. Neuropathol. Exptl. Neurol.*, 1999, **58**, 667–677.
38. B. Allen, E. Ingram, M. Takao, M. J. Smith, R. Jakes, K. Virdee, H. Yoshida, M. Holzer, M. Craxton, P. C. Emson, C. Atzori, A. Migheli, R. A. Crowther, B. Ghetti, M. G. Spillantini and M. Goedert, *J. Neurosci.*, 2002, **22**, 9340–9351.
39. J. Götz, F. Chen, R. Barmettler and R. M. Nitsch, *J. Biol. Chem.*, 2001, **276**, 529–534.
40. J. Lewis, E. McGowan, J. Rockwood, H. Melrose, P. Nacharaju, M. Van Slegtenhorst, K. Gwinn-Hardy, M. P. Murphy, M. Baker, X. Yu, K. Duff, J. Hardy, A. Corral, W.-L. Lin, S.-H. Yen, D. Dickson, P. Davies and M. Hutton, *Nature Genet.*, 2000, **25**, 402–405.
41. N. S. Honson, J. R. Jensen, A. Abraha, G. F. Hall and J. Kuret, *Neurotox. Res.*, 2009, **15**, 274–283.
42. G. F. Hall, S. Lee and J. Yao, *J. Mol. Neurosci.*, 2002, **19**, 253–260.
43. M. Pickhardt, Z. Gazova, M. von Bergen, I. Khlistunova, Y. Wang, A. Hascher, E.-M. Mandelkow, J. Biernat and E. Mandelkow, *J. Biol. Chem.*, 2005, **280**, 3628–3635.
44. A. Crowe, C. Ballatore, E. Hyde, J. Q. Trojanowski and V. M. Y. Lee, *Biochem. Biophys. Res. Commun.*, 2007, **358**, 1–6.
45. M. Pickhardt, G. Larbig, I. Khlistunova, A. Coksezen, B. Meyer, E. M. Mandelkow, B. Schmidt and E. Mandelkow, *Biochemistry*, 2007, **46**, 10016–10023.

46. S. Taniguchi, N. Suzuki, M. Masuda, S.-i. Hisanaga, T. Iwatsubo, M. Goedert and M. Hasegawa, *J. Biol. Chem.*, 2005, **280**, 7614–7623.
47. C. Chirita, M. Necula and J. Kuret, *Biochemistry*, 2004, **43**, 2879–2887.
48. A. Crowe, W. Huang, C. Ballatore, R. L. Johnson, A.-M. L. Hogan, R. Huang, J. Wichtermann, J. McCoy, D. M. Huryn, D. S. Auld, A. B. Smith, J. Inglese, J. Q. Trojanowski, C. P. Austin, K. R. Brunden and V. M. Y. Lee, *Biochemistry*, 2009, **48**, 7732–7745.
49. J. E. Kristiansen, *Dan. Med. Bull.*, 1989, **36**, 178–185.
50. A. Mansouri and A. A. Lurie, *Am. J. Hematol.*, 1993, **42**, 7–12.
51. A. Kupfer, C. Aeschlimann, B. Wermuth and T. Cerny, *Lancet*, 1994, **343**, 763–764.
52. G. J. Naylor, B. Martin, S. E. Hopwood and Y. Watson, *Biol. Psychiatry*, 1986, **21**, 915–920.
53. C. M. Wischik, P. Bentham, D. J. Wischik and K. M. Seng, *Alzheimer's and Dementia*, 2008, **4**, T167.
54. B. Vellas, S. Andrieu, C. Sampaio, N. Coley and G. Wilcock, *Lancet Neurol.*, 2008, **7**, 436–450.
55. R. T. Staff, T. S. Ahearn, A. D. Murray, P. Bentham, K. M. Seng and C. Wischik, *Alzheimer's and Dementia*, 2008, **4**, T775.

GLUCOSE METABOLISM AS TARGET

Insulin Resistance and Neurodegeneration: Type-2 Versus Type-3 Diabetes Mellitus

SUZANNE M. DE LA MONTE AND JACK R. WANDS

Divisions of Neuropathology and Gastroenterology, Departments of Medicine, Pathology, and Neurology, and the Liver Research Center, Rhode Island Hospital and the Warren Alpert Medical School of Brown University, Providence, RI, USA

12.1 Overview

Alzheimer's disease (AD) is a neurodegenerative disease associated with cell loss, abundant neurofibrillary tangles, dystrophic neurites, amyloid precursor protein-amyloid-β (APP-Aβ) deposits, activation of prodeath cascades, impaired energy balance, oxidative stress, and DNA damage. Understanding AD pathogenesis requires that we develop a mechanistic framework, which interlinks all of these phenomena. Over the past several years, growth in knowledge has been robust and diversified, leading to mounting evidence the insulin deficiency and insulin resistance mediate neurodegeneration, particularly in AD. One of the major questions being interrogated is what causes the brain-insulin resistance and insulin deficiency. Is it the amyloid, the

RSC Drug Discovery Series No. 2
Emerging Drugs and Targets for Alzheimer's Disease
Volume 1: Beta-Amyloid, Tau Protein and Glucose Metabolism
Edited by Ana Martinez

Published by the Royal Society of Chemistry, www.rsc.org

hyperphosphorylated *tau*, insulin-degrading enzyme, gene polymorphisms or mutations, or yet unknown factors? At the same time, the arguments have grown in complexity with respect to the potential contributions of Type-II diabetes mellitus (T2DM), metabolic syndrome, and obesity in the pathogenesis of AD. Herein, we review evidence that: 1) extensive disturbances in brain insulin and insulin-like growth factor (IGF) signalling mechanisms represent early and progressive abnormalities that could account for the majority of molecular, biochemical, and histopathological lesions in AD; 2) experimental brain diabetes produced by intracerebral administration of streptozotocin shares many features in common with AD, including cognitive impairment and disturbances in acetylcholine homeostasis; 3) T2DM causes brain-insulin resistance, oxidative stress, and cognitive impairment, but its aggregate effects fall far short of mimicking AD; and 4) brain-insulin resistance may be treatable with insulin-sensitiser agents, *i.e.* drugs currently used to treat T2DM. We conclude that the term, "Type-III diabetes" accurately reflects the fact that AD represents a form of diabetes that selectively involves the brain, and has molecular and biochemical features that overlap with both Type-1 and Type-II diabetes mellitus.

12.1.1   Alzheimer's Disease

AD can only be diagnosed with certainty by postmortem demonstration of abundant neurofibrillary tangles and neuritic plaques, accompanied by deposition of amyloid precursor protein-amyloid-beta (APP-Aβ) in plaques and vessel walls in specific brain regions. Dementia-associated structural lesions are caused by neuronal cytoskeletal collapse and accumulation of hyperphosphorylated and poly-ubiquitinated microtubule-associated proteins, such as *tau*, resulting in the formation of neurofibrillary tangles, dystrophic neuritis, and neuropil threads.[1-3] Progressive loss of fibers and cells, and disconnection of synaptic circuitry mediate the cerebral atrophy that worsens over time. The biochemical, molecular, and cellular abnormalities that precede or accompany AD neurodegeneration, including increased activation of prodeath genes and signalling pathways, impaired energy metabolism, mitochondrial dysfunction, chronic oxidative stress, and DNA damage are stereotypical,[4-11] yet they lack a clear etiology. For at least three decades of intense AD research, the inability to interlink these abnormalities under a single primary pathogenic mechanism resulted in the emergence of several heavily debated theories, each focused on how one particular component of AD triggers the development of all other abnormalities. However, re-evaluation of the older literature revealed that deficits in cerebral glucose utilisation and energy metabolism represent very early abnormalities that precede or accompany the initial stages of cognitive dysfunction.[12-14] These observations led to the concept that impaired insulin signalling has an important role in the pathogenesis of AD.[5]

12.1.2 Insulin and Insulin-Like Growth-Factor Signalling

Insulin and insulin-like growth factor type 1 (IGF-I) mediate their effects by activating complex intracellular signalling pathways starting with ligand binding to cell surface receptors, and followed by autophosphorylation and activation of their receptor tyrosine kinases.[15–17] Insulin/IGF-I receptor tyrosine kinases then phosphorylate insulin receptor substrate (IRS) molecules,[15,18–20] which transmit signals downstream by activating the Erk MAPK and PI3 kinase/Akt pathways, and inhibiting glycogen synthase kinase-3β (GSK-3β). IRS signalling increases cell growth, survival, energy metabolism, and cholinergic gene expression, and it inhibits oxidative stress and apoptosis.[20–27] These very same signalling pathways are utilised by many cell types throughout the body, and therefore are practically universal. Moreover, these pathways regulate development, growth, and survival, and are impaired with senescence, carcinogenesis, and neurodegeneration.

12.2 Human Studies Support the Concept that AD is Type-3 Diabetes

Analysis of postmortem human brains with advanced AD revealed striking reductions in the levels of insulin and IGF-I polypeptide and receptor genes in the brain.[5] All of the signalling pathways that mediate insulin and IGF-I stimulated neuronal survival, *tau* expression, energy metabolism, and mitochondrial function were found to be perturbed in AD. In addition to establishing that, like all other pancreatic and intestinal polypeptide genes, insulin is expressed in the adult human brain, the results taught us that endogenous brain deficiencies in insulin, IGF-I and IGF-II and their corresponding receptors, in the absence of T2DM or obesity, were linked to the most common form of dementia-associated neurodegeneration in the Western hemisphere. Since the abnormalities identified in the brain mimicked the effects of both T1DM and T2DM, yet none of the patients had these diseases, we proposed that AD be regarded as a brain-specific form of diabetes mellitus, and coined the term, "Type-III diabetes".

To determine whether brain insulin/IGF resistance and deficiency were causal in the pathogenesis of AD, we extended our investigations by analysing brains with different severities, *i.e.* Braak stages[28,29] of AD.[10] To accomplish this, we measured genes encoding insulin, IGF-I, and IGF-II polypeptides and their corresponding receptors, *tau*, and APP, and we used competitive binding assays to characterise the degree to which growth-factor-transmitted signalling was impaired in brains with AD. In addition, we examined indices of energy metabolism and cholinergic function. We demonstrated progressive AD Braak-stage-dependent reductions in insulin, IGF-I, and IGF-II receptor expression, with more pronounced deterioration in insulin and IGF-I compared with IGF-II receptors. The lowest levels of gene expression were in brains with AD Braak stage 6 (late or end-stage). Therefore, loss of insulin and IGF-I receptor-bearing neurons begins early, and progresses with the disease such that in the advanced stages, the deficits are severe and global. These results provided further evidence that the signalling abnormalities in AD are not restricted to insulin pathways, since they also involve IGF-I and IGF-II stimulated mechanisms. Analysis of growth factor polypeptide genes revealed AD Braak-stage-dependent impairments in insulin, IGF-I, and IGF-II expression, corresponding with progressive trophic factor withdrawal. Again, the results support the hypothesis that abnormalities in insulin and IGF signalling mechanisms begin early in the course of AD, and therefore are likely have an important role in its pathogenesis.

The eventual paucity of local growth-factor gene expression could substantially impair growth-factor signalling and produce a state of growth-factor withdrawal, which is a well-established mechanism of neuronal death. Therefore, to complement the molecular data, we performed competitive equilibrium and saturation binding assays to determine if reduced levels of growth-factor receptor expression were associated with and perhaps mediated by impaired ligand–receptor binding as occurs with insulin/IGF resistance. Those investigations demonstrated progressive declines in top-level binding (BMAX) to the insulin, IGF-I, and IGF-II receptors, but either no change or increased binding affinity, suggesting that impaired insulin/IGF actions in AD brains were mediated by decreased polypeptide and receptor gene expression due to cell loss.

In vitro and *in vivo* experiments demonstrated that neuronal and oligodendrocyte survival and function are closely tied to the integrity of insulin and IGF signalling mechanisms in brain.[10,30–35] Impairments in insulin/IGF signalling lead to deficits in energy metabolism with attendant increases in oxidative stress, mitochondrial dysfunction, proinflammatory cytokine activation, and APP expression.[4,10,32,34] Correspondingly, in AD, the reductions in neuronal and oligodendrocyte-specific biomarkers, and increases in astrocyte and microglial inflammatory/injury-associated genes were also attributed to progressive brain insulin/IGF deficiency and resistance. The fact that microglial, astrocytic, and APP gene expression are all increased in the early stages of neurodegeneration supports the inflammatory hypothesis of AD.[6] Previous studies demonstrated that microglial activation promotes APP-Aβ accumulation,[36–38] and that APP gene expression and cleavage increase with oxidative stress.[39] Therefore, the mechanism we propose is that impaired insulin/IGF signalling leads to increased oxidative stress and mitochondrial dysfunction,[40–42] which induce APP gene expression and APP cleavage.[39] Attendant accumulations of APP-Aβ cause local neurotoxicity,[43–45] exacerbating oxidative-stress-induced APP expression and APP-Aβ deposition.

A critical goal in these investigations was to draw connections between brain insulin/IGF deficiency and resistance and the major dementia-associated structural and biochemical abnormalities in AD. In this regard, postmortem studies demonstrated that the Braak stage associated declines in *tau* mRNA paralleled progressive reductions in insulin and IGF-I receptor expression in AD. In addition, the studies demonstrated AD Braak-stage-associated deficits in choline acetyltransferase (ChAT) expression and reduced colocalisation of

ChAT and insulin or IGF-I receptor immunoreactivity in cortical neurons. These results correspond with experimental data demonstrating that neuronal *tau* and ChAT expression are regulated by IGF-I and insulin.[30] Therefore, brain insulin and IGF deficiency and resistance could account for the cytoskeletal collapse, neurite retraction, synaptic disconnection, loss of neuronal plasticity, and deficiencies in acetylcholine production, all of which correlate with cognitive decline and dementia in AD. Altogether, the studies utilising postmortem human brain tissue provided solid evidence that AD is associated with fundamental abnormalities in insulin/IGF signalling mechanisms that are highly correlated with development and progression of structural, molecular, and biochemical lesions that correlate with dementia. Although the abnormalities noted in AD share features in common with Types 1 and 2 diabetes mellitus, they are, nonetheless, distinguished by the dual presence of trophic-factor deficiencies and trophic-factor-receptor resistance, ergo the term, "Type-III diabetes".

12.2.1 Experimental Animal Model Results

The human postmortem brain studies linked many of the characteristic molecular and pathological features of AD to reduced expression of the insulin and IGF genes and their corresponding receptors. However, without direct experimentation leading to cause–effect data, conclusions drawn from human studies would be correlative rather than mechanistic. Therefore, we utilised experimental models to determine if diabetes-mellitus-type molecular and biochemical abnormalities could be produced in central nervous system (CNS) neurons and brain by exposing them to Streptozotocin (STZ). STZ is 2-Deoxy-2{[methyl-(nitrosoamino)carbonyl]amino}D-glucopyranose ($C_8H_{15}N_3O_7$), *i.e.* a nitrosamide methylnitrosourea linked to the C2 position of D-glucose. Once metabolised, the N-nitrosoureido is liberated and causes DNA damage through generation of reactive oxygen species such as superoxide, hydrogen peroxide, and nitric oxide.[46,47] STZ causes diabetes because the compound is taken up by insulin-producing cells, such as beta cells in pancreatic islets.

We treated rats with a single intracerebral injection of STZ (ic-STZ), and allowed them to grow older for 4 to 8 weeks. The rats were subjected to Morris water maze tests of spatial learning and memory, and their brains were examined for histopathological, biochemical, and molecular indices of AD-type neurodegeneration. Although a similar model had been generated much earlier by other investigators,[48–51] and it was noted that the ic-STZ treatments reduced cerebral glucose utilisation[51] and oxidative metabolism,[48] inhibited insulin receptor function,[42] and caused progressive deficits in learning, memory, cognitive behaviour, and cerebral energy balance,[41,50] efforts were not made to connect these effects to AD by characterising the neuropathology, molecular pathology, abnormalities in genes expression pertinent to insulin and IGF-1 signalling in brain, or evaluate the integrity of the pancreas. Our goal in generating the model was to demonstrate that AD-type neurodegeneration with features of Type-III diabetes could be produced in the absence of either Type 1 or Type 2 DM.

The ic-STZ-injected rats did not have elevated blood glucose or insulin levels, and pancreatic architecture and insulin immunoreactivity were similar to the control, yet their brains were atrophied and they had striking evidence of neurodegeneration with cell loss, gliosis, and increased immunoreactivity for p53, activated glycogen synthase kinase 3β, phospho*tau*, ubiquitin, and APP-Aβ.[32,33] Moreover, qRT-PCR studies demonstrated that the ic-STZ-treated brains had significantly reduced expression of genes corresponding to neurons (Hu), oligodendroglia (MAG-1), and choline acetyltransferase (ChAT), and increased expression of genes encoding glial fibrillary acidic protein (GFAP), microglia-specific proteins (AIF-1), acetylcholinesterase (AChE), *tau*, and APP[32,33]. Increased p53, and decreased Hu and MAG-1 expression in ic-STZ-treated brains suggest that neuronal and oligodendroglial cell loss and cerebral atrophy were mediated by apoptosis. These findings correspond well with previous studies demonstrating increased expression of various pro-apoptosis molecules, including p53,[52,53] colocalisation of increased p53 immunoreactivity in neurons and white-matter glia, and reduced levels of Hu and MAG-1 mRNA in human brains with AD. Loss of oligodendroglia could contribute to the white-matter degeneration[54] and synaptic disconnection[55–58] that occur early in the course of AD.

The above-mentioned adverse effects of ic-STZ were associated reduced expression of genes encoding insulin, IGF-II, insulin receptor, IGF-I receptor, and IRS-1, and reduced ligand binding to the insulin and IGF-II receptors. Note that most of these same effects were detected in human brains with sporadic AD,[5] and worsened with AD progression.[10] The reduced levels of IRS-1 mRNA observed in both AD and ic-STZ treated rat brains were reminiscent of the murine IRS-1 and insulin receptor knockout models, which exhibit reduced brain and body weights due to impaired insulin stimulated growth and survival signalling.[59–61] The combined effects of decreased insulin/IGF polypeptide and receptor gene expression, insulin/IGF receptor binding, and IRS expression denote failure of brain insulin/IGF signalling mechanisms following ic-STZ treatment. Importantly, many of the molecular abnormalities that characterise AD and are produced by ic-STZ, including increased GSK-3β activation, increased *tau* phosphorylation, and decreased neuronal survival, could be mediated by downstream effects of impaired insulin and IGF signalling in the CNS. Therefore, the ic-STZ experimental animal model recapitulates many of the pathobiological features of AD-type neurodegeneration/Type-III diabetes.

Corresponding with the findings in AD,[5] the ic-STZ-treated brains had increased levels of activated GSK-3β, phospho*tau*, ubiquitin, APP and APP-Aβ, and decreased levels of *tau* protein. These results are consistent with previous studies, demonstrating that *tau* is regulated by insulin/IGF-I stimulation,[30,62] and that *tau* phosphorylation and ubiquitination increase with oxidative stress and activation of GSK-3β.[39] Similarly, APP mRNA increases with oxidative stress, and is a feature of sporadic AD.[5,10] Increased APP gene expression could account for APP-Aβ accumulation in AD and ic-STZ-treated brains. Potential sources of oxidative stress in AD and the ic-STZ model include: 1) mitochondrial dysfunction;[6,42,63] 2) microglial cell activation with increased cytokine release; and 3) impaired insulin/IGF signalling through PI3 kinase-Akt leading to increased levels of GSK-3β activity.

Finally, it was necessary to determine whether ic-STZ impaired acetylcholine homeostasis and cognitive function, as occur in AD. Molecular and bio-chemical studies revealed reduced levels of ChAT and increased levels of AChE mRNA and protein in ic-STZ-treated relative to control brains. Since ChAT is responsive to insulin and IGF-I stimulation, and energy metabolism is needed to produce acetyl-CoA for acetylcholine biosynthesis, deficits in insulin/IGF signalling and energy metabolism decrease ChAT expression and thereby impair cholinergic functions, which are key features of AD. Increased expression of AChE in ic-STZ brains promotes degradation of acetylcholine, thereby exacerbating the cholinergic deficits caused by reduced ChAT expression. The significance of these results is highlighted by the significant impairments in learning and memory detected in ic-STZ-treated rats.[32,33]

12.3   The Concept of Brain-Insulin Resistance

Interest in clarifying the roles of T2DM, peripheral insulin resistance, and hyperinsulinemia in relation to cognitive impairment, AD-associated neuronal cytoskeletal lesions, or APP-Aβ deposits in brain began between 5 and 10 years ago,[4,8,14,61,64–68] but within the last 2–3 years, this field has literally exploded with new information and a new concept, *i.e.* primary brain-insulin resistance and insulin deficiency mediate cognitive impairment and AD.[5,10,32,33,69–71] This idea was fuelled by evidence that *tau* expression and phosphorylation are regulated through insulin and IGF signalling cascades.[61,68] In addition, research performed in our laboratory demonstrated that many key aspects of the CNS degeneration that occurs in AD can be effectuated by impaired insulin signalling.[31,35,40,72]

12.3.1   Potential Roles of Obesity and Type-2 Diabetes Mellitus in AD Pathogenesis

There is on-going debate about the degree to which T2DM, and more recently, T1DM, contributes to AD pathogenesis. This concept has been fuelled by the rising prevalence rates of obesity, T2DM, and AD over the past several decades. Moreover, an interrelationship among these entities is suggested by the: 1) increased risk of developing mild cognitive impairment (MCI), dementia, or AD in individuals with T2DM[73,74] or obesity/dyslipidemic disorders;[75] 2) progressive brain-insulin resistance and insulin deficiency in AD;[5,10,70,71] 3) cognitive impairment in experimental animal models of T2DM and/or obesity;[76,77] 4) AD-type neurodegeneration and cognitive impairment in experimentally induced brain-insulin resistance and insulin deficiency;[33,63,78–80] 5) improved cognitive performance in experimental models and humans with

AD or MCI after treatment with insulin-sensitiser agents or intranasal insulin;[32,81–87] and 6) shared molecular, biochemical, and mechanistic abnormalities in T2DM and AD.[73,88–92] The urgency of this problem has been spot-lighted by the estimated 24 million people in the world with dementia, and the expectation that prevalence rates of AD are likely to double every 20 years in the future if current trends continue.[93] While aging is clearly the strongest risk factor for AD, emerging data suggest that T2DM and dyslidemic states can contribute substantially to the pathogenesis of AD, either directly or as cofactors.[93]

Epidemiologic studies provide strong evidence for a significant association between T2DM and MCI or dementia, and suggest that T2DM is a significant risk factor for developing AD.[73,94–98] However, those findings are not without controversy,[99] since other reports suggest that the risks of developing dementia or AD may be independent rather than linked to diabetes.[100] An alternative interpretation of the data is that, variability in rates and severities of insulin resistance, *i.e.* impaired ability to respond to insulin stimulation among different target organs, accounts for the limited overlap between T2DM and AD. Correspondingly, obesity (body mass index – BMI > 30) without T2DM produces a 3-fold increased risk for subsequently developing AD, whereas overweight but nonobese subjects (BMI 25–30) experience a two-fold increased risk for AD.[101] Although the risk is clearly higher among obese subjects, the difference is not striking, and therefore begs the question as to whether the adverse effects of obesity on cognitive function are specific, and truly contribute to the pathogenesis of AD.

Mechanistically, the increased risk of dementia in T2DM and obesity could be linked to chronic hyperglycemia, peripheral insulin resistance, oxidative stress, accumulation of advanced glycation end-products, increased production of proinflammatory cytokines, and/or cerebral microvascular disease.[98] The potential role of cerebral microvascular disease as a complicating, initiating, or accelerating component of AD has been recognised for years.[102] However, more recently, a magnetic resonance imaging (MRI) study demonstrated that older adults with T2DM have a moderately increased risk for developing lacunes and hippocampal atrophy, and that the severity of those lesions increases with duration and progression of T2DM.[103] Another study showed that T2DM and impaired fasting glucose occur significantly more frequently in AD than in non-AD controls.[104] However, since diffuse and neuritic plaques were similarly abundant in T2DM and control brains, and neurofibrillary tangles, one of the hallmarks and correlates of dementia in AD, were not increased in T2DM,[104] the results suggest that T2DM can enhance progression, but may not be sufficient to cause AD. Therefore, what remains unclear is the net contribution of T2DM or obesity to the pathogenesis of AD-type neurodegeneration. To address this question, we utilised an established experimental model of chronic high fat diet (HFD) feeding of C57BL/6 mice. We examined the degree to which obesity/T2DM was sufficient to produce histopathological, molecular, and/or biochemical brain abnormalities of AD-type neurodegeneration, *i.e.* Type-III diabetes.

HFD feeding for 16 weeks doubled mean body weight, caused T2DM, and marginally reduced mean brain weight.[105] Those effects were associated with significantly increased levels of *tau*, IGF-I receptor, insulin receptor substrate-1 (IRS-1), IRS-4, ubiquitin, glial fibrillary acidic protein, and 4-hydroxynonenal, and decreased expression of β-actin. Importantly, HFD feeding also caused brain-insulin resistance manifested by reduced top-level (BMAX) insulin-receptor binding, and modestly increased brain-insulin gene expression. However, the brains of HFD fed mice did not exhibit AD histopathology, or increased levels of APP-Aβ or phospho*tau*. Nor did they have significant impairments in IGF signalling, which typically occur in AD.[10] In essence, although the chronic obesity with T2DM model developed brain atrophy with insulin resistance, oxidative stress, and cytoskeleton degradation in the brain, the effects were modest compared with AD[5,10] and experimental models of Type-III diabetes,[32,33] and for the most part, the molecular, biochemical, and histopathological features that typify AD were not present. Therefore, T2DM and obesity may contribute to, *i.e.* serve as cofactors, but by themselves are probably not sufficient to cause AD.

12.3.2 Ceramides May Cause Brain-Insulin Resistance and Degeneration in Obesity and T2DM

Results of human and experimental animal model studies suggest that the relationship between obesity-T2DM-metabolic syndrome and cognitive impairment may be dictated by the coexistence and severity of hepatic insulin resistance caused by steatosis or steatohepatitis. In this regard, irrespective of etiology, steatohepatitis can be associated with cognitive impairment and neuropsychiatric dysfunction.[106–112] In experimental models of hepatic steatosis or steatohepatitis induced by various agents including, chronic HFD feeding, with or without diet-induced obesity, nitrosamine-mediated injury, or chronic alcohol exposure, hepatic insulin resistance was consistently found to be associated with brain-insulin resistance.[32–33,105,113,114] These observations collectively suggest that something related to dysregulated lipid metabolism in liver might cause cognitive impairment and neurodegeneration. The rationale for this concept stemmed from the knowledge that insulin normally stimulates lipogenesis, which increases triglyceride storage in the liver,[115,116] but if the process proceeds unchecked, hepatocytes begin to suffer ER stress, oxidative damage, mitochondrial dysfunction, and inflammatory reactions. Attendant alterations in membrane lipid composition shift the liver cells toward insulin resistance.[115,117] Insulin resistance promotes lipolysis,[118] and lipolysis generates toxic lipids, *i.e.* ceramides, which further impair insulin signalling, mitochondrial function, and cell viability.[117,119,120]

Ceramides are lipid signalling molecules that modulate positive or negative cellular responses such as, proliferation, motility, plasticity, inflammation, apoptosis, and insulin resistance.[121] In brain, ceramides impair insulin signalling[122–124] and activate proinflammatory cytokines.[121,125,126] These adverse effects of ceramides are mediated in part by inhibition of insulin-stimulated PI3 kinase-Akt.[127–130] In diet-induced obesity, the mechanisms of enhanced ceramide production in adipocytes and attendant insulin resistance have been well documented.[121,131–135] However, peripheral insulin resistance confined to liver, such as in nonalcoholic steatohepatitis (NASH), need not be associated with obesity. Since our experimental models with hepatic steatosis or steato-hepatitis were consistently associated with insulin resistance and oxidative stress in both liver and brain, even in the absence of obesity,[32,33,105,113,114,136,137] we considered the possibility that toxic lipids generated in liver mediate neu-rodegeneration and cognitive impairment.

The diet-induced obesity studies utilising the C57BL/6 HFD feeding model was associated with steatohepatitis, which emerged after 16 weeks of HFD feeding, and worsened after 20 weeks. Molecular and biochemical assays of liver demonstrated that progressive steatohepatitis was correlated with corre-sponding increases in the expression of proceramide genes in liver, and cer-amide in serum.[114] In contrast, proceramide gene expression was not increased in brains of HFD-fed mice, indicating that if ceramides had mediated the neurodegeneration and brain-insulin resistance, their sources would have been extra-CNS in origin, such as liver. We extended the investigations by measuring proceramide gene expression and ceramide levels in other disease states associated with steatohepatitis or hepatic steatosis, including chronic HFD feeding without obesity, low-level nitrosamine exposure, and chronic alcohol feeding.[32,33,105,113,114] In each of these conditions, we detected significantly increased levels of proceramide gene expression in liver and increased ceramide levels in liver and/or serum.

12.3.3   The Liver–Brain Axis of Neurodegeneration

Since ceramides, sphingosines, and other toxic lipids are lipid soluble and therefore can cross the blood/brain barrier, we hypothesised that ceramides generated in the context of hepatic steatosis or steatohepatitis mediate neuro-degeneration by precipitating a cascade leading to brain-insulin resistance. Correspondingly, we found that limited *in vitro* exposure to bioactive ceramides causes neuronal insulin resistance, reduced viability, neurotransmitter expres-sion, and mitochondrial function, and increased oxidative stress, DNA damage, and lipid peroxidation.[138] Further preliminary studies showed that intraperitoneal injection of bioactive ceramides impairs learning and causes brain-insulin resistance with mild neurodegeneration, similar to the effects of diet-induced obesity with T2DM and NASH. These observations support the hypothesis that ceramides generated in extra-CNS tissues, particularly liver, can cause brain-insulin resistance and thereby mediate neurodegeneration.

12.4   Type-3 Diabetes *Versus* Type-2 Diabetes and Neurodegeneration

The results from various studies, together provide strong evidence in support of the hypothesis that neurodegeneration is mediated by insulin resistance, and that there are dual mechanisms of brain-insulin resistance: one caused by brain-specific disease, and the other due to disease processes originating outside of the CNS (Figure 12.1). In AD, insulin resistance is confined to the CNS and associated with IGF resistance and insulin/IGF deficiency. Its cause is still under investigation, although genetic factors play a key role in some cases. Given its likeness and overlap with the molecular, biochemical, and physiological features of both T1DM and T2DM, we suggest that AD be regarded as Type-III diabetes.

On the other hand, cognitive impairment and neurodegeneration arising in the context of peripheral insulin resistance are far less severe than in AD. Our overarching hypothesis is that high fat intake and other exposures leading to

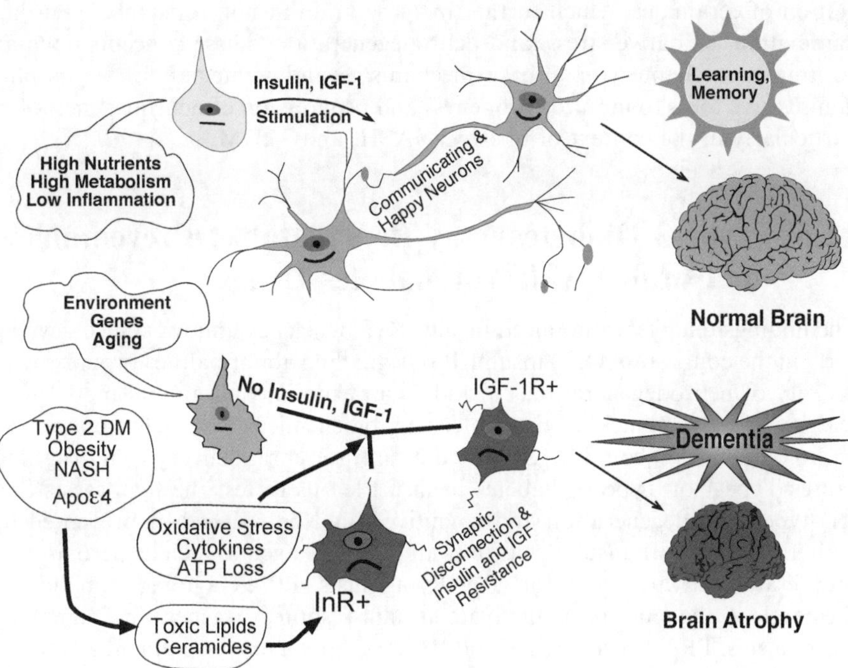

Figure 12.1 Brain-insulin-resistance-mediated neurodegeneration: Intact insulin and IGF signaling are essential for CNS neuronal survival and plasticity. Direct CNS insults stemming from environmental exposures, aging, and underlying genetic factors that threaten the integrity of insulin/IGF gene expression, result in trophic factor withdrawal and death or impaired function of insulin and IGF receptor-bearing cells, which have cholinergic function and mediate neuronal plasticity. CNS oxidative stress, inflammation, and metabolic insults exacerbate brain insulin resistance and neurodegeneration. Peripheral insulin resistance caused by T2DM, obesity, or NASH, and the presence of the Apoε4 gentotype, promote hepatic synthesis of toxic lipids, including ceramides, which cross the blood/brain barrier and cause insulin/IGF resistance. Therefore, peripheral insulin resistance diseases can precipitate, propagate, or exacerbate AD via a liver–brain axis of neurodegeneration.

hepatic insulin resistance, contribute to cognitive impairment via a liver–brain axis of neurodegeneration, as diagrammed in Figure 12.1. In brief, obesity, T2DM, and metabolic syndrome promote excess intracellular storage of lipids, eventually resulting in ER stress, inflammation, and local insulin resistance. Development of steatohepatitis with hepatic insulin resistance is permissive to the generation of cytotoxic lipids that can enter the circulation, penetrate the blood/brain barrier, and cause CNS insulin resistance, oxidative stress, and proinflammatory cytokine activation. Irrespective of cause, the long-term consequences of brain-insulin/IGF resistance include neuronal loss and reduced neurotransmitter functions required for plasticity, learning, and memory. In addition, brain-insulin resistance impairs oligodendrocyte survival and function, resulting in reduced myelin integrity, and increased local generation of ceramides, which further increase brain-insulin resistance, neuroinflammation, oxidative stress, and neurodegeneration. These concepts open an exciting new chapter on disease mechanisms and strategies for developing noninvasive tools to monitor proneness and progression of neurodegeneration, particularly in the context of obesity, NASH, and T2DM.

12.5   Type-3 Diabetes May Be Treatable, Preventable, or Curable with Antidiabetes Drugs

The findings that: 1) pronounced insulin/IGF deficiency and resistance develop early in the course of AD; 2) insulin/IGF signalling abnormalities progress with severity of neurodegeneration;[5,10] and 3) an experimental animal model with features closely mimicking the molecular, biochemical, and neuroanatomical pathologies of AD could be generated by intracerebral delivery of a drug that causes Type-1 or Type-II diabetes mellitus, led us to test the hypothesis that AD-type neurodegeneration and cognitive could be reduced or prevented by early treatment with insulin-sensitiser antidiabetes agents, such as peroxisome-proliferator-activated receptor (PPAR) agonists. PPAR agonists function at the level of the nucleus to activate insulin-responsive genes and signalling mechanisms. PPAR-α, PPAR-δ, and PPAR-ã are all expressed in adult human brains, including AD, but PPAR-δ is the most abundant of the 3 isoforms.[6] The experimental design involved treating rats with ic-STZ, followed by a single intraperitoneal injection of saline, a PPAR-α (GW7647; 25 µg/Kg), PPAR-δ (L-160,043; 2 µg/Kg), or PPAR-γ (F-L-Leu; 20 µg/kg) activator (CalBiochem, Carlsbad, CA).[32] The doses applied were considerably lower than those routinely used to treat T2DM. The major effects of the PPAR agonist treatments were to prevent brain atrophy, preserve insulin and IGF-II receptor bearing CNS neurons, and particularly with regard to the PPAR-δ agonist, prevent ic-STZ induced deficits in learning and memory.[32] Since the ic-STZ-mediated losses of insulin and IGF-expressing cells were not prevented by the PPAR agonist treatments, the PPAR agonists probably functioned by preserving insulin and IGF responsive (receptor-bearing) cells, including neurons and oligodendrocytes. In support of this concept was the finding insulin and IGF receptor expression and binding were increased by the PPAR agonist treatments.

PPAR-agonist-mediated preservation of insulin/IGF responsive neurons was associated with increased expression of ChAT, which has an important role in cognition, as cholinergic neuron deficits are a fundamental feature of AD.[139–142] This effect of the PPAR agonist treatments is consistent with the fact that ChAT expression is regulated by insulin/IGF,[30,62] and insulin/IGF resistance mediates cognitive impairment in AD. The PPAR agonist-associated increases in MAG-1 expression, corresponding to oligodendroglia, were of particular interest because previous research demonstrated that one of the earliest AD lesions was white-matter atrophy and degeneration with loss of oligodendroglial cells.[54] Within the context of the present discussion, white-matter atrophy in AD can now be interpreted as a manifestation of CNS insulin/IGF resistance since oligodendroglia require intact insulin/IGF signalling mechanisms for survival and function, including myelin synthesis.[143,144] Besides preserving insulin and IGF receptor-bearing CNS cells and signalling mechanisms germane to survival, energy metabolism, and neurotransmitter functions, the PPAR agonists rescued the ic-STZ model by lowering critical AD-associated indices of oxidative stress including, microglial and astrocyte activation, p53, nitric oxide synthase and NADPH oxidase gene expression, lipid peroxidation, DNA damage, APP expression, and *tau* phosphorylation.[6,32,33,37,38,145,146] In support of the hypothesis that AD/T3DM is caused by brain-insulin resistance, subjects with early AD exhibited cognitive improvement and/or stabilisation of cognitive impairment following treatment with intranasal insulin or a PPAR agonist.[83,85,147–150]

12.6 Conclusions

The postmortem human and experimental animal model data summarised above provide strong evidence favouring the hypothesis that AD is intrinsically a neuroendocrine disease caused by selective impairments in insulin and IGF signalling mechanisms, including deficiencies in local insulin and IGF production and insulin/IGF resistance. Referring to AD as "Type-III diabetes" is justified because the fundamental molecular and biochemical abnormalities overlap with Type-1 and Type-II diabetes, rather than mimic the effects of either one. Correspondingly, most of the biochemical and molecular abnormalities, neurocognitive deficits, and neuropathology associated with experimental Type-3 DM (produced by ic-STZ), can be prevented by early treatment with PPAR agonists, particularly the PPAR-δ subtype. On the other hand, solid epidemiological and experimental evidence has emerged supporting the concept that cognitive impairment can be caused by peripheral insulin resistance diseases, including obesity, T2DM, metabolic syndrome, and NASH. Although the severity of neurodegeneration is generally modest relative to AD/ T3DM, it is nonetheless associated with brain-insulin resistance. We concluded that the CNS structural and functional abnormalities produced by T2DM/

metabolic syndrome/obesity were not sufficient to cause AD, but they could serve as cofactors in its pathogenesis and progression. We linked brain-insulin-resistance-mediated neurodegeneration arising in the context of peripheral insulin resistance diseases to hepatic steatosis or steatohepatitis with increased production of cytotoxic lipids, including ceramides, which cause insulin resistance and can readily cross the blood/brain barrier. Detection of elevated levels of serum ceramide or proceramide gene expression in liver could help identify individuals with T2DM/metabolic syndrome/obesity who are at risk for developing cognitive impairment. The most exciting aspect of this thesis is that whether the brain-insulin resistance is caused by intrinsic CNS neurodegeneration or peripheral insulin resistance diseases, the treatments may be similar or possibly even identical.

Acknowledgements

Supported by AA02666, AA-11431, AA12908, and AA-16126 from the National Institutes of Health.

References

1. X. P. Wang and H. L. Ding, *Neurosci. Bull.*, 2008, **24**, 105.
2. K. A. Jellinger, *Neurodegener. Dis.*, 2008, **5**, 118.
3. J. J. Jalbert, L. A. Daiello, K. L. Lapane, *Epidemiol. Rev.*, 2008.
4. S. M. de la Monte and J. R. Wands, *J. Alzheimers. Dis.*, 2005, **7**, 45.
5. E. Steen, B. M. Terry, E. J. Rivera, J. L. Cannon, T. R. Neely, R. Tavares, X. J. Xu, J. R. Wands and S. M. de la Monte, *J. Alzheimers. Dis.*, 2005, **7**, 63.
6. S. M. de la Monte and J. R. Wands, *J. Alzheimers. Dis.*, 2006, **9**, 167.
7. P. I. Moreira, M. S. Santos, R. Seica and C. R. Oliveira, *J. Neurol. Sci.*, 2007, **257**, 206.
8. S. Hoyer, *J. Neural. Transm.*, 2002, **109**, 341.
9. R. A. Nixon, *Ageing Res. Rev.*, 2003, **2**, 407.
10. E. J. Rivera, A. Goldin, N. Fulmer, R. Tavares, J. R. Wands and S. M. de la Monte, *J. Alzheimers. Dis.*, 2005, **8**, 247.
11. P. Revill, M. A. Moral and J. R. Prous, *Drugs Today (Barc.)*, 2006, **42**, 785.
12. P. Iwangoff, R. Armbruster, A. Enz and W. Meier-Ruge, *Mech. Ageing Dev.*, 1980, **14**, 203.
13. N. R. Sims, D. M. Bowen, C. C. Smith, R. H. Flack, A. N. Davison, J. S. Snowden and D. Neary, *Lancet*, 1980, **1**, 333.
14. S. Hoyer, *Adv. Exp. Med. Biol.*, 2004, **541**, 135.
15. M. G. Myers, X. J. Sun and M. F. White, *Trends Biochem. Sci.*, 1994, **19**, 289.
16. T. O'Hare and P. F. Pilch, *Int. J. Biochem.*, 1990, **22**, 315.

17. A. Ullrich, J. R. Bell, E. Y. Chen, R. Herrera, L. M. Petruzzelli, T. J. Dull, A. Gray, L. Coussens, Y. C. Liao and M. Tsubokawa, *Nature*, 1985, **313**, 756.
18. X. J. Sun, P. Rothenberg, C. R. Kahn, J. M. Backer, E. Araki, P. A. Wilden, D. A. Cahill, B. J. Goldstein and M. F. White, *Nature*, 1991, **352**, 73.
19. M. F. White, R. Maron and C. R. Kahn, *Nature*, 1985, **318**, 183.
20. X. J. Sun, D. L. Crimmins, M. J. Myers, M. Miralpeix and M. F. White, *Mol. Cell. Biol.*, 1993, **13**, 7418.
21. B. M. Burgering and P. J. Coffer, *Nature*, 1995, **376**, 599.
22. M. Delcommenne, C. Tan, V. Gray, L. Rue, J. Woodgett and S. Dedhar, *Proc. Natl. Acad. Sci. USA.*, 1998, **95**, 11211.
23. H. Dudek, S. R. Datta, T. F. Franke, M. J. Birnbaum, R. Yao, G. M. Cooper, R. A. Segal, D. R. Kaplan and M. E. Greenberg, *Science*, 1997, **275**, 661.
24. G. Kulik, A. Klippel and M. J. Weber, *Mol. Cell. Biol.*, 1997, **17**, 1595.
25. K. Lam, C. L. Carpenter, N. B. Ruderman, J. C. Friel and K. L. Kelly, *J. Biol. Chem.*, 1994, **269**, 20648.
26. J. F. Mill, M. V. Chao and D. N. Ishii, *Proc. Natl. Acad. Sci. USA.*, 1985, **82**, 7126.
27. D. G. Puro and E. Agardh, *Science*, 1984, **225**, 1170.
28. H. Braak and E. Braak, *Neurobiol. Aging*, 1997, **18**, S85.
29. Z. Nagy, D. M. Yilmazer-Hanke, H. Braak, E. Braak, C. Schultz and J. Hanke, *Dement. Geriatr. Cogn. Disord.*, 1998, **9**, 140.
30. S. M. de la Monte, G. J. Chen, E. Rivera and J. R. Wands, *Cell. Mol. Life Sci.*, 2003, **60**, 2679.
31. S. M. de la Monte, T. R. Neely, J. Cannon and J. R. Wands, *Cell. Mol. Life Sci.*, 2001, **58**, 1950.
32. S. M. de la Monte, M. Tong, N. Lester-Coll, M. Plater Jr and J. R. Wands, *J. Alzheimers. Dis.*, 2006, **10**, 89.
33. N. Lester-Coll, E. J. Rivera, S. J. Soscia, K. Doiron, J. R. Wands and S. M. de la Monte, *J. Alzheimers Dis.*, 2006, **9**, 13.
34. S. J. Soscia, M. Tong, X. J. Xu, A. C. Cohen, J. Chu, J. R. Wands and S. M. de la Monte, *Cell. Mol. Life Sci.*, 2006, **63**, 2039.
35. J. Xu, J. Eun Yeon, H. Chang, G. Tison, G. Jun Chen, J. R. Wands and S. M. De La Monte, *J. Biol. Chem.*, 2003, **278**, 26929.
36. I. Blasko, M. Stampfer-Kountchev, P. Robatscher, R. Veerhuis, P. Eikelenboom and B. Grubeck-Loebenstein, *Aging Cell.*, 2004, **3**, 169.
37. P. Eikelenboom and W. A. van Gool, *J. Neural Transm.*, 2004, **111**, 281.
38. E. E. Tuppo and H. R. Arias, *Int. J. Biochem. Cell. Biol.*, 2005, **37**, 289.
39. G. J. Chen, J. Xu, S. A. Lahousse, N. L. Caggiano and S. M. de la Monte, *J. Alzheimers Dis.*, 2003, **5**, 209.
40. S. M. de la Monte and J. R. Wands, *CMLS, Cell. Mol. Life. Sci.*, 2002, **59**, 882.
41. S. Hoyer and H. Lannert, *Ann. N Y Acad. Sci.*, 1999, **893**, 301.

42. S. Hoyer, S. K. Lee, T. Loffler and R. Schliebs, *Ann. N Y Acad. Sci.*, 2000, **920**, 256.

43. A. Lorenzo and B. A. Yankner, *Ann. N Y Acad. Sci.*, 1996, **777**, 89.

44. T. Niikura, Y. Hashimoto, H. Tajima and I. Nishimoto, *J. Neurosci. Res.*, 2002, **70**, 380.

45. E. Tsukamoto, Y. Hashimoto, K. Kanekura, T. Niikura, S. Aiso and I. Nishimoto, *J. Neurosci. Res.*, 2003, **73**, 627.

46. A. D. Bolzan and M. S. Bianchi, *Mutat. Res.*, 2002, **512**, 121.

47. T. Szkudelski, *Physiol. Res.*, 2001, **50**, 537.

48. R. Duelli, H. Schrock, W. Kuschinsky and S. Hoyer, *Int. J. Dev. Neurosci.*, 1994, **12**, 737.

49. S. Hoyer, D. Muller and K. Plaschke, *J. Neural Transm. Suppl.*, 1994, **44**, 259.

50. H. Lannert and S. Hoyer, *Behav. Neurosci.*, 1998, **112**, 1199.

51. K. Plaschke and S. Hoyer, *Int. J. Dev. Neurosci.*, 1993, **11**, 477.

52. S. M. de la Monte, Y. K. Sohn and J. R. Wands, *J. Neurol. Sci.*, 1997, **152**, 73.

53. J. H. Su, G. Deng and C. W. Cotman, *J. Neuropathol. Exp. Neurol.*, 1997, **56**, 86.

54. S. M. de la Monte, *Ann. Neurol.*, 1989, **25**, 450.

55. A. Brun, X. Liu and C. Erikson, *Neurodegeneration*, 1995, **4**, 171.

56. H. Lassmann, P. Fischer and K. Jellinger, *Ann. N Y Acad. Sci.*, 1993, **695**, 59.

57. H. Lassmann, R. Weiler, P. Fischer, C. Bancher, K. Jellinger, E. Floor, W. Danielczyk, F. Seitelberger and H. Winkler, *Neuroscience*, 1992, **46**, 1.

58. X. Liu, C. Erikson and A. Brun, *Dementia*, 1996, **7**, 128.

59. S. Doublier, C. Duyckaerts, D. Seurin and M. Binoux, *Growth Horm. IGF Res.*, 2000, **10**, 267.

60. T. Nishiyama, T. Shirotani, T. Murakami, F. Shimada, M. Todaka, S. Saito, H. Hayashi, Y. Noma, K. Shima and H. Makino, *et al., Gene.*, 1994, **141**, 187.

61. M. Schubert, D. P. Brazil, D. J. Burks, J. A. Kushner, J. Ye, C. L. Flint, J. Farhang-Fallah, P. Dikkes, X. M. Warot, C. Rio, G. Corfas and M. F. White, *J. Neurosci.*, 2003, **23**, 7084.

62. M. Hong and V. M. Lee, *J. Biol. Chem.*, 1997, **272**, 19547.

63. S. Hoyer, H. Lannert, M. Noldner and S. S. Chatterjee, *J. Neural. Transm.*, 1999, **106**, 1171.

64. S. Craft, S. Asthana, D. G. Cook, L. D. Baker, M. Cherrier, K. Purganan, C. Wait, A. Petrova, S. Latendresse, G. S. Watson, J. W. Newcomer, G. D. Schellenberg and A. J. Krohn, *Psychoneuroendocrinology*, 2003, **28**, 809.

65. S. Craft, S. Asthana, G. Schellenberg, L. Baker, M. Cherrier, A. A. Boyt, R. N. Martins, M. Raskind, E. Peskind and S. Plymate, *Ann. N Y Acad. Sci.*, 2000, **903**, 222.

66. W. Farris, S. Mansourian, M. A. Leissring, E. A. Eckman, L. Bertram, C. B. Eckman, R. E. Tanzi and D. J. Selkoe, *Am. J. Pathol.*, 2004, **164**, 1425.

67. S. Hoyer, *Eur. J. Pharmacol.*, 2004, **490**, 115.
68. M. Schubert, D. Gautam, D. Surjo, K. Ueki, S. Baudler, D. Schubert, T. Kondo, J. Alber, N. Galldiks, E. Kustermann, S. Arndt, A. H. Jacobs, W. Krone, C. R. Kahn and J. C. Bruning, *Proc. Natl. Acad. Sci. USA.*, 2004, **101**, 3100.
69. S. Craft, *Arch. Neurol.*, 2005, **62**, 1043.
70. S. Craft, *Alzheimer Dis. Assoc. Disord.*, 2006, **20**, 298.
71. S. Craft, *Curr. Alzheimer Res.*, 2007, **4**, 147.
72. S. M. de la Monte, N. Ganju, K. Banerjee, N. V. Brown, T. Luong and J. R. Wands, *Alcohol Clin. Exp. Res.*, 2000, **24**, 716.
73. F. Pasquier, A. Boulogne, D. Leys and P. Fontaine, *Diabetes Metab.*, 2006, **32**, 403.
74. A. Verdelho, S. Madureira, J. M. Ferro, A. M. Basile, H. Chabriat, T. Erkinjuntti, F. Fazekas, M. Hennerici, J. O'Brien, L. Pantoni, E. Salvadori, P. Scheltens, M. C. Visser, L. O. Wahlund, G. Waldemar, A. Wallin and D. Inzitari, *J. Neurol. Neurosurg. Psychiatry.*, 2007, **78**, 1325.
75. I. J. Martins, E. Hone, J. K. Foster, S. I. Sunram-Lea, A. Gnjec, S. J. Fuller, D. Nolan, S. E. Gandy and R. N. Martins, *Mol. Psychiatry.*, 2006, **11**, 721.
76. G. Winocur and C. E. Greenwood, *Neurobiol. Aging*, 2005, **26**(Suppl 1), 46.
77. G. Winocur, C. E. Greenwood, G. G. Piroli, C. A. Grillo, L. R. Reznikov, L. P. Reagan and B. S. McEwen, *Behav. Neurosci.*, 2005, **119**, 1389.
78. M. P. Biju and C. S. Paulose, *Biochem. Mol. Biol. Int.*, 1998, **44**, 1.
79. A. Nitta, R. Murai, N. Suzuki, H. Ito, H. Nomoto, G. Katoh, Y. Furukawa and S. Furukawa, *Neurotoxicol. Teratol.*, 2002, **24**, 695.
80. M. Weinstock and S. Shoham, *J. Neural. Transm.*, 2004, **111**, 347.
81. M. N. Haan, *Nature Clin. Pract. Neurol.*, 2006, **2**, 159.
82. G. Landreth, *Exp. Neurol.*, 2006, **199**, 245.
83. G. Landreth, *Curr. Alzheimer Res.*, 2007, **4**, 159.
84. W. A. Pedersen, P. J. McMillan, J. J. Kulstad, J. B. Leverenz, S. Craft and G. R. Haynatzki, *Exp. Neurol.*, 2006, **199**, 265.
85. M. A. Reger, G. S. Watson, W. H. Frey 2nd, L. D. Baker, B. Cholerton, M. L. Keeling, D. A. Belongia, M. A. Fishel, S. R. Plymate, G. D. Schellenberg, M. M. Cherrier and S. Craft, *Neurobiol. Aging*, 2006, **27**, 451.
86. M. A. Reger, G. S. Watson, P. S. Green, C. W. Wilkinson, L. D. Baker, B. Cholerton, M. A. Fishel, S. R. Plymate, J. C. Breitner, W. Degroodt, P. Mehta and S. Craft, *Neurology*, 2008, **70**, 440.
87. G. S. Watson, T. Bernhardt, M. A. Reger, B. A. Cholerton, L. D. Baker, E. R. Peskind, S. Asthana, S. R. Plymate, L. Frolich and S. Craft, *Neurobiol. Aging*, 2006, **27**, 38.
88. G. Marchesini and R. Marzocchi, *Clin. Liver Dis.*, 2007, **11**, 105.
89. M. R. Nicolls, *Curr. Alzheimer Res.*, 2004, **1**, 47.
90. D. Papandreou, I. Rousso and I. Mavromichalis, *Clin. Nutr.*, 2007, **26**, 409.

91. D. Pessayre, *J. Gastroenterol. Hepatol.*, 2007, **22**(Suppl 1), S20.
92. M. M. Yeh and E. M. Brunt, *Am. J. Clin. Pathol.*, 2007, **128**, 837.
93. C. Qiu, D. De Ronchi and L. Fratiglioni, *Curr. Opin. Psychiatry*, 2007, **20**, 380.
94. M. N. Haan and R. Wallace, *Annu. Rev. Public Health*, 2004, **25**, 1.
95. L. J. Launer, *Curr. Diab. Rep.*, 2005, **5**, 59.
96. J. A. Luchsinger and R. Mayeux, *Curr. Atheroscler. Rep.*, 2004, **6**, 261.
97. J. A. Luchsinger, C. Reitz, B. Patel, M. X. Tang, J. J. Manly and R. Mayeux, *Arch. Neurol.*, 2007, **64**, 570.
98. R. A. Whitmer, *Curr. Neurol. Neurosci. Rep.*, 2007, **7**, 373.
99. P. Cicconetti, N. Riolo, C. Priami, L. Tafaro and E. Ettore, *Recenti. Prog. Med.*, 2004, **95**, 535.
100. W. Xu, C. Qiu, B. Winblad and L. Fratiglioni, *Diabetes*, 2007, **56**, 211.
101. R. A. Whitmer, E. P. Gunderson, C. P. Quesenberry Jr, J. Zhou and K. Yaffe, *Curr. Alzheimer Res.*, 2007, **4**, 103.
102. D. Etiene, J. Kraft, N. Ganju, T. Gomez-Isla, B. Gemelli, B. T. Hyman, E. T. Hedley-Whyte, J. R. Wands and S. M. De La Monte, *J. Alzheimers Dis.*, 1998, **1**, 119.
103. E. S. Korf, L. R. White, P. Scheltens and L. J. Launer, *Diabetes Care*, 2006, **29**, 2268.
104. J. Janson, T. Laedtke, J. E. Parisi, P. O'Brien, R. C. Petersen and P. C. Butler, *Diabetes*, 2004, **53**, 474.
105. N. Moroz, M. Tong, L. Longato, H. Xu and S. M. de la Monte, *J. Alzheimers Dis.*, 2008, **15**, 29.
106. R. Ramasamy, S. J. Vannucci, S. S. Yan, K. Herold, S. F. Yan and A. M. Schmidt, *Glycobiology*, 2005, **15**, 16R.
107. M. D. Kopelman, A. D. Thomson, I. Guerrini and E. J. Marshall, *Alcohol Alcohol*, 2009, **44**, 148.
108. J. E. Elwing, P. J. Lustman, H. L. Wang and R. E. Clouse, *Psychosom. Med.*, 2006, **68**, 563.
109. J. M. Loftis, M. Huckans, S. Ruimy, D. J. Hinrichs and P. Hauser, *Neurosci. Lett.*, 2008, **430**, 264.
110. W. Perry, R. C. Hilsabeck and T. I. Hassanein, *Dig. Dis. Sci.*, 2008, **53**, 307.
111. K. Karaivazoglou, K. Assimakopoulos, K. Thomopoulos, G. Theocharis, L. Messinis, G. Sakellaropoulos and C. Labropoulou-Karatza, *Liver Int.*, 2007, **27**, 798.
112. J. J. Weiss and J. M. Gorman, *Curr. HIV/AIDS Rep.*, 2006, **3**, 176.
113. M. Tong, M. Lawton, A. Neusner, L. Longato, J. R. Wands, S. M. de la Monte, *J. Alzheimers Dis.*, 2009, (in press).
114. L. E. Lyn-Cook, M. Lawton, M. Tong, E. Silbermann, L. Longato, P. Jiao, P. Mark, J. R. Wands, H. Xu, S. M. de la Monte. *J. Alzheimers Dis.* 2009, **16**, 715.
115. J. Capeau, *Diabetes. Metab.*, 2008, **34**, 649.
116. B. L. Leonard, R. N. Watson, K. M. Loomes, A. R. Phillips and G. J. Cooper, *Acta. Diabetol.*, 2005, **42**, 162.

117. E. W. Kraegen and G. J. Cooney, *Curr. Opin. Lipidol.*, 2008, **19**, 235.
118. Y. Kao, J. H. Youson, J. A. Holmes, A. Al-Mahrouki and M. A. Sheridan, *Gen. Comp. Endocrinol.*, 1999, **114**, 405.
119. W. L. Holland and S. A. Summers, *Endocr. Rev.*, 2008, **29**, 381.
120. M. Langeveld and J. M. Aerts, *Prog. Lipid. Res.*, 2009.
121. S. A. Summers, *Prog. Lipid. Res.*, 2006, **45**, 42.
122. G. Arboleda, T. J. Huang, C. Waters, A. Verkhratsky, P. Fernyhough and R. M. Gibson, *Eur. J. Neurosci.*, 2007, **25**, 3030.
123. C. E. Chalfant, K. Kishikawa, M. C. Mumby, C. Kamibayashi, A. Bielawska and Y. A. Hannun, *J. Biol. Chem.*, 1999, **274**, 20313.
124. B. Liu, L. M. Obeid and Y. A. Hannun, *Semin. Cell. Dev. Biol.*, 1997, **8**, 311.
125. L. Bryan, T. Kordula, S. Spiegel and S. Milstien, *Biochim. Biophys. Acta*, 2008, **1781**, 459.
126. J. R. Van Brocklyn, *Mini. Rev. Med. Chem.*, 2007, **7**, 984.
127. N. A. Bourbon, L. Sandirasegarane and M. Kester, *J. Biol. Chem.*, 2002, **277**, 3286.
128. E. Hajduch, A. Balendran, I. H. Batty, G. J. Litherland, A. S. Blair, C. P. Downes and H. S. Hundal, *Diabetologia*, 2001, **44**, 173.
129. T. C. Nogueira, G. F. Anhe, C. R. Carvalho, R. Curi, S. Bordin and A. R. Carpinelli, *Pancreas*, 2008, **37**, 309.
130. D. J. Powell, E. Hajduch, G. Kular and H. S. Hundal, *Mol. Cell. Biol.*, 2003, **23**, 7794.
131. L. A. Consitt, J. A. Bell and J. A. Houmard, *IUBMB Life*, 2009, **61**, 47.
132. W. L. Holland, J. T. Brozinick, L. P. Wang, E. D. Hawkins, K. M. Sargent, Y. Liu, K. Narra, K. L. Hoehn, T. A. Knotts, A. Siesky, D. H. Nelson, S. K. Karathanasis, G. K. Fontenot, M. J. Birnbaum and S. A. Summers, *Cell. Metab.*, 2007, **5**, 167.
133. W. L. Holland, T. A. Knotts, J. A. Chavez, L. P. Wang, K. L. Hoehn and S. A. Summers, *Nutr. Rev.*, 2007, **65**, S39.
134. B. Vistisen, L. I. Hellgren, T. Vadset, C. Scheede-Bergdahl, J. W. Helge, F. Dela and B. Stallknecht, *Eur. J. Endocrinol.*, 2008, **158**, 61.
135. J. R. Zierath, *Cell. Metab.*, 2007, **5**, 161.
136. M. Tong, L. Longato, S. M. de la Monte, *BMC Cell. Biol.*, 2009, (under review).
137. M. Tong, A. Neusner, L. Longato, M. Lawton, J. R. Wands, S. M. de la Monte, *J. Alzheimers Dis.*, 2009, **17**(4), 827.
138. M. Tong and S. M. de la Monte, *J. Alzheimers Dis.*, 2009, **16**, 705.
139. D. S. Auld, T. J. Kornecook, S. Bastianetto and R. Quirion, *Prog. Neurobiol.*, 2002, **68**, 209.
140. S. B. Dunnett and H. C. Fibiger, *Prog. Brain Res.*, 1993, **98**, 413.
141. O. Felician and T. A. Sandson, *J. Neuropsychiatry Clin. Neurosci.*, 1999, **11**, 19.
142. D. L. Price, *Annu. Rev. Neurosci.*, 1986, **9**, 489.
143. D. Chesik, J. De Keyser and N. Wilczak, *J. Mol. Neurosci.*, 2008, **35**, 81.

144. M. Lopes-Cardozo, J. E. Sykes, R. H. Van der Pal and L. M. van Golde, *J. Dev. Physiol.*, 1989, **12**, 117.
145. W. R. Markesbery and J. M. Carney, *Brain Pathol.*, 1999, **9**, 133.
146. L. M. Sayre, D. A. Zelasko, P. L. Harris, G. Perry, R. G. Salomon and M. A. Smith, *J. Neurochem.*, 1997, **68**, 2092.
147. C. Benedict, M. Hallschmid, A. Hatke, B. Schultes, H. L. Fehm, J. Born and W. Kern, *Psychoneuroendocrinology*, 2004, **29**, 1326.
148. C. Benedict, M. Hallschmid, K. Schmitz, B. Schultes, F. Ratter, H. L. Fehm, J. Born and W. Kern, *Neuropsychopharmacology*, 2007, **32**, 239.
149. C. Benedict, M. Hallschmid, B. Schultes, J. Born and W. Kern, *Neuroendocrinology*, 2007, **86**, 136.
150. G. S. Watson, B. A. Cholerton, M. A. Reger, L. D. Baker, S. R. Plymate, S. Asthana, M. A. Fishel, J. J. Kulstad, P. S. Green, D. G. Cook, S. E. Kahn, M. L. Keeling and S. Craft, *Am. J. Geriatr. Psychiatry*, 2005, **13**, 950.

Insulin-Like Growth Factor I as a Disease-Modifying Therapy in Alzheimer's Dementia

ANA M. FERNANDEZ AND IGNACIO TORRES-ALEMAN

Cajal Institute, CSIC and CIBERNED, Avda Dr. Arce 37, Madrid, 28002, Spain

13.1 Introduction

After several decades of intense search, the ultimate causes of Alzheimer's pathology remain uncharacterised. While the amyloid cascade is currently the favoured hypothesis, whether it is the origin of the disease or merely a consequence of a prior derangement is still under debate.[1] The most favoured candidate for a disease-modifying therapy has been for a long time to reduce the amyloid load, but recent findings do not support this view. Thus, patients subjected to amyloid vaccination showed full erasure of the plaques without modification of the cognitive pathology and subsequent death.[2] The proposed explanation that the lack of effect was due to a late therapeutic intervention[3] seems more an attempt to hold the amyloid hypothesis on its formerly prominent position. Along this vein, it was recently proposed that Aβ oligomers rather than plaques are causing neuronal derangement and dementia,[4] or that N-APP, another peptide generated together with Aβ from the amyloid precursor peptide (APP), is the actual cause of neuronal demise.[5] An alternative

RSC Drug Discovery Series No. 2
Emerging Drugs and Targets for Alzheimer's Disease
Volume 1: Beta-Amyloid, Tau Protein and Glucose Metabolism
Edited by Ana Martinez

Published by the Royal Society of Chemistry, www.rsc.org

view, among many others, that is gaining momentum is that in late-onset AD, amyloidosis is probably a late event that is triggered by metabolic impairment. Metabolic disturbances will in turn originate inflammation, calcium dysregulation, mitochondrial derangement, oxidative stress, *etc.*, or a combination thereof that eventually leads to aberrant amyloid accumulation and a vicious cycle sets off. That is, amyloidosis forms part of the pathologic cascade but does not initiate it. Mutations leading to amyloidosis in familial AD would be directly pathogenic not requiring a triggering mechanism as in late-onset AD. But this would be the exception rather than the rule. Thus, familial and sporadic AD share a common pathological pathway[6] while their triggering causes are distinct; mutations directly causing amyloid overproduction in the former, metabolic derangement leading to amyloidosis in the latter. The fact that dementia is still present in the absence of amyloidosis or that aged individuals may present a relatively large amyloid burden without any signs of dementia has supported still an alternative explanation that tangles, rather than amyloid plaques are the biological culprits.[7] This would inevitably lead to the assumption that tauopathy is downstream of amyloidosis, as the initial pathogenic event in familial AD is increased amyloid burden, but this issue is far from being settled[8] (*i.e.* tauopathy precedes amyloid deposits in late-onset AD) and is beyond the scope of this review.

13.2   IGF-I and Development of AD Pathology

The notion that metabolic impairment is the cause of amyloidosis is appealing. Alzheimer's disease, as other diseases with rising prevalence may not merely be due to an increasingly older population. Rather, an inadequate life style that dominates modern societies could be a key additional trait. It is now widely accepted that the so-called metabolic syndrome is the direct cause of bad life-style habits and probably favours many of the modern diseases that plague our society, such as atherosclerosis, heart disease, diabetes or even cancer.[9] The metabolic syndrome defines a set of diffuse disturbances including systemic inflammation, dyslipidemia, abdominal obesity, hypertension, and insulin resistance, all apparently arising from visceral fat accumulation. The endocrine activity of visceral fat appears to ultimately trigger inflammation and oxidative stress, two processes possibly favouring development of diabetes and probably also AD.[10,11] A link between obesity, inevitably associated to the metabolic syndrome, or diabetes and AD has been shown in several epidemiological studies.[12] In view of the epidemic proportions of obesity and diabetes in developed countries, the constant rise in AD cases may turn even worse than the most pessimistic predictions should these pathological links prove correct.

But how may a general metabolic impairment lead to a specific brain disease such as AD? Is AD ultimately due to "bad fat", or what is almost the same; wrong dietary habits? First, although not generally recognised, we may consider that AD is a systemic disease, as previously suggested,[13] whose most prominent clinical characteristics are neurological. Many, if not all neurodegenerative conditions usually present with pathological perturbations in other organs. For instance, AD patients show peripheral amyloidosis and weight loss.[13,14] A striking connection between brain diseases in general and diabetes is also known,[15] a further hint that metabolic impairment associates with brain disease. As reviewed elsewhere (Nishijima *et al.*, in press), there are many other observations supporting the view that insulin function and AD are probably related. Is it then that what we so far have considered as peripheral resistance to insulin is also affecting brain cells? As glucose handling by brain cells was thought to be insulin independent this possibility was not generally accepted. However, the recognition of an ample set of effects of insulin on brain function other than glucose metabolism has changed this view.[16] Moreover, direct evidence of brain-insulin resistance in subjects with peripheral resistance has been documented.[17,18] While the incidence of AD in diabetic patients may be higher than the normal population,[19] although this is still disputed,[20] and insulin resistance increases the risk of AD,[21] there still remains the fact that many AD patients ($\sim$75%) are not diabetic. Thus, if still these AD patients have insulin resistance at brain level, why only this organ? Is it that AD is truly "brain diabetes", as recently suggested?[22]

Specific vulnerability of brain cells to metabolic impairment may be related to another member of the insulin family of peptides, insulin-like growth factor I (IGF-I). This growth factor/hormone displays two characteristics particularly relevant to brain disease; 1) IGF-I has a wide neuroprotective activity; and, 2) IGF-I is directly involved in insulin sensitivity[23] in ways that we still do not fully understand (see Figure 13.1). In previous work we found that defective serum IGF-I input at the blood/brain interfaces located between vessels and brain parenchyma, and vessels and cerebrospinal fluid (CSF) set in motion an AD-like neuropathology in rodents, including cognitive derangement, amyloidosis, tauopathy and inflammation.[24,25] Elaborating on these and other observations we postulated that AD is triggered by disturbed serum IGF-I input to the brain as an initial lesion.[26] Following disrupted IGF-I function, brain input of serum insulin is compromised through a compound reduced entrance of insulin (AD patients show reduced insulin levels in CSF at initial stages of the disease; A. Minguez-Cedazo, personal communication) together with reduced brain sensitivity to insulin (Figure 13.1).

The combined reduced trophic IGF-I input and resultant brain metabolic impairment starts the pathological cascade that ultimately leads to increased amyloid levels, aberrant tau phosphorylation, cognitive impairment and inflammation (see below). Major pathological disturbances inevitably associated to deranged metabolism, such as oxidative stress,[27] mitochondrial dysfunction or calcium dysregulation will be actually driving progress of the disease (Figure 13.2).

Because IGF-I is a universal cyto-protectant one could argue that any type of damaged cell, including affected neurons in Alzheimer's disease would benefit from it. This might be perfectly true. However, we consider that the beneficial effects of IGF-I in AD are due to its specific actions in the pathological disturbances involved in this disease, as already explained. Should this proposal

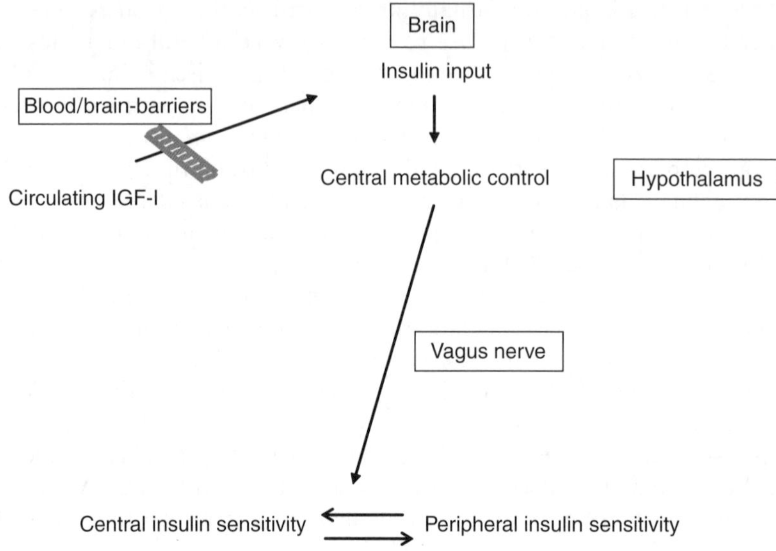

Figure 13.1 Circulating IGF-I controls peripheral insulin sensitivity. Proposed mechanism: serum IGF-I enters into the brain through the blood/brain barriers (BBB) and modulates insulin sensitivity in brain cells. In turn, insulin-sensitive neurons within hypothalamic circuits participate in the control of peripheral insulin sensitivity through the parasympathetic nervous system (vagus nerve) connected to the liver and pancreas. It is now recognized that peripheral insulin resistance associates to brain-insulin resistance.[135] Reduced serum IGF-I function derived either from serum IGF-I deficiency, loss of sensitivity to IGF-I at the BBB, or both, leads to decreased entrance of serum IGF-I into the brain. This results in compromised insulin signalling at the brain level. Subsequently, the metabolic circuitry will adapt to reduced insulin input and react consequently (increased food intake, increased adiposity) eventually leading to peripheral insulin insensitivity.[136] Accordingly, during healthy aging, serum IGF-I levels decline and insulin resistance ensues.[38] Should this proposal turn out to be correct, IGF-I targets within the hypothalamus may provide a key mechanism in the control of peripheral insulin sensitivity.

turn out to be correct, administration of IGF-I would constitute an actual disease-modifying therapy. If impaired serum IGF-I input to the brain initiates all the pathological changes found in AD,[26] those mechanisms leading to reduced serum IGF-I levels as well as reduced brain sensitivity to this growth factor could explain the origin of this devastating disease (Nishijima *et al.*, in press). New evidence allows us now to elaborate this proposal in greater detail.

It is important to remember that sporadic AD, the most frequent type of the disease, is associated to old age, a period where serum IGF-I levels are considerably reduced as compared to those found in young individuals.[28] The reason for this gradual decline is unknown but is probably originated in the

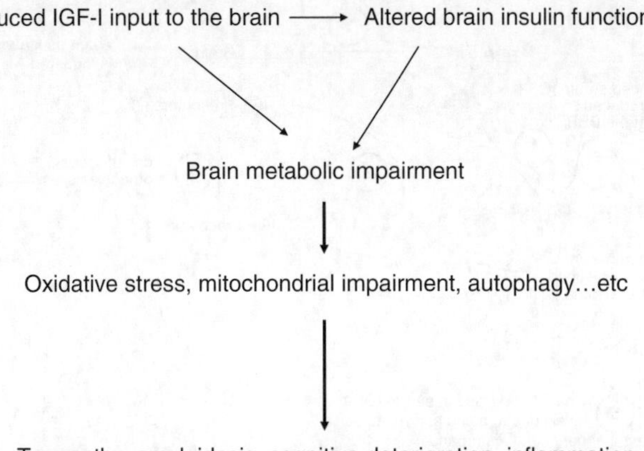

Figure 13.2 Hierarchy of pathological changes in AD. At the onset of the disease serum IGF-I input to the brain is reduced and as already explained, impaired insulin action at the brain ensues. The compound deficit of insulin and IGF-I signalling leads to brain metabolic impairment that originates oxidative stress, mitochondrial dysfunction, autophagy and dysregulated cell calcium handling. These pathological events lead to typical perturbations of the AD brain: cognitive loss, amyloidosis, tauopathy and inflammation.

liver, that produces around 70% of total IGF-I.[29] We reason that development of Alzheimer's disease along aging is a gradual process due to gradually declining serum IGF-I levels that progressively compromise brain IGF-I function (Figure 13.3). At the same time, in the first stages of the disease, the cells making up the blood/brain barriers (BBBs) would develop diminished sensitivity to IGF-I. Reduced sensitivity to IGF-I is a common trend in normal aging,[30] so in the case of AD, this reduced sensitivity should be more pronounced. Otherwise, AD will be an inevitable consequence of aging. We consider that interactions between specific environmental and genetic factors originate this abnormally high IGF-I resistance at the BBBs.[31]

Hence, either a normal age-associated serum IGF-I deficiency combined with an exaggerated resistance to IGF-I at the BBBs or an abnormally low serum deficiency together with a normal resistance to IGF-I at the BBBs will eventually compromise IGF-I input to the brain at the onset of AD. Of course, individual variations in these disturbances should be expected, being abnormally low serum IGF-I input the resulting final condition. In accordance with this proposal we observed that in early stages of the disease APP/PS2 mice had increased brain levels of IGF-I receptors that were corrected by IGF-I treatment.[32] Increased brain IGF-I receptors were also reported in AD, an observation interpreted by the authors as a resistant state to IGF-I in AD brain.[33] The opposite observation was reported by other authors who found reduced IGF-I mRNA levels in AD brains.[22] In both cases, reduced efficacy of IGF-I in the AD brain was argued to be the outcome of these apparently opposing

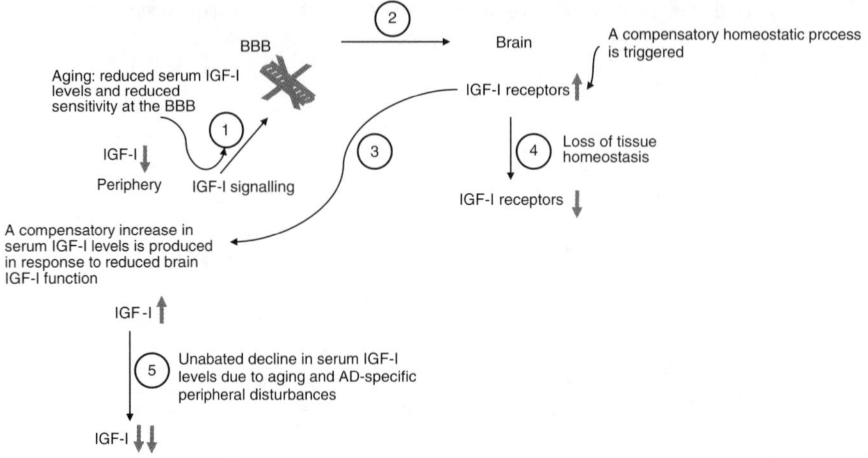

Figure 13.3 Chronology of changes in IGF-I function in Alzheimer's disease. 1: Aging reduces serum IGF-I levels producing in this way IGF-I deficiency. Aging is also associated to IGF-I resistance in at least several tissues. We propose development of IGF-I resistance at the BBBs as a primary pathogenic event in AD. This resistance may be due to impaired signal transduction without changes in the levels of IGF-I receptors at these barriers. 2: As a consequence of reduced IGF-I input to the BBBs less serum IGF-I enters into the brain and reduced Aβ clearance and degradation will also be elicited.[24] Lower amounts of IGF-I in the brain will produce a compensatory increase in brain IGF-I receptors in AD. We propose that this will be seen in the relatively early stages of the disease. 3: A compensatory increase in serum IGF-I levels takes place as a result. 4: At later stages the whole IGF-I system collapses as a consequence of even greater reduction in serum IGF-I, and brain IGF-I receptors are reduced. Tissue homeostasis is irreversibly altered.

observations. A possible way to reconcile this contradiction is that changes in brain IGF-I receptors in AD are stage dependent (Figure 13.3). Thus, as a result of reduced input of IGF-I to the brain, an initial homeostatic response would be to express more IGF-I receptors. Later, when brain function becomes severely disrupted, reduced brain IGF-I receptors may reflect a general loss of tissue homeostasis. For instance, protracted oxidative stress reduces IGF-I receptors.[34] And, as indicated, oxidative stress is one of the favourite candidate mechanisms involved in the pathological cascade in AD. So, in the initial stages we can expect high receptor levels, while in later stages the opposite may be true.

This would mean that in the first stages of the disease, IGF-I/insulin sensitisers should be beneficial. There are several on-going trials with insulin sensitisers that show promising results.[35,36] It is now increasingly recognised that early diagnosis of AD will significantly increase the efficacy of any potentially beneficial treatment. In fact, it has been argued that the failure of all clinical trials so far conducted may be because the patients were already too advanced in the pathology for any treatment to be effective.[37] In the case of insulin/IGF-I sensitisers this would be particularly true. Of note, insulin sensitisers are effective in Type-2 diabetics only in early stages of the disease. At this point we should emphasise that although no IGF-I sensitisers have been developed yet, by increasing insulin sensitivity, which, not coincidentally is also lowered in old age[38] and even more in AD patients,[39] IGF-I sensitivity would also increase. While still hypothetical, the assumption is based on the view that both hormones share the same intracellular pathways. For instance, a typical insulin sensitiser such as rosiglitazone, increases insulin sensitivity through various processes including decreased protein tyrosine phosphatase 1B (PTP1B)[40] or increased IRS-1 levels.[41] Both effects will increase sensitivity to IGF-I as PTP1B also inhibits IGF-I signalling, while increased levels of IRS-1 increase IGF-I signalling.

We hypothesised that reduced IGF-I efficacy in the brain in the initial stages of AD will lead to a compensatory increase in serum IGF-I levels[26] (Figure 13.3). This would be a process reminiscent of that seen in Type-2 diabetics. When insulin resistance in Type-II diabetes unfolds, a typical compensatory response is increased production of insulin by the pancreas leading to hyperinsulinemia. Recent observations in an animal model of brain IGF-I resistance, the brain-specific IGF-I receptor knockout mice confirm this postulate; these mice showed increased serum IGF-I levels.[42] Of course, although IGF-I directly influences pituitary GH secretion, one could argue that the absence of IGF-I feedback onto hypothalamic neurons governing GH secretion will produce a compensatory increase in the GH/IGF-I axis. If this is the case, high GH levels should be found in early stages of AD. But GH levels have been found largely unaltered in AD patients,[43] while GH responses to various secretagogues were highly variable.[44] More importantly, and in agreement with our proposal, AD patients show increased serum IGF-I levels in the early stages of the disease.[45] Furthermore, AD patients show decreased sensitivity to insulin as measured by Akt phosphorylation in lymphocytes.[46] Although not yet directly tested, we predict a similar attenuated response to IGF-I in AD subjects. Collectively, from these observations we could infer a somewhat unexpected conclusion. Levels of IGF-I in serum appear to be controlled by brain IGF-I sensitivity to this growth factor not only at the hypothalamic level but also at some as yet undetermined site(s). Although this proposal would require further studies, peripheral levels of insulin are also governed by a central mechanism.

In a subsequent stage of the disease, AD patients will develop IGF-I insufficiency probably due to exhaustion of the liver, its main producer. Indeed, during normal aging the liver produces lower amounts of IGF-I. The reason for this is not entirely clear. A combination of lower GH input, partially impaired liver GH receptor transduction, senescence of IGF-I-producing hepatocytes, altered liver IGF-I gene epigenetic regulation and other unknown processes (various hormonal age-associated derangements, age-associated diseases, *etc.*) may probably contribute.[47,48] In addition, a disruption in the brain mechanism controlling serum IGF-I levels would also contribute to decreased liver IGF-I output. It is known that liver IGF-I production is not fully dependent on GH input; *i.e.* at perinatal stages,[49] so liver IGF-I output could depend on other as yet unknown mechanisms. In AD, disrupted brain control of liver IGF-I output may be the cause of a specific greater reduction of serum IGF-I compared to already reduced levels in normal aged individuals. This is merely a theoretical proposal, but AD produces important changes in specific brain circuits,[50] and one of those affected may be related to the control of liver IGF-I output. In summary, a compound resistance/deficiency to IGF-I gradually unfolds, making the situation worse as less IGF-I trophic input will reach its targets in the brain. Importantly, as our proposal predicted, AD patients show reduced serum IGF-I levels at later stages in the disease; *i.e.* with high MME scores.[51] At this point, treatment of AD patients should include IGF-I or a mimetic.

But recent data suggests that brain IGF-I resistance, or at least hippocampal IGF-I resistance in brain-specific IGF-I receptor knockouts protects from AD amyloidosis and premature death in the Tg2576 mouse model of AD.[52] How can this be reconciled with our hypothesis? Disregarding essential drawbacks such as the fact that no IGF-I resistance could actually be detected in double IGF-IR$^{-/-}$/Tg2576 transgenic mice,[52] the key issue is the time and cell context where reduced IGF-I receptor levels are present. In the transgenic model used by Freude *et al.*,[52] IGF- I receptors are absent throughout life in a specific subset of forebrain neurons in combination with overproduction of a mutant human APP. In wild-type mice, IGF-I receptors are expressed by all types of brain cells, and no excess APP is produced. Although transgenic models are very useful in furthering our understanding of human disease, translating the results obtained in these animal models into a physiological context is not straightforward. For instance, major disturbances in hippocampal cytoarchitecture and development in brain IGF-IR$^{-/-}$ mice have been reported, together with early lethality[53] or behavioural impairments (unpublished observations). Conversely, no such effects were reported in similar mice by other authors.[42] Compensatory mechanisms in response to absence of IGF-I receptors, such as increased insulin receptor signalling, should also be considered.[54] Collectively, the data of Freude *et al.* and our own indicate that IGF-I is involved in AD, although the apparently opposite effects (beneficial *vs.* detrimental) require further clarification.

13.3   Pathological Traits in AD and IGF-I Function

Regardless of this controversial information, an additional, albeit parsimonious piece of evidence that IGF-I has a role in AD is the fact that all the pathological characteristics of Alzheimer's disease can be explained by disrupted brain IGF-I function. We will first briefly explore the three main pathological traits of AD and its relation to IGF-I.

13.3.1   IGF-I and Cognitive Deterioration

As discussed in detail elsewhere,[55] there is now enough evidence to directly relate IGF-I with cognitive functions, at least in experimental animals.

In humans, a correlation between levels of IGF-I and cognitive status during adulthood and into aging have been shown in various studies.[56] This may appear as an unusual trait for a growth factor, but several hormones and growth factors participate in higher brain functions.[57] A general characteristic of the modulatory effects of intercellular messengers on cognitive operations is their ability to affect underlying synaptic plasticity processes. IGF-I is no exception, as it is able to influence several of these, most prominently long-term depression.[58] IGF-I acts on cognitive processes at multiple levels, not only because of its influence on general cell function as a key homeostatic gauge, but also in cellular and molecular mechanisms essential to cognitive function such as glutamatergic transmission,[59–63] adult neurogenesis,[64–66] or neuronal excitability.[67,68] Based on findings on serum-IGF-I-deficient transgenic mice and those from several other models we may conclude that IGF-I is a trophic signal for glutamatergic synapses through at least three pathways: regulation of glutamate receptor trafficking,[62,69] neuronal Ca^{++} levels,[70–73] and Aβ metabolism.[74,75] Aβ has been shown to directly interfere with glutamatergic neurotransmission.[76] In turn, glutamate transmission is essential for long-term potentiation,[77] a cellular substrate for memory (Figure 13.4).

13.3.2 IGF-I and Amyloid Physiology

Despite recent advances,[78] the precise biological role of amyloid or its precursor APP is not well understood yet. APP has been associated to neuronal excitability,[79] modulation of the p75 neurotrophin receptor,[80] inhibition of Aβ

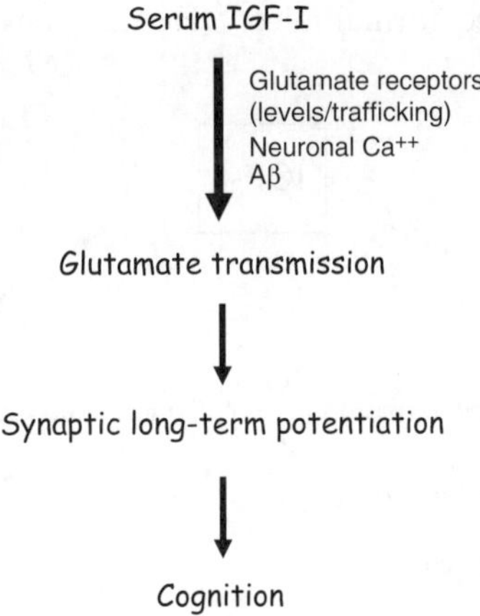

Figure 13.4 A possible functional relationship between serum IGF-I and cognition. Serum IGF-I controls glutamatergic tone in the brain through at least three complementary mechanisms: glutamate receptor numbers, Ca^{++} handling by neurons and Aβ levels. In turn, glutamate transmission is essential for mechanisms of synaptic plasticity such as long-term potentiation (LTP). LTP underlies cognitive operations.[137]

toxicity,[81] and many other processes, while Aβ has been shown to modulate neuronal excitability,[82] memory formation[83] or insulin/IGF-I signalling.[84] In turn, IGF-I appears to modulate brain amyloid through a variety of actions. A first one is by modulating its production, as IGF-I favours the so-called nonamyloidogenic pathway of APP processing whereby Aβ levels are kept low.[74] Alternatively, proamyloidogenic actions of IGF-I have also been reported by several authors, indicating increased Aβ production in response to IGF-I through the beta-secretase pathway.[85,86] Secondly, IGF-I lowers brain Aβ levels through enhanced clearance by increasing brain levels of Aβ carriers,[24] and together with insulin, stimulating cellular uptake of deleterious Aβ aggregates.[87] Insulin and IGF-I also protect neurons against the detrimental actions of Aβ.[88,89] Collectively, and not withstanding important discrepancies on the effect of IGF-I in Aβ production, the majority of data indicate that the role of IGF-I is to maintain low brain Aβ levels, and when Aβ accumulates, to protect neurons from its detrimental actions (Figure 13.5).

13.3.3   IGF-I and Tauopathy

Whether intracellular neurofibrillary tangles, rather than amyloid are the cause of dementia is hotly debated.[90] Intracellular deposits of aberrantly phosphorylated tau protein together with several other constituents abound in damaged brain regions of AD patients and develop prior to plaques. Because IGF-I inhibits the tau-kinase GSK-3, reduced levels of phosphorylated tau may be expected as a result of IGF-I actions on neurons, reducing in this way tauopathy.[91] This has been shown to be true in mouse models of the disease.[92] The opposite has also been documented, reduced brain IGF-I signalling leads to tauopathy in otherwise normal mice.[93] Hence, dysregulated GSK-3 activity,

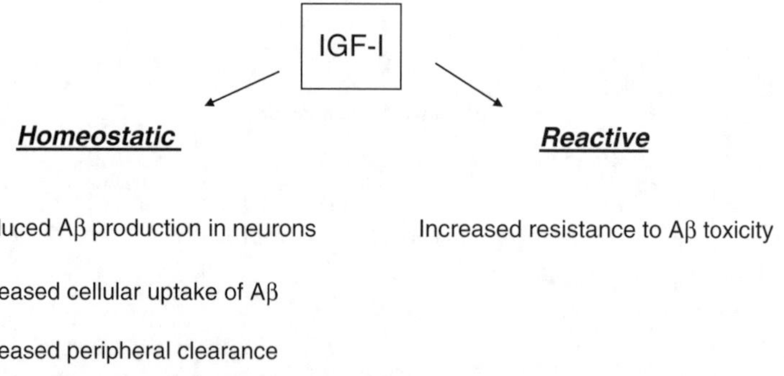

Figure 13.5   Mechanisms whereby IGF-I protects neurons from Aβ exposure. Under normal physiology, IGF-I uses at least three mechanisms to keep low levels of Aβ in the brain: 1) Reduces production of Aβ from its precursor APP; 2) Increases cellular uptake of Aβ; and 3) Stimulates clearance of Aβ through its carriers. Under conditions where these mechanisms are overcome (due, for example, to lower IGF-I input), IGF-I protects neurons from the neurotoxic action of excess Aβ.

a proposed mechanism for the disease, may be secondary to insufficient IGF-I tonic inhibition. Again, a major AD trait is related to IGF-I function.

13.4 IGF-I and the Pathological Cascade in AD

Our proposal is that prior to these 3 pathological derangements stands impaired IGF-I signalling.[26] Further elaborating on this idea we now emphasise two points not previously envisaged: 1) the important cooperative action of insulin dysfunction on this pathological cascade; and 2) the essential role of inflammation. Without a concomitant reduction in insulin signalling, AD will not develop. Without a protracted or maladaptive inflammatory response, AD will not progress.

13.4.1 IGF-I and Insulin Signalling in AD

A functional relationship between these two members of the insulin family is supported by ample evidence, but the underpinnings are not yet understood. We know that impaired serum IGF-I input to the brain will lead to reduced serum insulin levels.[94,95] In turn, low serum insulin input to the brain, together with ongoing reduced IGF-I input will further deteriorate peripheral insulin regulation.[95] Together with progressing central insulin resistance as a consequence of disturbed IGF-I signalling, insulin function as a whole will be deteriorated, as central insulin resistance leads to peripheral insulin resistance.[96] We propose that this is in part due to reduced insulin receptors in brain as a consequence of IGF-I-dependent regulation of these receptors. But the latter is hypothetical and experimental confirmation is pending. At any rate, brain-insulin receptors and insulin signalling are deteriorated in AD.[97,98] After this primary pathogenic event unfolds, the precise steps that lead to AD pathology cannot be placed in a hierarchically accurate order as yet, although major steps starting at the BBBs and eventually leading to neuronal death had already been drawn by us previously.[99] These pathological processes include oxidative stress, mitochondrial dysfunction and calcium dysregulation as necessary intermediate steps towards cell death (Figure 13.2).

13.4.2 IGF-I and Neuroinflammation

A major, not yet fully recognised driving force of AD pathology would be inflammation. Although the precise role of inflammation in AD needs further clarification, a neuroinflammatory lesion is probably very important in the process of neuronal loss. Any insult to the brain produces a homeostatic inflammatory response involving glial cells that ultimately target surrounding neurons.[100] Formerly based on indirect epidemiological evidence of protection against AD in subjects treated with anti-inflammatory drugs,[101] direct proof of inflammatory processes in AD brains is now accumulating.[10,102–104] Although trials with anti-inflammatory compounds in AD patients largely failed,[105,106] the rationale to target inflammatory pathways in this disease is still strong. For instance, blockade of inflammatory cytokines such as TNFα has been shown to delay pathology in the triple AD mouse model.[107] The use of PPARγ agonists with anti-inflammatory properties reduces microglial activation[108] and are promising new drugs for AD.

The mainstream idea of the role of inflammation in AD is that excess Aβ stimulates microglia in response to ongoing neuronal injury,[109] although this assumption has recently been questioned.[110] Reactive microglia positions close to amyloid lesions in AD brains.[111] Microglia activation is initially neuroprotective but as neuronal death continues unabated, reactive microgliosis eventually turns detrimental as in this later phase microglia changes into a neurotoxic phenotype.[112] Within this general scheme, the role of astrocytes is not clearly elucidated. Usually astrocytes are considered mostly passive bystanders of the inflammatory cascade triggered by active microglia, although they are recognised to be involved in subsequent steps that eventually feeds into chronic inflammation.[113] However, we have gathered evidence that astrocytes are primary players in brain inflammatory cascades partially independent of microglia responses.[114]

Initial *in vivo* studies suggested, although did not prove, a connection between IGF-I and neuroinflammation. Treatment of LPS-treated mice with IGF-I attenuated the ensuing brain-inflammatory response.[115] IGF-I also antagonised the blocking effects of inflammatory mediators on synaptic plasticity[116] or their deleterious role in axonopathy.[117] IGF-I is produced by stimulated microglia in a response that is interpreted as neuroprotective.[112] Indeed, active microglia produces *de novo* synthesis of IGF-I in all types of insults.[118] IGF-I also directly blocks proinflammatory signalling in astrocytes[119] through a complex ambivalent cascade involving the phosphatase calcineurin and the proinflammatory transcription factor NFκB. This pathway, altered also in AD brains (unpublished observations) probably acts as a master switch in responses towards and away from inflammation.[114] This astrocyte inflammatory/anti-inflammatory route when gone astray appears as a driving force in AD pathology (Fernandez *et al.*, in preparation). Genetic and pharmacological manipulation of this cascade in astrocytes abrogates progress of the disease at any stage, even at advanced phases (unpublished observations). Hopefully, different constituents of this pathway could become druggable targets for new disease-modifying therapies in AD.

13.4.3 Pathological Disturbances Arising from Brain Metabolic Impairment

Oxidative stress,[120] mitochondrial dysfunction,[121] autophagy[122] or altered cholesterol metabolism[123] can be linked to an initial lesion of cellular energy metabolism as a result of deteriorated insulin/IGF-I function. Indeed, energy imbalance will produce excess oxygen radicals that will compromise mitochondrial function and eventually lead to autophagic processes and impaired cholesterol handling. In turn, at least oxidative stress and mitochondrial derangement will aggravate metabolic impairment and decisively contribute to an inflammatory response. More hypothetically, and although cellular calcium dysregulation has been ascribed to excess Aβ,[124] as IGF-I and insulin affect neuronal calcium channels[70–73] and astrocytes are important cellular targets of IGF-I,[125] it is possible that early dysregulation of calcium function in astrocytes[126] can also be related to altered IGF-I/insulin input to brain tissue. Therefore, these disturbances previously associated to development of AD independently may all stem from impaired cell metabolism.

13.5 Mechanisms Leading to Reduced Serum IGF-I Input to the Brain

From the above-discussed data, it follows that processes involved in reducing IGF-I input to the brain are ultimately responsible of development of AD. The major factor leading to reduced IGF-I input to the brain is aging,[28] which is also the major risk factor for sporadic AD.[127] However, many other factors, discussed in detail elsewhere (Nishijima *et al.*, in press), including genetic and environmental events may powerfully impinge on IGF-I function. For instance, obesity has been linked to increased risk of AD[12], and we have shown that unbalanced diets leading to obesity produce reduced serum IGF-I input to the brain.[128] A sedentary life style, that alone does not affect IGF-I levels, is deleterious to the brain because it is associated with reduced serum IGF-I uptake by the brain.[129] More intriguingly, life factors supporting an active mind (social engagement, higher education) may also affect serum IGF-I input to the brain. In this regard, we have recently found that active brain areas specifically accumulate serum IGF-I (unpublished). In summary, low serum IGF-I input constitutes a unifying explanation of the relationship of these risk factors with AD.

13.6 Summary and Perspectives

We have discussed current evidence supporting the concept that late-onset Alzheimer's dementia may be caused by metabolic impairment in brain tissue as a whole ultimately compromising neuronal function, the more sensitive cell type in the brain. We further propose that metabolic impairment arises from an initial lesion at the periphery/brain interface that hinders appropriate signalling of serum IGF-I onto brain targets. This leads in the first instance to insulin dysregulation at brain level that together with defective IGF-I function originates a first wave of disturbances represented by oxidative stress and mitochondrial derangement that eventually lead to AD pathology through inflammation as a driving mechanism. If this scheme of events is correct, addressing IGF-I and insulin pathways will truly constitute a disease-modifying strategy in AD.

However, much work is still needed to prove this proposal as a whole. First, unequivocal evidence of the nature of insulin/IGF-I perturbations in AD brain is still lacking. Either increased, decreased or unaltered levels of these factors and their receptors have been variously reported, making it difficult to interpret the significance of these hormones in the disease. Secondly, their role in pathogenic events is not firmly established and at times controversial. Lastly, while several studies suggest a therapeutic effect of insulin/IGF-I in AD and experimental models,[92,130–132] others do not support this view.[133,134] While this lack of coherent information may be unavoidable in a complex disease such as AD, where firm diagnosis can not be established until death, it also points to a possible variety of pathological triggers in AD, some related to IGF-I/insulin, others unrelated. Alternatively, as knowledge of AD is sadly incomplete, we may still be describing a patchy picture of the disease that does not allow us to fill the gaps. Hence, a detailed longitudinal analysis of populations at risk (*i.e.* healthy elderly people in their early years) is very much needed. This analysis should be conducted with all available tools including genomic, proteomic and metabolomic approaches and medical image.

References

1. Q. Chen, A. Nakajima, S. H. Choi, X. Xiong and Y. P. Tang, *J. Neurosci. Res.*, 2008, **86**, 1615.
2. C. Holmes, D. Boche, D. Wilkinson, G. Yadegarfar, V. Hopkins, A. Bayer, R. W. Jones, R. Bullock, S. Love, J. W. Neal, E. Zotova, J. A. Nicoll, *The Lancet*, 2008, **372**, 216.
3. A. Abbott, *Nature*, 2008, **456**, 161.
4. J. P. Cleary, D. M. Walsh, J. J. Hofmeister, G. M. Shankar, M. A. Kuskowski, D. J. Selkoe and K. H. Ashe, *Nature Neurosci.*, 2005, **8**, 79.
5. A. Nikolaev, T. McLaughlin, D. D. M. 'Leary and M. Tessier-Lavigne, *Nature*, 2009, **457**, 981.
6. C. F. Lippa, A. M. Saunders, T. W. Smith, J. M. Swearer, D. A. Drachman, B. Ghetti, L. Nee, D. Pulaski-Salo, D. Dickson, Y. Robitaille, C. Bergeron, B. Crain, M. D. Benson, M. Farlow, B. T. Hyman, S. P. George-Hyslop, A. D. Roses and D. A. Pollen, *Neurology*, 1996, **46**, 406.
7. C. Ballatore, V. M. Y. Lee and J. Q. Trojanowski, *Nature Rev. Neurosci.*, 2007, **8**, 663.
8. G. Xu, V. Gonzales and D. R. Borchelt, *Alzheimer Dis. Assoc. Disord.*, 2002, **16**, 196.
9. P. L. Huang, *Dis. Model. Mech.*, 2009, **2**, 231.
10. H. Akiyama, S. Barger, S. Barnum, B. Bradt, J. Bauer, G. M. Cole, N. R. Cooper, P. Eikelenboom, M. Emmerling, B. L. Fiebich, C. E. Finch, S. Frautschy, W. S. Griffin, H. Hampel, M. Hull, G. Landreth, L. Lue, R. Mrak, I. R. Mackenzie, P. L. McGeer, M. K. O'Banion, J. Pachter, G. Pasinetti, C. Plata-Salaman, J. Rogers, R. Rydel, Y. Shen, W. Streit, R. Strohmeyer, I. Tooyoma, F. L. Van Muiswinkel, R. Veerhuis,

D. Walker, S. Webster, B. Wegrzyniak, G. Wenk and T. Wyss-Coray, *Neurobiol. Aging*, 2000, **21**, 383.
11. A. Nunomura, R. J. Castellani, X. Zhu, P. I. Moreira, G. Perry and M. A. Smith, *J. Neuropathol. Exp. Neurol.*, 2006, **65**, 631.
12. D. Gustafson, E. Rothenberg, K. Blennow, B. Steen and I. Skoog, *Arch. Intern. Med.*, 2003, **163**, 1524.
13. C. L. Joachim, H. Mori and D. J. Selkoe, *Nature*, 1989, **341**, 226.
14. H. White, C. Pieper and K. Schmader, *J. Am. Geriatr. Soc.*, 1998, **46**, 1223.
15. M. Ristow, *J. Mol. Med.*, 2004, **82**, 510.
16. W. Q. Zhao, H. Chen, M. J. Quon and D. L. Alkon, *Eur. J. Pharmacol.*, 2004, **490**, 71.
17. K. Anthony, L. J. Reed, J. T. Dunn, E. Bingham, D. Hopkins, P. K. Marsden and S. A. Amiel, *Diabetes*, 2006, **55**, 2986.
18. C. G. Jolivalt, C. A. Lee, K. K. Beiswenger, J. L. Smith, M. Orlov, M. A. Torrance and E. Masliah, *J. Neurosci. Res.*, 2008, **86**, 3265.
19. A. Ott, R. P. Stolk, A. Hofman, H. F. van, D. E. Grobbee and M. M. Breteler, *Diabetologia*, 1996, **39**, 1392.
20. I. Alafuzoff, L. Aho, S. Helisalmi, A. Mannermaa and H. Soininen, *Neuropathol. Appl. Neurobiol.*, 2009, **35**, 60.
21. E. Ronnemaa, B. Zethelius, J. Sundelof, J. Sundstrom, M. German-Gunnarsson, C. Berne, L. Lannfelt and L. Kilander, *Neurology.*, 2008, **71**, 1065.
22. E. Steen, B. M. Terry, E. J. Rivera, J. L. Cannon, T. R. Neely, R. Tavares, X. J. Xu, J. R. Wands and S. M. De La Monte, *J. Alzheimers Dis.*, 2005, **7**, 63.
23. D. R. Clemmons, *Horm. Res.*, 2004, **62**(Suppl 1), 77.
24. E. Carro, J. L. Trejo, T. Gomez-Isla, D. LeRoith and I. Torres-Aleman, *Nat. Med.*, 2002, **8**, 1390.
25. E. Carro, J. L. Trejo, C. Spuch, D. Bohl, J. M. Heard and I. Torres-Aleman, *Neurobiol. Aging*, 2006, **27**, 1618.
26. E. Carro and I. Torres-Aleman, *Eur. J. Pharmacol.*, 2004, **490**, 127.
27. R. Quiroz-Baez, E. Rojas, C. Arias, *Neurochem. Int.*, 2009, **55**, 662.
28. C. R. Breese, R. L. Ingram and W. E. Sonntag, *J. Gerontol.*, 1991, **46**, B180.
29. S. Yakar, J. L. Liu, B. Stannard, A. Butler, D. Accili, B. Sauer and D. LeRoith, *Proc. Natl. Acad. Sci. USA*, 1999, **96**, 7324.
30. J. J. Cao, P. Kurimoto, B. Boudignon, C. Rosen, F. Lima, B. P. Halloran, *J. Bone Miner. Res.*, 2007, **22**, 1271.
31. E. Carro and I. Torres-Aleman, *Keio. J. Med.*, 2006, **55**, 59.
32. C. Lopez-Lopez, M. O. Dietrich, F. Metzger, H. Loetscher and I. Torres-Aleman, *J. Neurosci.*, 2007, **27**, 824.
33. A. M. Moloney, R. J. Griffin, S. Timmons, R. O'Connor, R. Ravid, C. O'Neill, *Neurobiology of Aging*, 2010, **31**, 224.
34. T. L. Schmitt, H. Klein and W. Droge, *Redox. Rep.*, 2006, **11**, 105.
35. M. E. Risner, A. M. Saunders, J. F. Altman, G. C. Ormandy, S. Craft, I. M. Foley, M. E. Zvartau-Hind, D. A. Hosford and A. D. Roses, *Pharmacogenomics. J.*, 2006, **6**, 246.

36. G. S. Watson, B. A. Cholerton, M. A. Reger, L. D. Baker, S. R. Plymate, S. Asthana, M. A. Fishel, J. J. Kulstad, P. S. Green, D. G. Cook, S. E. Kahn, M. L. Keeling and S. Craft, *Am. J. Geriatr. Psychiatry*, 2005, **13**, 950.
37. M. A. Smith, C. S. Atwood, J. A. Joseph and G. Perry, *J. Neurosci. Res.*, 2002, **69**, 285.
38. A. M. Abbatecola, G. Paolisso, M. Lamponi, S. Bandinelli, F. Lauretani, L. Launer and L. Ferrucci, *J. Am. Geriatr. Soc.*, 2004, **52**, 1713.
39. M. N. Haan, *Nature Clin. Pract. Neurol.*, 2006, **2**, 159.
40. Y. Wu, J. P. Ouyang, K. Wu, S. S. Wang, C. Y. Wen and Z. Y. Xia, *Br. J. Pharmacol.*, 2005, **146**, 234.
41. Y. Miyazaki, H. He, L. J. Mandarino and R. A. DeFronzo, *Diabetes*, 2003, **52**, 1943.
42. L. Kappeler, C. M. De Magalhaes Filho, J. Dupont, P. Leneuve, P. Cervera, L. Perin, C. Loudes, A. Blaise, R. Klein, J. Epelbaum, B. Y. Le and M. Holzenberger, *PLoS. Biol.*, 2008, **6**, e254.
43. C. Proto, D. Romualdi, R. M. Cento, R. S. Spada, M. G. Di, R. Ferri and A. Lanzone, *Gynecol. Endocrinol.*, 2006, **22**, 213.
44. C. B. Nemeroff, K. R. Krishnan, B. M. Belkin, J. C. Ritchie, C. Clark, W. W. Vale, J. E. Rivier and M. O. Thorner, *Neuroendocrinology*, 1989, **50**, 663.
45. E. R. Vardy, P. J. Rice, P. C. Bowie, J. D. Holmes, P. J. Grant and N. M. Hooper, *J. Alzheimers Dis.*, 2007, **12**, 285.
46. P. Castri, L. Iacovelli, B. A. De, F. Giubilei, A. Moretti, F. T. Capone, F. Nicoletti and F. Orzi, *Eur. J. Neurosci.*, 2007, **26**, 2469.
47. X. Xu, S. A. Bennett, R. L. Ingram and W. E. Sonntag, *Endocrinology*, 1995, **136**, 4551.
48. B. Velasco, L. Cacicedo, E. Melian, G. Fernandez-Vazquez and F. Sanchez-Franco, *Eur. J. Endocrinol.*, 2001, **145**, 73.
49. P. Rotwein, D. P. Bichell and K. Kikuchi, *Mol. Reprod. Dev.*, 1993, **35**, 358.
50. M. A. Busche, G. Eichhoff, H. Adelsberger, D. Abramowski, K. H. Wiederhold, C. Haass, M. Staufenbiel, A. Konnerth and O. Garaschuk, *Science*, 2008, **321**, 1686.
51. T. Watanabe, A. Miyazaki, T. Katagiri, H. Yamamoto, T. Idei and T. Iguchi, *J. Am. Geriatr. Soc.*, 2005, **53**, 1748.
52. S. Freude, M. M. Hettich, C. Schumann, O. Stohr, L. Koch, C. Kohler, M. Udelhoven, U. Leeser, M. Muller, N. Kubota, T. Kadowaki, W. Krone, H. Schroder, J. C. Bruning, M. Schubert, *FASEB J.*, 2009, **23**, 3315.
53. W. Liu, P. Ye, J. R. O'Kusky, A. J. D'Ercole, *J. Neurosci. Res.*, 2009, **87**, 2821.
54. J. P. Liu, J. Baker, A. S. Perkins, E. J. Robertson and A. Efstratiadis, *Cell*, 1993, **75**, 59.
55. A. Aleman and I. Torres-Aleman, *Prog. Neurobiol.*, 2009, **89**, 256.
56. L. I. Arwert, J. B. Deijen and M. L. Drent, *Growth Horm. IGF. Res.*, 2005, **15**, 416.
57. E. C. McNay, *Curr. Opin. Pharmacol.*, 2007, **7**, 628.

58. Y. T. Wang and D. J. Linden, *Neuron.*, 2000, **25**, 635.
59. M. A. Castro-Alamancos and I. Torres-Aleman, *Proc. Natl. Acad. Sci. USA*, 1993, **90**, 7386.
60. d. l. Gonzalez, V. W. Buno, S. Pons, M. S. Garcia-Calderat, E. Garcia-Galloway and I. Torres-Aleman, *Neuroreport*, 2001, **12**, 1293.
61. W. E. Sonntag, S. A. Bennett, A. S. Khan, P. L. Thornton, X. Xu, R. L. Ingram and J. K. Brunso-Bechtold, *Brain. Res. Bull.*, 2000, **51**, 331.
62. Y. T. Wang and D. J. Linden, *Neuron*, 2000, **25**, 635.
63. C. Zona, M. T. Ciotti and P. Calissano, *Neurosci. Lett.*, 1995, **186**, 75.
64. M. A. Aberg, N. D. Aberg, H. Hedbacker, J. Oscarsson and P. S. Eriksson, *J. Neurosci.*, 2000, **20**, 2896.
65. M. Llorens-Martin, I. Torres-Aleman and J. L. Trejo, *Neuroscientist.*, 2009, **15**, 134.
66. J. L. Trejo, E. Carro and I. Torres-Aleman, *J. Neurosci.*, 2001, **21**, 1628.
67. E. Carro, A. Nunez, S. Busiguina and I. Torres-Aleman, *J. Neurosci.*, 2000, **20**, 2926.
68. A. Nunez, E. Carro and I. Torres-Aleman, *J. Neurophysiol.*, 2003, **89**, 3008.
69. R. C. Carroll, E. C. Beattie, M. von Zastrow and R. C. Malenka, *Nat. Rev. Neurosci.*, 2001, **2**, 315.
70. L. A. Blair and J. Marshall, *Neuron*, 1997, **19**, 421.
71. L. Gao, L. A. C. Blair, G. D. Salinas, L. A. Needleman and J. Marshall, *J. Neurosci.*, 2006, **26**, 6259.
72. M. Kanzaki, Y. Q. Zhang, H. Mashima, L. Li, H. Shibata and I. Kojima, *Nat. Cell Biol.*, 1999, **1**, 165.
73. T. Kleppisch, F. J. Klinz and J. Hescheler, *Brain Res.*, 1992, **591**, 283.
74. L. Adlerz, S. Holback, G. Multhaup and K. Iverfeldt, *J. Biol. Chem.*, 2007, **282**, 10203.
75. D. Shineman, A. Daina, M. Kim, V. Lee, *Biochemistry*, 2009, **48**, 3787.
76. H. Hsieh, J. Boehm, C. Sato, T. Iwatsubo, T. Tomita, S. Sisodia and R. Malinow, *Neuron*, 2006, **52**, 831.
77. S. J. Kim and D. J. Linden, *Neuron*, 2007, **56**, 582.
78. Y. Chen and C. Dong, *Cell Death Differ.*, 2009, **16**, 386.
79. S. F. Santos, N. Pierrot, N. Morel, P. Gailly, C. Sindic and J. N. Octave, *J. Neurosci.*, 2009, **29**, 4708.
80. C. Matrone, R. Marolda, S. Ciafre, M. T. Ciotti, D. Mercanti and P. Calissano, *Proc. Natl. Acad. Sci USA.*, 2009, **106**, 11358.
81. M. Gralle, M. G. Botelho and F. S. Wouters, *J. Biol. Chem.*, 2009, **284**, 15016.
82. P. R. Turner, K. O'Connor, W. P. Tate and W. C. Abraham, *Prog. Neurobiol.*, 2003, **70**, 1.
83. A. Garcia-Osta and C. M. Alberini, *Learn. Mem.*, 2009, **16**, 267.
84. M. Townsend, T. Mehta and D. J. Selkoe, *J. Biol. Chem.*, 2007, **282**, 33305.
85. C. Costantini, H. Scrable and L. Puglielli, *EMBO J.*, 2006, **25**, 1997.

86. S. Freude, K. Schilbach and M. Schubert, *Curr. Alzheimer Res.*, 2009, **6**, 213.

87. W. Q. Zhao, P. N. Lacor, H. Chen, M. P. Lambert, M. J. Quon, G. A. Krafft and W. L. Klein, *J. Biol. Chem.*, 2009, **284**, 18742.

88. F. G. De Felice, M. N. N. Vieira, T. R. Bomfim, H. Decker, P. T. Velasco, M. P. Lambert, K. L. Viola, W. Q. Zhao, S. T. Ferreira and W. L. Klein, *Proc. Natl. Acad. Sci.*, 2009, **106**, 1971.

89. S. Dore, S. Bastianetto, S. Kar and R. Quirion, *Ann. N.Y. Acad. Sci.*, 1999, **890**, 356.

90. J. Avila, J. J. Lucas, M. Perez and F. Hernandez, *Physiol. Rev.*, 2004, **84**, 361.

91. M. Hong and V. M. Lee, *J. Biol. Chem.*, 1997, **272**, 19547.

92. E. Carro, J. L. Trejo, A. Gerber, H. Loetscher, J. Torrado, F. Metzger and I. Torres-Aleman, *Neurobiol. Aging*, 2006, **27**, 1250.

93. C. M. Cheng, V. Tseng, J. Wang, D. Wang, L. Matyakhina and C. A. Bondy, *Endocrinology*, 2005, **146**, 5086.

94. L. A. Foster, N. K. Ames and R. S. Emery, *Physiol. Behav.*, 1991, **50**, 745.

95. R. H. Muzumdar, X. Ma, S. Fishman, X. Yang, G. Atzmon, P. Vuguin, F. H. Einstein, D. Hwang, P. Cohen and N. Barzilai, *Diabetes*, 2006, **55**, 2788.

96. S. Obici, Z. Feng, G. Karkanias, D. G. Baskin and L. Rossetti, *Nat. Neurosci.*, 2002, **5**, 566.

97. L. Frolich, D. Blum-Degen, H. G. Bernstein, S. Engelsberger, J. Humrich, S. Laufer, D. Muschner, A. Thalheimer, A. Turk, S. Hoyer, R. Zochling, K. W. Boissl, K. Jellinger and P. Riederer, *J. Neural. Transm.*, 1998, **105**, 423.

98. E. J. Rivera, A. Goldin, N. Fulmer, R. Tavares, J. R. Wands and S. M. De La Monte, *J. Alzheimers Dis.*, 2005, **8**, 247.

99. I. Torres-Aleman, *Expert Opinion on Therapeutic Targets*, 2007, **11**, 1535.

100. I. Pineau and S. Lacroix, *Glia*, 2009, **57**, 351.

101. P. L. McGeer, M. Schulzer and E. G. McGeer, *Neurology*, 1996, **47**, 425.

102. B. Brugg, Y. L. Dubreuil, G. Huber, E. E. Wollman, N. Delhaye-Bouchaud and J. Mariani, *Proc. Natl. Acad. Sci. U.S.A.*, 1995, **92**, 3032.

103. P. L. McGeer and E. G. McGeer, *Brain Res. Brain Res. Rev.*, 1995, **21**, 195.

104. D. Pratico and J. Q. Trojanowski, *Neurobiol. Aging*, 2000, **21**, 441.

105. P. Pasqualetti, C. Bonomini, F. G. Dal, L. Paulon, E. Sinforiani, C. Marra, O. Zanetti and P. M. Rossini, *Aging Clin. Exp. Res.*, 2009, **21**, 102.

106. P. S. Aisen, K. A. Schafer, M. Grundman, E. Pfeiffer, M. Sano, K. L. Davis, M. R. Farlow, S. Jin, R. G. Thomas and L. J. Thal, *JAMA: The J. Am. Med. Asso.*, 2003, **289**, 2819.

107. F. E. McAlpine, J. K. Lee, A. S. Harms, K. A. Ruhn, M. Blurton-Jones, J. Hong, P. Das, T. E. Golde, F. M. LaFerla, S. Oddo, A. Blesch and M. G. Tansey, *Neurobiol. Dis.*, 2009, **34**, 163.

108. J. C. Roberts, S. L. Friel, S. Roman, M. Perren, A. Harper, J. B. Davis, J. C. Richardson, D. Virley, A. D. Medhurst, *Exp. Neurol.*, 2009, **216**, 459.

109. S. E. Hickman, E. K. Allison and K. J. El, *J. Neurosci.*, 2008, **28**, 8354.
110. W. J. Streit, H. Braak, Q. S. Xue, I. Bechmann, *Acta Neuropathol.*, 2009, **118**, 475.
111. T. Bolmont, F. Haiss, D. Eicke, R. Radde, C. A. Mathis, W. E. Klunk, S. Kohsaka, M. Jucker and M. E. Calhoun, *J. Neurosci.*, 2008, **28**, 4283.
112. S. Jimenez, D. Baglietto-Vargas, C. Caballero, I. Moreno-Gonzalez, M. Torres, R. Sanchez-Varo, D. Ruano, M. Vizuete, A. Gutierrez and J. Vitorica, *J. Neurosci.*, 2008, **28**, 11650.
113. D. Rossi and A. Volterra, *Brain Res. Bull.*, 2009, **80**, 224.
114. A. M. Fernandez, S. Fernandez, P. Carrero, M. Garcia-Garcia and I. Torres-Aleman, *J. Neurosci.*, 2007, **27**, 8745.
115. R. Dantzer, G. Gheusi, R. W. Johnson and K. W. Kelley, *Neuroreport*, 1999, **10**, 289.
116. F. O. Maher, R. M. Clarke, A. Kelly, R. E. Nally and M. A. Lynch, *J. Neurochem.*, 2006, **96**, 1560.
117. A. Wilkins and A. Compston, *J. Neurochem.*, 2005, **92**, 1487.
118. I. Torres-Aleman, *Mol. Neurobiol.*, 2000, **21**, 153.
119. S. Pons and I. Torres-Aleman, *J. Biol. Chem.*, 2000, **275**, 38620.
120. S. Boudina, H. Bugger, S. Sena, B. T. O'Neill, V. G. Zaha, O. Ilkun, J. J. Wright, P. K. Mazumder, E. Palfreyman, T. J. Tidwell, H. Theobald, O. Khalimonchuk, B. Wayment, X. Sheng, K. J. Rodnick, R. Centini, D. Chen, S. E. Litwin, B. E. Weimer and E. D. Abel, *Circulation*, 2009, **119**, 1272.
121. S. M. De La Monte and J. R. Wands, *J. Alzheimers Dis.*, 2006, **9**, 167.
122. M. Bains, M. L. Florez-McClure, K. A. Heidenreich, *J. Biol. Chem.*, 2009, **284**, 20398.
123. I. Kadish, O. Thibault, E. M. Blalock, K. C. Chen, J. C. Gant, N. M. Porter and P. W. Landfield, *J. Neurosci.*, 2009, **29**, 1805.
124. M. P. Mattson, B. Cheng, D. Davis, K. Bryant, I. Lieberburg and R. E. Rydel, *J. Neurosci.*, 1992, **12**, 376.
125. S. Fernandez, A. M. Fernandez, C. Lopez-Lopez and I. Torres-Aleman, *Growth Hormone & IGF Research*, 2007, **17**, 89.
126. K. V. Kuchibhotla, C. R. Lattarulo, B. T. Hyman and B. J. Bacskai, *Science*, 2009, **323**, 1211.
127. R. Mayeux, *Annu. Rev. Neurosci.*, 2003, **26**, 81.
128. M. O. Dietrich, A. Muller, M. Bolos, E. Carro, M. L. Perry, L. V. Portela, D. O. Souza and I. Torres-Aleman, *Neuromolecular Med.*, 2007, **9**, 324.
129. J. L. Trejo, E. Carro, A. Nunez and I. Torres-Aleman, *Rev. Neurosci.*, 2002, **13**, 365.
130. M. A. Reger, G. S. Watson, P. S. Green, L. D. Baker, B. Cholerton, M. A. Fishel, S. R. Plymate, M. M. Cherrier, G. D. Schellenberg, W. H. Frey Ii and S. Craft, *J. Alzheimers Dis.*, 2008, **13**, 323.
131. E. Carro, J. L. Trejo, T. Gomez-Isla, D. LeRoith and I. Torres-Aleman, *Nature Med.*, 2002, **8**, 1390.
132. D. Aguado-Llera, E. Arilla-Ferreiro, A. Campos-Barros, L. Puebla-Jimenez and V. Barrios, *J. Neurochem.*, 2005, **92**, 607.

133. T. A. Lanz, C. T. Salatto, A. R. Semproni, M. Marconi, T. M. Brown, K. E. Richter, K. Schmidt, F. R. Nelson, J. B. Schachter, *Biochem.-Pharmacol.*, 2008, **75**, 1083 .

134. J. J. Sevigny, J. M. Ryan, C. H. van Dyck, Y. Peng, C. R. Lines and M. L. Nessly, *Neurology*, 2008, **71**, 1702.

135. K. Gerozissis, *Eur. J. Pharmacol.*, 2008, **585**, 38.

136. M. S. Kim, Y. K. Pak, P. G. Jang, C. Namkoong, Y. S. Choi, J. C. Won, K. S. Kim, S. W. Kim, H. S. Kim, J. Y. Park, Y. B. Kim and K. U. Lee, *Nature Neurosci.*, 2006, **9**, 901.

137. A. Gruart, M. D. Munoz and J. M. Gado-Garcia, *J. Neurosci.*, 2006, **26**, 1077.

Ketone Bodies as a Therapeutic for Alzheimer's Disease

SAMUEL T. HENDERSON

Accera, Inc., 380 Interlocken Crescent, Ste. 780, Broomfield, CO 80021, USA

14.1 Introduction

Alzheimer's disease (AD, OMIM 104300) is an age-associated, neurodegenerative disease characterised clinically by a progressive decline in memory and language,[1] and pathologically by accumulation of senile plaques and neurofibrillar tangles in the brain.[2] Another prominent feature of AD is regional cerebral hypometabolism. Hypometabolism is most prominent in the posterior cingulate, parietal, temporal, and prefrontal cortex, and occurs very early in the disease.[3] Hypometabolism may contribute to both the cognitive decline and the pathology associated with AD.[4,5]

The etiology of AD remains controversial, yet genetic links found in familial AD strongly implicate processing of the amyloid precursor protein (APP) as a central player in the disease. Cases of familial, or early onset, AD are associated with variation in several genes involved in APP processing, including: APP, presenilin 1 (PSEN1) and presenilin 2 (PSEN2).[6] Cases of sporadic, or late onset, AD are generally not associated with variation in APP, PSEN1 or PSEN2, and instead are associated with risk factors. The primary risk factors for sporadic AD are age and the carriage status of the epsilon 4 (E4) variant of the apolipoprotein E gene (APOE). Possession of one or more E4 alleles increases risk and lowers the age of onset. The link between APOE4 and APP

RSC Drug Discovery Series No. 2
Emerging Drugs and Targets for Alzheimer's Disease
Volume 1: Beta-Amyloid, Tau Protein and Glucose Metabolism
Edited by Ana Martinez
© Royal Society of Chemistry 2010
Published by the Royal Society of Chemistry, www.rsc.org

remains elusive, yet many mechanisms have been extensively studied and reviewed.[7–9]

Despite significant advances in AD research, there exists a tremendous need for new AD treatments. Most developed countries will experience a dramatic demographic shift toward an older population in the next 50 years, which is anticipated to greatly increase the prevalence of AD. The rise in the number of AD patients will place a tremendous social and economic burden on the developed world.[10] Throughout this book numerous promising new therapeutic targets are reviewed, and it is hoped that one, or a combination of approaches, can treat or prevent this terrible disease. In this chapter, the potential for targeting hypometabolism is discussed and, in particular, the use of ketone bodies to address the low metabolic rates of cerebral glucose metabolism is presented.

14.2 Brain-Energy Metabolism

The energy demands of the human brain are substantial. The brain is one of the most metabolically active organs in the body. While the brain is only about 2% of body weight, it receives 15% of cardiac output, accounts for 20% of whole-body oxygen consumption, and utilises 25% of total body glucose. Unlike many other organs in the body, the brain does not efficiently utilise fat as an energy source. Measurements of ateriovenous differences across the brain indicate minimal extraction of fat, yet considerable extraction of glucose and oxygen. Therefore, under normal conditions, the adult human brain relies almost exclusively on glucose as an energy substrate.[11] Glucose offers several advantages as a fuel substrate that makes it the preferential choice for the brain. One of the main advantages is the speed at which energy can be generated. Glucose is rich in potential energy and highly soluble in aqueous solutions, therefore it can be readily mobilised when energy demands are increased. In addition, glucose can yield energy both rapidly through glycolysis, and efficiently through oxidative phosphorylation. While glycolysis yields only 2 molecules of ATP per molecule of glucose, it can generate ATP almost twice as fast as through oxidative phosphorylation. Therefore, although a minor pathway for energy generation in the brain, the rapid generation of energy by glycolysis is strategically important for many activation tasks.[12]

Measurements of ateriovenous differences across the brain indicate that the brain extracts approximately 50% of the oxygen and 10% of the glucose in arterial blood. Oxygen consumption across the adult human brain is approximately 160 mmol/100 g min, and glucose consumption is approximately 31 mmol/100 g min. The extraction of fats is minor, and the vast majority of the extracted oxygen is used to metabolise glucose, with a ratio of oxygen/glucose utilisation of approximately 5.2. The theoretical ratio of oxygen/glucose use for the complete oxidation of glucose is 6. Hence, not all the glucose is oxidised and a portion is likely shunted toward synthesis pathways, to be incorporated into lipids, proteins, and glycogen, and also the precursor of certain neurotransmitters such as g-aminobutyric acid (GABA), glutamate, and acetylcholine.

Also, small amounts of oxygen are used in other pathways such as synthesis and degradation of monoamine neurotransmitters.[11]

14.2.1 Glucose Metabolism is Coupled to Neuronal Activity

While housekeeping functions of neurons and glial cells require energy in the form of ATP, much of the energy consumed by the brain appears to be used for neuronal signalling processes. The major excitatory neurotransmitter in the brain is glutamate. Glutamate is released by glutamatergic neurons into the synaptic cleft, and then cleared from the synapse by uptake into astrocytes. Within astrocytes, glutamate is converted to the synaptically inactive glutamine, which is released for uptake in neurons. Once in neurons, glutamine is converted back to glutamate for signalling (Figure 14.1). The process of glutamate signalling and cycling may use much of the energy generated by the metabolism of glucose.[13,14] For example, Sibson *et al.* used ^{13}C NMR spectroscopy in rats to calculate the rates of glucose metabolism and compared them to the rates of glutamate cycling between neurons and astrocytes. These studies indicate a near 1:1 stoichiometric ratio between glucose metabolism and glutamate cycling. The authors suggest that "This finding indicates that the majority of cortical energy production supports functional (synaptic) glutamatergic neuronal activity".[15] Using anatomic and physiologic data of rats and primates, Attwell and Laughlin reach a similar conclusion. The authors confirm that the brain is an "expensive tissue" with rates of ATP use (30 µmol ATP/g/min) equal to that of a human leg muscle running a marathon. Notably, based on their calculations most of the energy is used for action potentials and postsynaptic effects of glutamate.[16]

Several models have been proposed that couple energy metabolism and synaptic activity that explain the brains reliance on glucose as an energy source (Figure 14.1). Glucose is ideally suited to both rapidly generate ATP via glycolysis, which is required for the millisecond clearance of glutamate from the synapse, and to efficiently generate of ATP via oxidative phosphorylation, which is required for sustained high levels of ATP in neurons. The emerging model suggests that anaerobic metabolism of glucose occurs mostly in astrocytes. The rapid generation of energy by glycolysis provides energy for clearance of neurotransmitters (largely glutamate and GABA) from the synapse. A portion of the lactate generated in astrocytes is then release as substrate for neurons. Neurons are thought to generate most of their ATP by oxidative phosphorylation of both glucose and lactate.[17,18]

Despite the brain's reliance on glucose for energy, the brain stores little glucose in the form of glycogen, and instead relies on glucose in circulation for much of its function.[11] The reliance on circulating glucose and the coupling of glucose metabolism and synaptic function makes the brain especially vulnerable to dysfunction of either glucose transport or utilisation. This is well illustrated in cases of GLUT1 deficiency syndrome (GLUT1 DS, OMIM 606777).

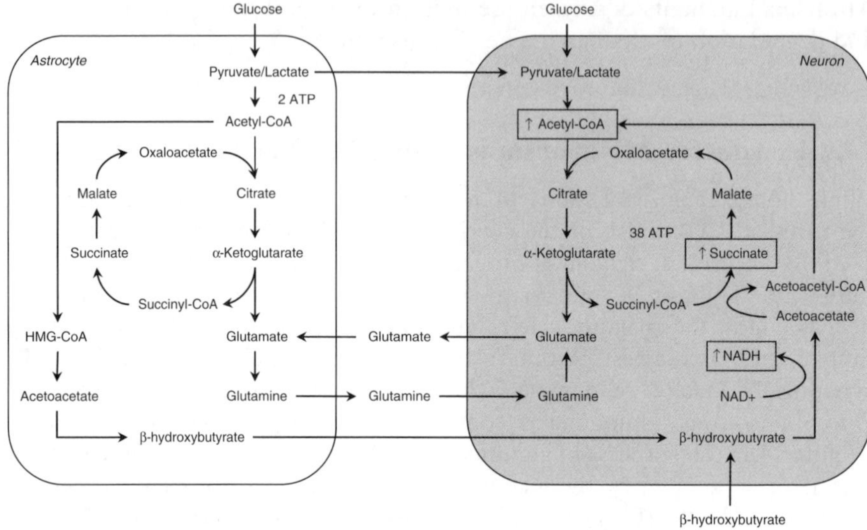

Figure 14.1 Coupling of brain-glucose metabolism and neuronal activity. During neuronal signalling, glutamatergic neurons release glutamate into the synaptic space that is then cleared from the synapse by astrocytes. The millisecond clearance of glutamate is fuelled by rapid ATP synthesis generated by glycolysis. Within astrocytes, glutamate is converted to glutamine that is released for uptake in neurons. Neurons convert glutamine back to glutamate for signaling. Neurons generate most of their ATP through complete oxidation of glucose. Ketone bodies, such as β-hydroxybutyrate can also serve as an energy-generating substrate and may be synthesized by astrocytes or transported from circulation. Boxed metabolites indicate metabolic changes in neurons upon metabolism of ketone bodies. Conversion of β-hydroxybutyrate to acetoacetate generates NADH, thereby reducing the mitochondrial NAD/NADH couple and increasing the oxidation of the coenzyme Q couple. Mitochondrial metabolism of acetoacetate increases the pool of succinate providing a substrate for complex II and by passing complex I. Metabolism of acetoacetate increases acetyl-CoA levels providing substrate for the citric acid cycle. The net effect of these changes is increased metabolic efficiency.

Glucose is transported into the brain by facilitative transporters encoded by the GLUT1 and GLUT3 genes. The GLUT1 protein is prominently expressed in the endothelial cells of the cerebral microvasculature and in the astrocyte footpads that contact the vasculature, and transports the majority of glucose into the brain.[19] Due to the substantial glucose requirements of brain, both copies of the GLUT1 gene must be present for proper function. In GLUT1 DS, one copy of the gene is nonfunctional and the cerebrospinal glucose concentrations are 30–40% of blood concentrations.[20] The clinical features of GLUT1 DS are serious and include: seizures, hypotonia, ataxia, language deficits, and microcephaly.[21]

14.3 Hypometabolism in Alzheimer's Disease

Examination of the rate of glucose utilisation in the brain can be accomplished using ^{18}F-2-deoxy-2-fluoro-D-glucose (18FDG), a positron-emitting tracer. When used in conjunction with positron emission tomography (PET), the FDG PET technique can provide *in vivo* quantitation of regional brain metabolism. FDG PET studies in the early 1980s compared AD subjects with normal controls and found significant reductions in the global cerebral metabolic rate of glucose (CMRglc). In these studies, significant correlations were found between reduced CMRglc and worsening performance on measures of cognitive function.[22] Subsequent studies have revealed that low CMRglc in AD is not simply a global decrease in glucose use across the brain, but rather maps to very specific areas of the brain. Low CMRglc in these areas results in a characteristic, reproducible pattern of hypometabolism in AD. Abnormally low rates of CMRglc are found in the posterior cingulated,[23] and parietal, temporal, and prefrontal cortices.[24] This pattern is robust and has been proposed as a diagnostic tool for AD (Figure 14.2).[25,26]

The characteristic pattern of hypometabolism is an early event in AD. The pattern has been noted in at-risk populations well before clinical symptoms of dementia or cognitive impairment become evident. For example, hypometabolism has been detected in cases of mild cognitive impairment (MCI). Mild cognitive impairment has been defined as scoring significantly lower on cognitive tests than an age-matched peer group, but the impairment is not severe enough to interfere with activities of daily living. Estimates on the prevalence vary, yet as many as 19% of those over 65 may have MCI. Mild cognitive impairment is frequently considered a risk state for later development of dementia. Some studies have found that as many as 50% of MCI patients will convert to dementia and in particular to AD.[27] Drzezga *et al.* used FDG PET to examine the pattern of hypometabolism in MCI patients who converted to AD over a one-year timeframe. Twenty two MCI patients were evaluated at baseline and one year later by cognitive testing and FDG PET. Eight of the patients converted to probable AD, termed MCI$_{AD}$, while 12 remained as MCI, termed MCI$_{MCI}$. Both groups demonstrated hypometabolism in the posterior cingulated cortex (PCC) compared to normal controls, and in addition, the MCI$_{AD}$ group showed the typical AD pattern with low glucose rates in temporal and parietal cortices as well as the PCC. Over the course of the year the MCI$_{AD}$ group demonstrated additional reductions in CMRglc in temporal and parietal cortices and in the PCC.[28]

14.3.1 Early Occurrence of Hypometabolism

Additional, more complete, longitudinal studies have confirmed the presence of the AD pattern of hypometabolism in MCI and even earlier in presymptomatic normal elderly. Mosconi *et al.* examined changes in brain-glucose metabolism that occurred in normal elderly as they progressed to MCI and eventually as they progressed to confirmed postmortem diagnosis of AD. Four normal

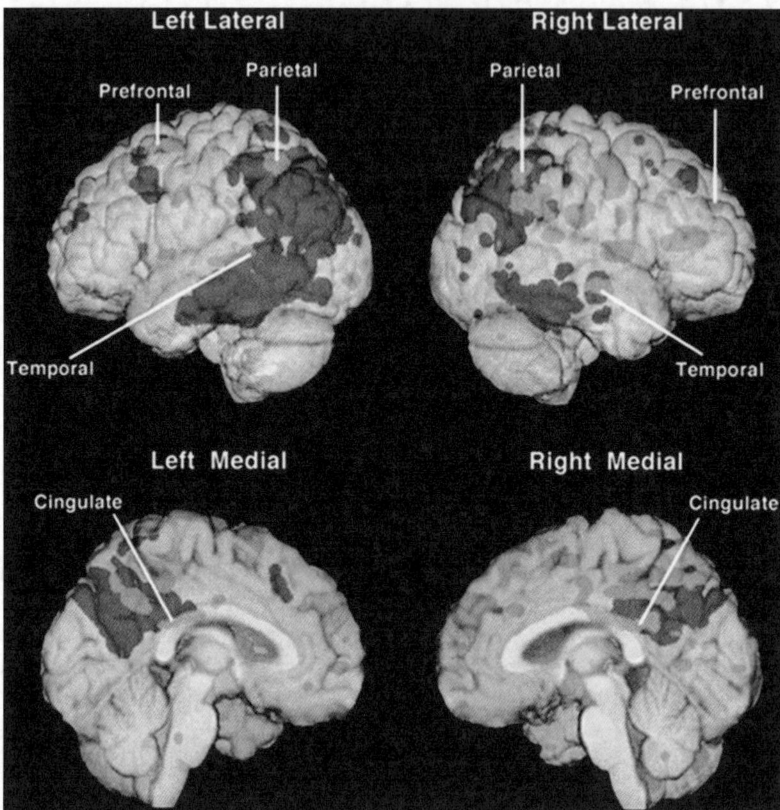

Figure 14.2 Regional hypometabolism in Alzheimer's disease. Shown is a 3D surface-projection map of abnormally low CMRglc as shown by FDG PET superimposed on a map of volume-rendered MRI of the brain. The dark grey areas show low CMRglc in subjects with probable AD, with notable decreases bilaterally in the posterior cingulate, parietal, temporal, and prefrontal cortex. The light grey regions show regions of low CMRglc in young-adult (mean age 30.7 years of age) E4 carriers. Young-adult E4 carriers demonstrate low CMRglc in a pattern similar to probable AD patients. Figure taken from Reiman *et al.*[3] Copyright (2004) National Academy of Sciences, USA.

healthy elderly were followed for 13 years (± 5 years). Two of the subjects declined to MCI and two to AD. The diagnosis of AD was confirmed by the postmortem presence of senile plaques and neurofibrillar tangles. The long-itudinal FDG PET scans revealed progressive declines in CMRglc in temporal and parietal cortices as well as the PCC.[29]

In a remarkable series of experiments, Reiman and coauthors examined how early the AD pattern of hypometabolism could be detected. In order to find subjects at risk of developing AD and who may have early nonclinical signs of the disease, Reiman and coauthors screened subjects between the ages of 50 and 65 years of age with a family history of AD, for those who were homozygous for the E4 allele. These subjects are at high risk of developing AD. Eleven E4/ E4 participants with a family history of AD (mean age 55.4) were compared to 22 normal controls (mean age 56.3). The groups did not differ in any cognitive testing. However, the E4/E4 homozygotes showed modest declines in glucose metabolism in the same regions identified in AD subjects. This finding demonstrated that hypometabolism is an early event in the disease and occurs before MCI.[25] To further trace the origin of the hypometabolism, Reiman *et al.* examined the pattern of glucose utilisation in young adults. Twelve E4 carriers (all E4/E3) were compared to 15 E4 noncarriers (all E3/E3). Similar to the previous study, E4 carriers (E4(+)) were found to have minor declines in CMRglc bilaterally in the posterior cingulate, parietal, temporal, and prefrontal cortices, despite showing no signs of cognitive impairment (Figure 14.2). Remarkably, the mean age of the E4 carriers in this study was 30.7 years.[3] Therefore, low regional CMRglc appears to be a very early event in the disease process, well before any clinical signs of dementia are evident, and may be present, in a limited form, decades before any outward signs of cognitive deficits are present.

The AD pattern of reduced CMRglc is not exclusive to E4 carriers. Corder *et al.* examined 31 patients diagnosed with probable AD, of these: 16 were E3/ E4, 11 E3/E3 and 4 E2/E3. These 31 subjects demonstrated a pattern of decline in CMRglc in the frontal region and the temporoparietal regions of the cortex. Individual genotypes were examined for differences in CMRglc and none were found.[30] However, studies using more advanced techniques have identified some differences between E4(+) and E4(−) participants. E4(+) subjects may have more global declines in CMRglc.[31]

The causes of the regional reductions in CMRglc in AD are not fully understood. They may simply be due to a reduction in the density or activity of terminal neuronal fields or glial cells, a metabolic defect within neurons or glia, or a combination these factors. Several authors have addressed this question with the use of kinetic FDG PET. Kinetic PET was used to examine rates of glucose transport and rates of glucose phosphorylation by hexokinase, the first committed step in glycolysis. Piert *et al.* examined 10 normal control subjects with 8 subjects with probable AD (mean MMSE score of 18.1) using dynamic FDG PET. Significant declines in glucose transport and hexokinase activity were found in the parietal and temporal cortices in the AD subjects. In addition, hexokinase activity was reduced in the putamen, cerebellar and occipital cortex. As discussed in previous sections, most glucose use is coupled to synaptic activity. Therefore, loss of synaptic fields in AD would be predicted to show a decline in glucose uptake. However, if the remaining synapses are functioning properly there should not be a decrease in the hexokinase rate constant. Yet, this study found a decrease in the rate of hexokinase activity, which suggests that there may be loss of synaptic activity without substantial atrophy of neurons and glia. Also, the rate of hexokinase decrease was not sufficient to account for the large, 20–40%, decreases in CMRglc seen in AD, suggesting other defects in metabolism were present.[32]

14.3.2   Compensatory Activation

Given the brains dependence on glucose and the coupling of glucose metabolism with synaptic activity, how is it that the declines in CMRglc
and hexokinase activity occur without cognitive symptoms? The early occurrence of hypometabolism in E4(+) subjects and those diagnosed with MCI or
AD may be compensated for by the recruitment of increased regions of the
brain to accomplish tasks. Alzheimer's patients in several studies were found
to have greater activations of regions of the cortex than normal controls.[33,34]
For example, Grady *et al.* examined 12 healthy elderly controls (average age
71 ± 4 years; average MMSE score 28 ± 1) and 12 mild probable AD patients
(average age 74 ± 9; average MMSE score 22 ± 5) by PET to examine neural
activity during memory tasks. Subjects with probable AD exhibited increased
recruitment of prefrontal regions in an attempt to complete the semantic and
episodic memory tasks.[35] Similar to the presymptomatic findings of hypometabolism, this compensatory activation can be detected in cognitively
normal, at-risk carriers of the APOE4 allele. Bookheimer *et al.* examined 30
cognitively normal, right-handed, older adults (mean age 63 ± 8 years) by
functional MRI to measure regions of activation while subjects memorised
and recalled unrelated pairs of words. In addition, subjects were genotyped
for variation in APOE locus. Sixteen subjects were E4(+) (14 E3/E4; 2 E4/
E4). Fourteen subjects were E4(−) (14 E3/E3). Among E4(+) subjects, both
the magnitude and the extent of brain activation during verbal memory
challenge were greater than those among similar subjects who were E4(−).
The authors conclude "Our results indicate that, as a group, older persons
with a genetic risk for Alzheimer's disease have alterations in brain function
without obvious morphologic or behavioural indications of impending disease".[36] Bondi *et al.* examined 20 normal, right-handed, older adults (mean
age 76 ± 5 years) who were part of a longitudinal aging study. Ten of the
subjects were E4(−) (all E3/E3) and ten were E4(+), 8 E4/E3, 2 E4/E4. The
goal of the study was to examine the role E4 carriage status played
during a picture-encoding task on blood-oxygen-level-dependent (BOLD)
brain response. Similar to the earlier study, in multiple brain regions, the
E4(+) group showed significantly greater BOLD brain response compared to
the E4(−) group while learning new pictures compared to viewing a repeated
picture.[37] These studies suggest that during periods of recall, elderly E4(+)
subjects recruit larger neuronal fields to accomplish the same task as E4(−)
subjects.

Similar to the early occurrence of hypometabolism in E4 carriers, the
increased compensatory activation can be detected in E4(+) subjects as young
adults. When 18 young (age 20–35 years), healthy E4(+) subjects were compared to age-matched E4(−) subjects, significant increased regions of activation
were present in the E4(+) subjects.[38] The mechanism behind this compensatory
activation is unclear, yet may be related to reduced metabolic capacity in E4(+)
neuronal tissue, which necessitates the activation of larger areas to accomplish
tasks.

14.4 Causes of Hypometabolism

The precise mechanism of the cause of hypometabolism and why it is present in certain regions of the brain remains unknown. Low CMRglc may be due to inhibition of metabolism mediated by alterations in the processing of either the amyloid precursor protein (APP) or the ApoE4 protein. Mahley and coauthors have proposed the structural difference between ApoE3 and ApoE4 proteins results in different handling of the protein during secretion. Chang *et al.* examined the toxicity of ApoE4 protein fragments in Neuro-2a cells, a mouse neuroblastoma line. When truncated versions of ApoE4 were expressed in the cells, they were frequently found outside of the endoplasmic reticulum and the golgi. Some versions of the fragments escaped the secretory pathway and were found localised in mitochondria and inhibited mitochondrial function.[39] The disturbances in mitochondrial function has been suggested to contribute to low CMRglc seen in AD and in presymptomatic E4(+) carriers.[40]

The Aβ peptide generated from the cleavage of the App protein may also play a role in hypometabolism. Several hypotheses on the function of Aβ have been proposed. Aβ may bind heme leading to a loss of complex IV activity.[41] Aβ may stimulate GSK3β activity leading to the phosphorylation of pyruvate dehydrogenase and the inhibition of energy metabolism.[42] The uncleaved APP protein may accumulate across mitochondrial import channels, causing mito-chondrial dysfunction.[43] These mechanisms have been the subject of several excellent reviews to which the interested reader is referred.[41,44–46]

Other authors have suggested that hypometabolism in AD may be directly related to reduced expression of genes involved in energy production in the affected areas. Liang *et al.* examined expression levels of 80 genes involved in metabolism and energy generation in 33 late-onset AD brain samples (15 males and 18 females) and compared the expression levels to healthy controls. Samples were taken from several brain regions by laser-capture microdissec-tion. Cells chosen were nontangle-bearing neurons from autopsied brains of AD cases and normal controls. Regions chosen were the PCC, parietotemporal cortices, and the visual cortex. The visual cortex typically does not demonstrate hypometabolism in AD and served as an internal control. In the PCC, sig-nificant declines in the expression of nuclear-encoded components of the elec-tron transport chain (ETC) were evident and these decreases were less evident in the visual cortex. The genes included components of all the major complexes of the ETC, including the F1 ATPase. This study suggests that the low CMRglc found in AD and presymptomatic cases may be directly due to low expression of these genes.[47]

14.5 Hypometabolism and APP Processing

Low rates of CMRglc appear early in the disease well before large amounts of Aβ are predicted to be present, and in AD subjects who are E4(−). Therefore, some authors have investigated if disturbances in energy metabolism may result in alterations in APP processing. Inhibition of energy production by treating cells with sodium azide leads to an 80-fold increase in the generation of an amyloidic C-terminal fragment of APP and accumulation of APP in the secretory pathway.[48] This increase in APP fragmentation may be due to increased β-secretase (BACE) activity, the rate-limiting step of Aβ generation. When wild-type mice were treated with agents that induce acute energy deprivation, such as insulin and 2-deoxygluose, a long lasting increase in Bace1 enzyme levels was detected. Elevated Bace1 levels were durable, lasting at least seven days after treatment. Furthermore, it was found that such treatments led to a 2-fold increase in Aβ40 levels seven days after treatment.[5] In a follow-up study, O'Conner *et al.* found that the mechanism of activation of BACE was post-transcriptional. Energy deprivation in cultured primary neurons caused phosphorylation of the translation initiation factor eIF2α and increased translation of the BACE mRNAs. In addition, chronic energy depletion in a transgenic mouse model of AD, led to increased levels of BACE, Aβ and amyloid load. These studies suggest that chronic, low-level, energy stress may increase the activities of APP processing enzymes ultimately leading to the pathology of AD.[49]

Importantly, these studies suggest that hypometabolism is an early event in AD that may give rise to generation of Aβ and amyloid deposition. As such, these findings have significant implications in developing therapeutic interventions for AD, suggesting that addressing the metabolic defects found in presymptomatic, at-risk individuals may offer protection from AD.

14.6   Hypometabolism and the Default Network

If hypometabolism is a precursor to the pathology of AD, it still leaves open the question of why specific areas of the brain would be susceptible to the development of metabolic disturbances as well as amyloid deposition. A clue to this question may be related to the observations that regions of the brain showing hypometabolism in AD significantly overlap with regions identified in the brain's "default network". The default network can be considered the baseline activity of the brain. One way to examine this baseline activity is by measuring oxygen consumption, and specifically the brain oxygenation extraction fraction (OEF). The OEF is defined as the ratio of the oxygen used by the brain to the amount of oxygen delivered by the blood. It is well known that during specific activation tasks local blood flow increases, glucose consumption increases, but oxygen consumption does not. The phenomenon has been interpreted as evidence of an increase in glycolysis in astroctyes to provide rapid generation of ATP for maintenance of synaptic function.[17] This increase in local oxygen content can be measured by fMRI and is referred to as the blood-oxygen-level-dependent (BOLD) signal. Hence, during task activation the BOLD signal increases as local circulation increases oxygen in those regions. As local oxygen levels increase without an increase in brain utilisation, the ratio of oxygen used relative to what is available in the circulation decreases. Hence, the OEF signal decreases. Therefore, unlike the BOLD signal, a decrease in OEF is associated with activation, and an increase in OEF is associated with deactivation.

Raichle *et al.* used PET to measure OEF during both resting and activation tasks in an attempt to measure the baseline neuronal activity of the brain. By measuring OEF they were able to identify regions of the brain that would preferentially be deactivated during task-specific activation, but would be active when the subject was resting quietly with their eyes closed. The consistent decreases in activity regardless of the task being performed suggested to the authors that, "We posit that areas decreasing their activity in this manner may be tonically active in the baseline state, as distinguished from areas that are transiently activated in support of varying goal-directed activities".[50]

Subsequently, these baseline, or default, areas were mapped to a network of brain regions. Several different methods have been used to determine the areas of the brain that compose the default network. A group of studies used methods similar to Raichle *et al.* and measured areas of deactivation after a variety of task activations, including verbal, auditory and visual tasks.[51] In addition, Greicius *et al.* used functional connectivity MRI to map the default network. Functional connectivity MRI (fcMRI) measures interregional correlations of BOLD signal changes and can map areas that share variations in BOLD signal under different conditions. Fourteen healthy, right-handed subjects (mean age 21.2 years) were measured in a resting state, during a test of working memory, and a visual pro-cessing task. The fcMRI mapping of default activity agreed well with deactivation mapping by PET.[52] These studies help establish the view that the default network represents a brain system involving physically and functionally connected brain areas. The mapping techniques suggest an interconnected network comprising prefrontal cortex, posterior cingulate, parietal lobule, temporal cortex, and pre-frontal cortex (see Figure 14.3).

The observation that the default network is most active when one is resting quietly with eyes closed has given consideration to the idea that the default network functions in spontaneous cognition; thoughts that do not require outside stimuli and are sometimes referred to as stimulus-independent thought (SITs). These activities include, planning for the future, day dreaming, and importantly for the present discussion, memory retrieval. A review of the function and activities of the default network are beyond the scope of this chapter and the reader is referred to a more complete review of the subject.[51]

The default network is also unique in its energy requirements. As noted in previous sections, greater than 90% of the glucose taken up by the brain is oxidatively degraded to generate ATP, resulting in 36 mol of ATP per mol of glucose. Task-specific activation relies extensively on glycolysis to rapidly generate ATP, resulting in 2 mol of ATP per mol of glucose. In addition, the relatively minor increases in blood flow do not account for the significant amounts of energy being utilised by specific tasks. Instead, it has been proposed that the majority of energy utilised by the brain is in the absence of any par-ticular task, hence being used by default activity.[12] Also, the default network may use more energy than other networks in the brain. Minoshima *et al.* compared the metabolic activity of 66 subjects with probable AD with 22

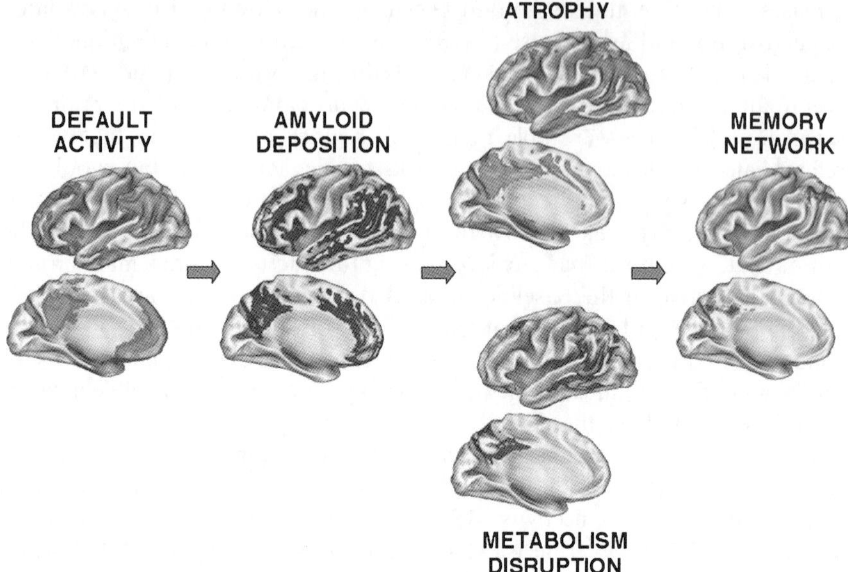

Figure 14.3 Convergence of the default network and Alzheimer's disease pathology. Regions associated with the default network show remarkable overlap with those areas affected in Alzheimer's disease. Each image represents a projection of findings on surface of left hemisphere. Dark areas represent: Default activity is a representation of activity in young adults as measured by $^{15}OH_2O$ PET. Amyloid deposition is a representation of deposition of amlyoid in older adults measured by ^{11}C PIB PET. Atrophy is a representation of longitudinal atrophy in older adults measured by structural MRI. Metabolic disruption is a representation of low CMRglc as measured by FDG PET in probable AD patients. Memory network is a representation of retrieval success in young adults measured by fMRI. Figure taken from Buckner *et al.*[53] Copyright 2005 The Society for Neuroscience.

normal controls by FDG PET. In the normal controls it was noted that glucose metabolism in PCC was about 20% higher than in other regions of brain. Also, it was noted that the AD subjects had a 21–22% reduction in the PCC,[23] consistent with many other findings. Therefore, the default network is unique in both its continuous activity and high rates of glucose use.

14.7   Metabolism Hypothesis of AD

The regions of the default network seem to be preferential areas that demonstrate hypometabolism, amyloid deposition, and atrophy in AD. In establishing baseline activity, which would become central to the default network, Raichle and coauthors noted that the PCC was vulnerable to several insults, such as carbon-monoxide poisoning, ischemia, and AD. The authors proposed "We wonder whether the exceptionally high metabolic rate exhibited by the posterior cingulate and precuneus adds to their vulnerability".[50] By examining five imaging techniques, Buckner and coauthors greatly expanded this finding and demonstrate the remarkable overlap between the default network and pathology of AD. Buckner *et al.* combined data on default activity by H_2O PET, amyloid deposition by PIB PET, structural atrophy by MRI, glucose metabolism by FDG PET, and memory-retrieval effects by fMRI to coincidently map default-network activity to areas affected by AD (see Figure 14.3). The overlap between regions of default activity and AD, led the authors to speculate on the existence of a link between the metabolic characteristics of the network creating an environment conducive to AD. This model was referred to as the "metabolism hypothesis" of AD (see Figure 14.3).[51,53] The authors speculate that high continuous levels of glucose utilisation in the default network early in life may be conducive to the future development of hypometabolism, plaque deposition, and cell atrophy that characterises AD.[53]

14.8 Environmental Factors in Hypometabolism

It is unclear why high rates of continuous glucose metabolism would predispose the default network to be susceptible to the specific hypometabolism and amyloid deposition seen in AD. However, an important clue to the susceptibility of the default network may be provided by the genetic risk factors associated with late onset AD. Greater than 90% of the cases of AD are late onset (also known as sporadic AD). The two major risk factors for the late onset AD are increasing age and the carriage status of E4 allele of the APOE gene. Importantly, possession of an E4 allele is not 100% penetrant. E4 only increases the risk of developing AD and many non-E4 carriers develop AD. Therefore, other genetic and environmental factors play a role in influencing the risk profile. Hence, it is crucial to ask what the environmental factor or factors might be, and how they are related to APOE4 and the metabolic rate of the default network.

A clue to the environmental factors may be found by looking at the distribution of the APOE4 allele in different populations around the globe. There are three common variants of APOE, commonly referred to as: E2, E3, and E4. The distribution of these alleles varies in different ethnic populations. Corbo and Sacchi made a key observation that the frequency of APOE4 is low in populations with long historical exposure to agriculture, and high in populations with little exposure to agriculture. For example, low frequencies of E4 are found in Moroccans (0.085), Turks (0.068), and Sardinians (0.052); while high frequencies are found in Nigerians (0.296), Sudanese (0.291) and Papuans (0.368). Frequencies of E4 tend to be high in Native American populations, for example Cayapas in Ecuador have an E4 frequency of 0.28. Yet, an exception in Native American populations is found in descendents of Mayans, who possess an E4 frequency of 0.089, which is similar to Moroccans. Mayans are unique in that they developed an extensive corn-based agricultural society. This suggested to Corbo and Sacchi that there was something discordant between possession of E4 and development of agriculture. They proposed that APOE4 was an example of a "thrifty gene". The authors conclude "The comparison confirms that to carry APOE*4 may become disadvantageous in a Westernised environment; therefore, the aboriginal populations with high APOE*4 frequency could experience increased rates of CAD and AD when exposed to western lifestyle".[54]

The thrifty genotype hypothesis was proposed by Neel in 1962 to explain high rates of Type-II diabetes in populations recently exposed to Western lifestyle, in particular Native Americans. The basic idea was that long-time hunter-gatherer populations possessed a genetic make-up that was exceptionally efficient at the utilisation and storage of food stuffs, such that they would readily store food as fat in preparation for inevitable food shortages due to the nature of the hunter-gatherer lifestyle. When these populations are exposed to the abundant, fat-rich, foods characteristic of Western lifestyle, the propensity to store the food as fat gives rise to obesity and to an increased risk of Type-II diabetes.[55,56]

One criticism of the thrifty gene hypothesis is that all organisms are "thrifty". It is evolutionarily advantageous for long-lived species to store resources when conditions are abundant in preparation for lean times. A modification of the thrifty gene hypothesis has been proposed wherein, rather than a change in the availability of food, the type of food changed. Some analyses of hunter-gatherer economics suggests that humans were extremely effective at food gathering and had limited wants.[57] One change that has been suggested to contribute to Type-II diabetes in populations not accustomed to agricultural diets, was the change in dietary carbohydrate levels. Several authors have suggested that the primary change in diet that occurred, during the transition to grain-based agriculture, was a switch from a low-carbohydrate hunter-gatherer diet, to a high-carbohydrate agricultural diet.[58] This switch, and the inherent thrifty nature of humans, likely explains the high rates of obesity and Type-II diabetes in modern times, and may be related to the development of AD.

14.8.1   High-Carbohydrate Diets and Alzheimer's Disease

It has been proposed that the change in macronutrient content that humans underwent in the switch to agriculture is central to the etiology of AD, as evidenced by the selection against E4 carriers in those populations. In brief, the change in diet would induce alterations in lipid/glucose metabolism that may lead to inappropriate cleavage of APP, and poor functioning of other susceptible proteins such as glucose transporters. The alterations may be evident as hypometabolism, accumulation of amyloid, and the generation of reactive oxygen species. Interestingly, this viewpoint suggests that there is nothing inherently deleterious about APOE4, it is only deleterious in the context of Western diet and lifestyle.[8]

While this model remains speculative, a relatively simple corollary to this hypothesis would be that low-carbohydrate or ketogenic diets would be protective against AD as well as other diseases or disorders that may be due to this dietary shift. A growing body of evidence seems to support the idea that high-carbohydrate diets may be deleterious and low-calorie or ketogenic diets are neuroprotective (for reviews see refs. 59–61). In particular, the presence of ketone bodies may be beneficial to the AD patient.

14.9 Ketogenic Diets

Ketogenic diets (KDs) became of interest in the 1920s where they were developed to reduce the occurrence of seizures in epileptics. The rationale for this type of diet was based on the ancient observation that fasting reduced seizures. In the 5th century, Hippocrates noted that a man suffering from epilepsy was cured by abstaining from all food and drink. KDs were developed to mimic the physiological changes seen in extended fasting. KDs are very low in carbohydrates and proteins and very high in fat, and have been used successfully for many years to treat refractive childhood epilepsy.[62] Importantly, KDs have also been used successfully to mitigate the symptoms of GLUT1 DS. The rationale for this treatment is to provide a substitute metabolic fuel when adequate amounts of glucose cannot be transported to the brain.[63] Thus, as will be explained in greater detail below, the induction of ketosis in GLUT1 DS parallels to the use of ketone bodies to address hypometabolism in AD.

A hallmark of KD is the elevation of circulating ketone bodies (ketosis), which also occurs during extended fasting. One function of ketosis is to provide an alternative to glucose for the brain during periods of food deprivation or low glucose availability. Owen *et al.* examined brain metabolism in obese subjects who were undergoing a complete fast for 5–6 weeks. During the fast, the subjects were restricted to one multivitamin capsule, 17 mEq of NaCl, and 1500 ml of water per day. Under these fasting conditions, glucose is produced predominantly from catabolism of proteins and glycerol from fat stores. Rates of glucose synthesis were calculated during the fast based on nitrogen excretion rates. It was estimated that the total glucose synthesised and available to the body during the fast was approximately 33 g of glucose per day, well short of the brain's normal requirement of $>$ 110 grams/day of glucose.[64]

If the brain is dependent on glucose, then why doesn't the body simply make more glucose? The primary sources of *de novo* synthesised glucose are amino acids. To provide the high levels of glucose required for the brain during a fast would require a significant breakdown of tissue, predominantly muscle. Such loss of lean body mass during periods of food scarcity would make it difficult to hunt or gather food and would likely be maladaptive over the course of evolution. Instead, the utilisation of ketone bodies allows for the tapping of abundant fat stores. During the 5–6 week fast, the concentration of the major ketone body β-hydroxybutyrate was elevated to the range of 4–8 mM, 10- to 20-fold over normal 12-h fasting values. Cerebral use of ketone bodies was found to be significant. Arteriovenous differences across the brain measured rates of β-hydroxybutyrate extraction of 0.23 to 0.43 mmoles per litre of whole blood, similar to rates of glucose extraction. It was estimated that under these conditions ketone bodies supply about 60% of the brain's energy requirements.[64] Therefore, a normal physiological response to low food and glucose availability is the provision of ketone bodies from fat stores for brain metabolism.

14.10 Regulation of Ketogenesis

Ketone bodies (KBs) are produced mainly by the liver from fatty acids (FA) during periods of fasting or when on a low-carbohydrate, high-fat diet, such as during mammalian neonatal development. KBs consist of β-hydroxybutyrate (BHB), acetoacetate (ACA) and acetone. Ketogenesis requires both an abundant source of FA in circulation and conditions in the liver favourable for FA oxidation. Ketogenesis can be considered to be regulated at 3 key steps. First, is the release of FA from adipocytes. Conditions that favour storage of fat, such as high-carbohydrate diets, limit adipocyte release of FFA and hamper KB production. Interestingly, KB may regulate FFA release as part of a feedback loop. β-hydroxybutyrate may be a ligand for the HM74a receptor leading to reduced FFA secretion from adipocytes.[65] HM74a is better known as the niacin receptor, a potent lipid-lowering agent. Second, is the decision between oxidation or esterification. In hepatocytes, carnitine palmitoyltransferase-I (CPT-1) acts as the major regulator of FA oxidation. CPT-1 functions to transfer activated FFA into the mitochondria. Conditions that favour FA synthesis, such as high carbohydrate diets, inhibit CPT-I and promote esterification of FA to phospholipids or triacylglycerols, and prevent oxidation of FA and the production of KB. Third, is the fate of acetyl-CoA generated from FA oxidation. Acetyl-CoA may enter the citric acid cycle, be used for synthesis or FA or cholesterol, or be used for ketogenesis. The generation of ketone bodies is promoted by high rates of FA oxidation when rapid accumulation of acetyl-CoA exceeds the capacity of the citric acid cycle.

Importantly, the regulation of KB production is strongly influenced by carbohydrate in the diet. This has consequences in relation to modern dietary practices. A typical modern diet, is very rich in carbohydrates (> 50% of total calories consumed) and hence high rates of ketogenesis rarely occur. Significant ketosis is only encountered when dysregulation of insulin signalling has occurred, such as in diabetes. Therefore, ketosis has historically been viewed as a deleterious, abnormal condition. This is likely an artifact of modern diets. Throughout much of human evolution, ketosis likely served as a valuable survival mechanism to fuel brain metabolism during times of food scarcity and to preserve muscle mass.[64]

One impact of modern carbohydrate-rich diets, is that under these conditions, the brain, and in particular the default network, would be fuelled continuously and at high metabolic rates by glucose only. Only very rarely would sufficiently high levels of ketone bodies be produced to have any metabolic or protective effects. Without the protective effects of bouts of ketosis, the highly metabolically active default network may be particularly vulnerable.

14.11 Neuroprotective Effects of Ketogenic Diets and Ketone Bodies

Is there evidence that KDs and KBs are neuroprotective? KDs have demonstrated potential in treating several neurological conditions (see Table 14.1), including, AD,[66] amyotrophic lateral sclerosis,[67] traumatic brain injury[68] and Parkinson's disease.[69] The change in macronutrient content in a KD diet from a standard diet induces a set of changes that could contribute to neuroprotective effects, such as reduced glucose, low insulin/IGF signalling, and increased levels of uncoupling proteins.[70] For a complete review the reader is referred to a review of this topic, see ref. 60.

Several authors have asked if KBs alone have neuroprotective qualities. When KBs are used in cell-culture systems or infused into animal models, numerous reports have indicated neuroprotection (Table 14.2). For example, infusion of KB into rodents protects them from glutatmate toxicity,[71] ischemia[72] and 1-methyl-4-phenyl-1,2,3,6-tetrahydropyridine (MPTP) toxicity.[73] For a complete review the reader is referred a review of this topic, see ref. 61.

Ketone bodies have several properties that make them advantageous to conditions of low food availability. First, they are derived from abundant fat stores in body and can be produced in large amounts. Secondly, they are a very efficient fuel, requiring only 3 steps for conversion to acetyl-CoA. Thirdly, they are metabolised without the use of carnitine. Fourthly, the metabolism of KBs spare muscle catabolism of proteins. Fifthly, KB may increase autophagy, thereby removing damaged components from the cell.

Ketone bodies are a readily metabolised substrate for energy generation and many studies have shown that KB can lead to increased metabolic efficiency.

Table 14.1 Neuroprotection of ketogenic diets.

Injury	Lesion	Species	Outcome	Reference
Amyotrophic lateral sclerosis	Transgenic SOD1 mouse	Mice	Increased motor neuron counts	67
Traumatic brain injury	Controlled cortical impact	Rats	Reduced contusion volume	68
Alzheimer's disease	Transgenic APP expression	Mice	Reduced Aβ levels	66
KA-induced seizures	Kainic acid	Mice	Increased cell survival	100
GLUT1 haploinsufficiency	Glucose deprivation	Human	Decreased seizure frequency	101
Parkinson's disease	Human PD patients	Human	Improved motor function	69
Ischemia	Cardiac arrest induced ischemia	Rats	Protection from neurodegeneration	102

Table 14.2 Neuroprotective effects of ketone bodies.

Intervention	Injury	Lesion	Species	Outcome	Reference
Injection of acetoacetate	Glutamate toxicity	Inhibition of glycolysis by iodoacetate	Rat, cell culture	Neuroprotection	103
Infusion of 4 mM BHB, 5 mM ACA	Glutamate toxicity	Incubation with 5 mM glutamate	Cell culture	Increased cell survival	71
Infusion of BHB	Glutamate toxicity	Glutamate and iodoaceate treatment	Rats	Neuroprotection and reduced lipid peroxidation	104
1 mM BHB 1 mM ACA	Glutamate toxicity	Glutamate treatment	Cell culture	Increased mitochon-drial efficiency	105
4 mM BHB	Hypoxia	2-h exposure to hypoxia	Cell culture	Increased cell survival	106
Infusion BHB	Hypoxia	Carotid artery ligation	Mice	Maintained ATP and low lactate	107
Infusion BHB	Ischemia	Occulsion of middle cerebral artery	Mice	Reduced cerebral infarct area	72
Infusion BHB	Traumatic brain injury	Controlled cortical impact	Rats	Restored ATP levels after CCI	108
Ketogenic agent	Alzheimer's disease	Memory problems in Alzheimer's disease	Human	Improved cognitive performance	90
BHB treatment	Alzheimer's disease	Aβ in cell-culture model of AD	Cell culture	Increased cell survival	83
BHB infusion	Parkinson's disease	MPTP lesioning	Mice	Improved neuronal survival, improved mitochondrial efficiency	73
BHB treatment	Parkinson's disease	Rotenone treatment of cells	Cell culture	Increased cell survival	109

Early studies in spermatozoa revealed that both ACA and BHB increased sperm motility, while at the same time decreasing oxygen consumption by 10–29%.[74,75] Veech and coauthors have extensively studied the metabolic efficiency of KBs. Working with Veech, Sato *et al.* treated perfused rat hearts with levels of KBs found during starvation (4 mM BHB and 1 mM ACA) and compared the rates of oxygen consumption with heart cells treated with glucose alone. Ketone body treatment was found to increase cardiac efficiency by 13% compared to control hearts. Sato *et al.* further examined changes in metabolites during administration of KB in the hearts. By this method, the authors attributed the increase in metabolic efficiency to several key factors, including; increasing acetyl-CoA pools and increasing the strength of the NADH/NAD redox couple (see Figure 14.1). When BHB is converted to ACA it generates NADH, thereby reducing the mitochondrial NAD/NADH couple and increasing the oxidation of the coenzyme Q couple. This increase in redox potential was proposed as a key source of improved mitochondrial efficiency.[76] In addition, other authors have demonstrated that KB metabolism increases the mitochondrial pool of succinate (Figure 14.1) allowing bypass of complex-I deficiencies.[73]

Studzinski *et al.* examined mitochondrial efficiency after treating aged beagle dogs (age 9–11 years) with a ketogenic agent for two months. This study utilised unique fats called medium-chain triglycerides (MCTs) to induce ketosis in test animals. MCTs are triglycerides comprised of fatty-acid chains between 5–12 carbons. Due to the short chain lengths of medium chain fatty acids, MCT are not subject to the regulation imposed on long-chain fatty acids and are well known for their ketogenic properties.[77] Importantly, the oxidation of medium-chain fatty acids occurs regardless of carbohydrate in the diet, and differs substantially from a ketogenic diet in that no restriction of carbohydrate or protein intake is required. The ketogenic treatment led to modest levels of ketosis, in the range of 0.1 to 0.3 mM. Mitochondria were isolated from parietal and frontal lobes, washed free of KBs, and measured for oxygen consumption and metabolic efficiency. Mitochondria isolated from the parietal lobe of the ketogenic dogs demonstrated both improved respiration rates and an increased ability to drive electrons through complex I of the mitochondrial ETC. The net effect of these changes is increased metabolic efficiency. In addition, the authors examined signs of oxidative damage in both cytoplasmic and mitochondrial fractions from the parietal lobe. Significant reductions were found in both protein carbonyls and 3-nitrotyrosine specifically in the mitochondrial fraction. No changes were seen in the cytoplasmic fraction, suggesting that KBs act in the mitochondria to improve efficiency and reduce the generation of reactive oxygen species.[78] The reader is referred to several reviews that extensively cover the metabolic efficiency of KB metabolism.[42,61,79]

Since KBs are normally produced during periods of food deprivation they may have other survival properties that confer beneficial properties to neurons. Incubating cells with KBs has been demonstrated to increase rates of chaperone-mediated autophagy. Autophagy (self-eating) is a mechanism where large cellular structures, such as organelles and foreign bodies, can be delivered to lysosomes for degradation and recycling. Autophagy is increased when cells are starved of nutrients, in particular amino acids. Human embryonic fibroblasts treated with 1 to 10 mM BHB demonstrated increased degradation of long-lived proteins and increased autophagic activity. This was attributed to increased oxidation of cellular components by BHB and activation of recycling pathways.[80] Autophagy may function under normal conditions to rid cells of damaged or accumulated proteins. This may be particularly useful in neuronal cells. Inhibition of autophagy in several model systems promotes neurodegeneration.[81] Autophagy may be important mechanism in AD, and may function to clear axonal obstructions that result from misprocessing of APP.[82] However, this model has yet to be tested.

14.12 Ketosis and Alzheimer's Disease

Few studies have directly tested the effects of a KD or KBs in AD. Yet, the studies from cell culture, mice, and humans are encouraging. In cell-culture models of AD, exposure of cultured hippocampal cells from 18-day embryonic rats to 5 μM Aβ42 resulted in a 50% decrease in cell number. When the cells were simultaneously exposed to 4 mM BHB, a doubling of cell survival was observed, suggesting the BHB offers protection from Aβ toxicity.[83] Van de Auwera *et al.* tested a KD in a transgenic line of mice expressing the "London" APP mutation (V717I) driven by the THY1 promoter. These animals exhibit significant levels of Aβ at as little as 3 months of age and extensive plaque deposition by 12–14 months.[84] Sixteen, 3-month-old, female mice were fed either a ketogenic chow or standard chow for 43 days. The ketogenic chow consisted primarily of lard and butter fat. It was 79% fat (mostly saturated fats) and less than 1% carbohydrate. KB levels were significantly elevated during the treatment period ranging from 2–9 mM. Among the animals on the KD, a significant 25% reduction in levels of both Aβ40 and Aβ42 was noted (Figure 14.4).[66] This result may seem to contradict other findings in transgenic mouse models that found that high fat diets increased Aβ loads.[85,86] However, in the earlier studies fat was added to the diet without decreasing carbohydrate content, therefore these were not KDs, instead they were high-fat, high-carbohydrate diets more similar to a Western diet. Thus, when evaluating studies that examine dietary effects on disease, it is important to look at the entire macronutrient content of diet. Diets that restrict carbohydrates result in a physiologic profile that is very similar to fasting, regardless of the fat content of the diet.[87]

The cognitive effects of a KD in human AD patients have not been reported. One factor hampering such a study may be the difficulty in compliance in AD patients to a low carbohydrate KD. Changes in food selection toward sweet foods have been noted multiple times in AD. Mungas *et al.* examined food preference using a telephone survey comparing 31 patients with probable AD with 43 normal elderly controls. Participants with probable AD showed a preference for both sweet high-fat foods and sweet low-fat foods. The authors

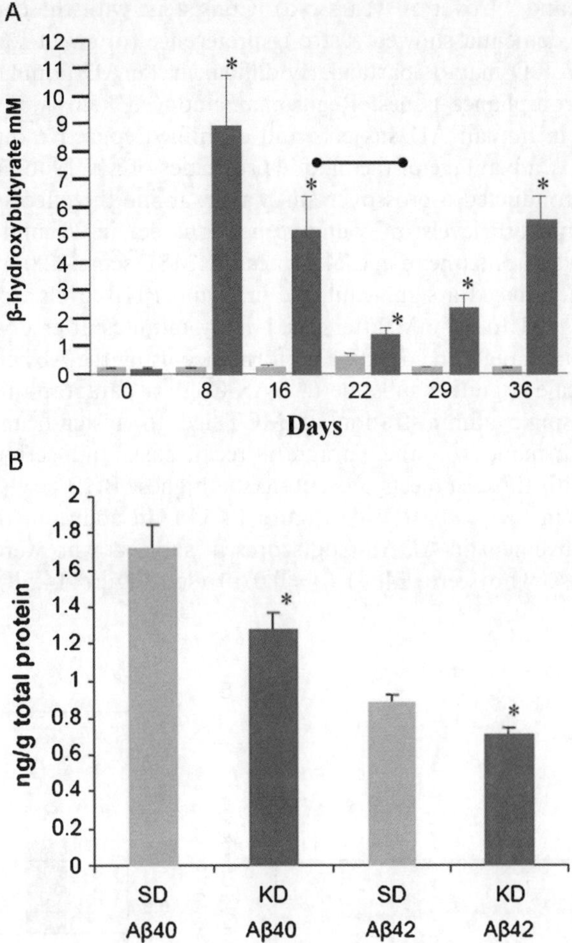

Figure 14.4   Ketogenic diet in a mouse model of AD. Ketogenic diet increases circulating ketone bodies and reduces total Aβ load in mouse model of AD. (A) Ketogenic diet led to rapid and substantial elevation of serum β-hydroxybutyrate. Ketogenic diet (KD) shown in dark grey, standard diet (SD) shown in light grey, error bars represent standard error of the mean. (B) After 43 days on diet, levels of both Aβ40 and Aβ42 were reduced 25% in mice fed ketogenic chow. * represent significant p-values ($p < 0.05$) Graphs from Van de Auwera *et al.*[66] Copyright 2005.

conclude: "Results provide preliminary evidence that craving for sweet food may be a significant part of the clinical syndrome of dementia, but further research is needed to delineate the psychological and biological mechanisms accounting for it".[88] Keene and Hope investigated the reports from caregivers that those suffering from dementia showed a marked increase in the desire for sweet foods. The authors examined 17 patients with dementia, 14 patients with dementia and who were hyperphagic, and 32 nondemented controls (18 under

50 years old and 14 over 50 years old). Consistent with the earlier findings, subjects with dementia showed a strong preference for sweet foods.[89] Hence, compliance to KD may be particularly difficult in an AD population.

To avoid compliance issues, Reger *et al.* induced ketosis without dietary modification in human AD subjects and examined cognitive outcomes. This study also took advantage of the unique properties of MCTs to induce ketosis. Reger *et al.* conducted a crossover study to examine the effects of acute elevation of serum KB levels on cognitive performance in 20 mild to moderate probable AD subjects (mean age 74.7; mean MMSE score 22.0). A single 40-g dose of MCT induced a significant rise in serum BHB levels from a baseline value of 0.04 mM to 0.5 mM after 2 h. Ninety minutes after dosing, subjects were tested for changes in cognitive performance using the Alzheimer's disease assessment scale-cognitive subscale (ADAS-Cog), a paragraph-recall test and others. The single administration of MCT led to a significant correlation between performance on the paragraph-recall task and serum BHB concentration, with those subjects presenting the highest BHB levels showing the most improvement ($p = 0.02$) (see Figure 14.5A). In addition, there was significant improvement in ADAS-Cog scores in subjects who were E4(−) compared to those who were E4(+) ($p = 0.039$) (see Figure 14.5B). The rapid,

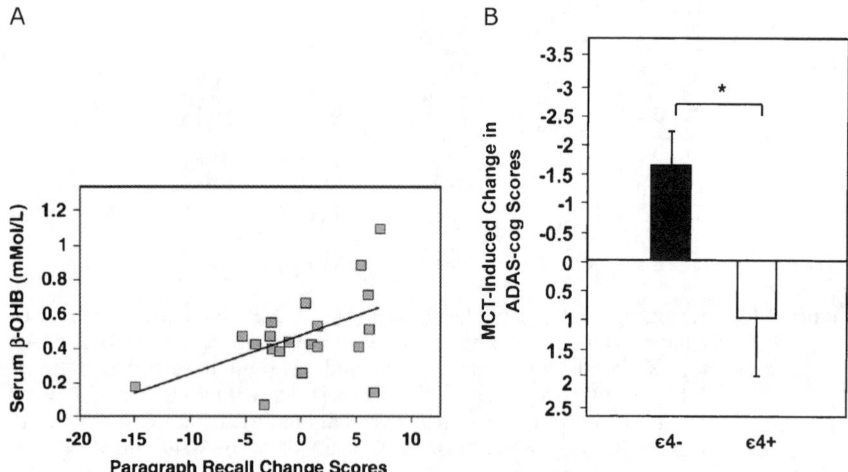

Figure 14.5   Acute ketosis in Alzheimer's disease patients improves cognitive performance. Ketosis induced by consumption of medium-chain triglycerides led to rapid improvement in cognitive tests in a mild to moderate Alzheimer's disease population. (A) Serum β-hydroxybutyrate levels correlate with improvement in the paragraph recall task $r = 0.50$, $p = 0.02$. (B) Bars represent average change from baseline in ADAS-Cog score, black bar represents APOE4(−) subjects, white bar represents APOE4(+) subjects. 1.5 point improvement in ADAS-Cog scores were observed in APOE4(−) subjects 90 minutes after dosing, $p = 0.039$. Graphs from Reger *et al.*[90] Copyright (2004) Elsevier Publishing.

single-dose, improvement seen in cognitive tasks suggests that the effect is driven by enhanced neuronal metabolism.[90]

In a follow-up study, Henderson *et al.* examined the effects on cognition of chronic induction of ketosis in mild to moderate AD patients (mean MMSE of 23). Ketosis was induced by daily dosing of 20 g of MCTs for 90 days, in a randomised, double-blind, placebo-controlled, multicentre trial conducted at 23 clinical sites within the United States. When examined on days 45 and 90, 20 g of MCTs successfully induced significant rises in serum BHB levels. Two hours after administration, mean BHB levels were 0.36 mM on day 45 and 0.39 mM on day 90. After 45 days of treatment, a significant 1.9-point difference in mean change from baseline in ADAS-Cog scores between placebo and the MCT groups was noted. As in the earlier study, a significant interaction was found between cognitive outcomes and APOE4 carriage status. Subjects who lacked the E4 allele, performed significantly better in ADAS-Cog scores compared to placebo at both days 45 and 90. In addition, those subjects who were E4(–) and were administered greater than 80% of the total required dosage showed a significant 5.3 point difference from placebo in change from baseline ADAS-Cog scores at day 90. As with the earlier study, postdose serum BHB levels correlated with improvement in ADAS-Cog scores, suggesting the induction of ketosis may be beneficial to AD patients, particularly if they lack an APOE4 allele (Figure 14.6).[91]

Importantly, the level of ketosis achieved in both the Reger *et al.* and the Henderson *et al.* study are considerably lower than those used in infusion studies, and lower than levels predicted for efficacy based on theoretical arguments.[42] This suggests that easily achievable levels of ketosis may be beneficial to the AD patient in the absence of dietary changes.

14.13   The Pharmacogenomics of APOE4 and Induced Ketosis

In the Reger *et al.* and Henderson *et al.* studies, efficacy was most notable in E4(–) subjects.[90,91] Interestingly, APOE4 effects have been noted in other studies targeting the insulin pathway as a treatment for AD. In a series of studies, Craft and coauthors found that E4(–) subjects rapidly responded to treatment with glucose and insulin[92] and when exposed to nasal insulin.[93] E4 effects were also seen in a larger study with the insulin sensitising agent rosiglitazone. E4(–) subjects taking an 8 mg dose of rosiglitazone demonstrated a significant difference in ADAS-Cog when compared to placebo scores after 24 weeks. E4(+) subjects show no difference or got slightly worse.[94]

Why would these treatments work preferentially in participants lacking the AD risk factor APOE4? In the rosiglitazone study it was proposed that fragments of ApoE4 protein interfered with mitochondrial function and prevented E4 carriers from responding to PPARγ activation.[94] It is possible that such damaged mitochondria also fail to metabolise KBs.

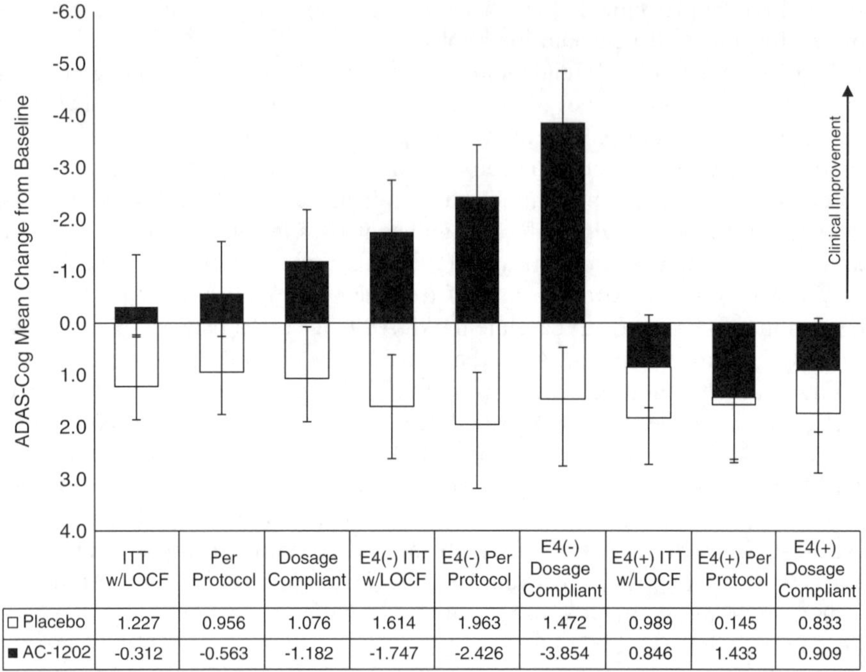

	ITT w/LOCF	Per Protocol	Dosage Compliant	E4(-) ITT w/LOCF	E4(-) Per Protocol	E4(-) Dosage Compliant	E4(+) ITT w/LOCF	E4(+) Per Protocol	E4(+) Dosage Compliant
□ Placebo	1.227	0.956	1.076	1.614	1.963	1.472	0.989	0.145	0.833
■ AC-1202	-0.312	-0.563	-1.182	-1.747	-2.426	-3.854	0.846	1.433	0.909

Figure 14.6 Chronic ketosis in Alzheimer's disease patients improves cognitive performance. Summary graph of mean change in ADAS-Cog score from baseline after 90 days of treatment with MCT or Placebo. The MCT used in this study was a caprylic triglyceride referred to as AC-1202. Graph shows analysis of different population groups in the study including, the intention-to-treat with last observation carried forward (ITT w/LOCF), per protocol, and dosage-compliant groups, each stratified by APOE4 carriage status. Black columns represent subjects receiving AC-1202. White columns represent subjects receiving placebo. Error bars represent standard error of the mean. Table displays mean change from baseline for each group. Mean changes from baseline was largest in APOE4(−) subjects who were dosage compliant. Graph from Henderson *et al.*[91] Copyright 2005.

An alternative hypothesis relates to the role of APOE4 in AD. In a study of 25 AD subjects, E4(−) carriers were found to have higher fasting insulin levels than normal controls.[95] In a follow-up study of 31 subjects with mild to moderate probable AD, E4(−) subjects were found to have significantly lower glucose disposal rates.[92] High fasting insulin and low glucose disposal rates both suggest that E4(−) subjects are insulin resistant. The insulin resistance of E4(−) subjects may explain their responsiveness to KBs. KBs are relatively impermeable to the blood/brain barrier. KBs are transported into the brain by the monocarboxylate transporter (MCTs) carrier proteins. These carrier proteins transport short-chain monocarboxylic acids such as ketone bodies and lactate.[96] MCT1 is widely expressed and found in endothelial cells of the

BBB.[97] The levels of monocarboxylate transporters decrease after weaning but are known to be elevated during fasting and in other conditions where insulin resistance occurs.[98] Mild insulin resistance in E4(−) subjects may increase the levels of MCT1 transporter proteins and allow for better transport of KBs into the brain.

It is tempting to speculate that the improved insulin sensitivity and unresponsiveness to KBs in E4(+) subjects may be related to their increased risk of developing AD and to the targeting of the default network. This model is summarised in Figure 14.7. Studies from the default network suggest that

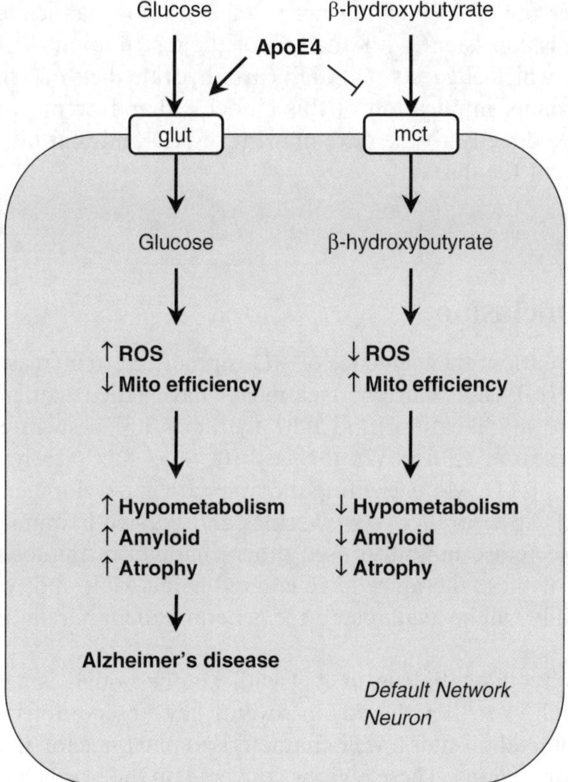

Figure 14.7 Influence and consequences of ApoE4 on neuronal metabolism. The differential binding of ApoE4 protein to triglyceride rich lipoprotein particles influences insulin sensitivity and affects transport of glucose and ketone bodies. By suppressing fat metabolism, ApoE4 increases insulin sensitivity, promoting the transport of glucose, and suppressing the transport of β-hydroxybutyrate, resulting in an increased reliance on glucose by the default network. Metabolism of ketone bodies reduces reactive oxygen species (ROS) and improves mitochondrial efficiency. Overreliance on glucose in APOE4 carriers leads to generation of ROS resulting in a cascade of hypometabolism, amyloid deposition and atrophy, ultimately leading to the pathology of Alzheimer's disease.

continuous high metabolic rates of glucose use predispose those regions to metabolic disruption, amyloid accumulation, and cell atrophy.[53] Metabolism of KBs may reduce oxidative damage and improve mitochondrial efficiency, thereby lessening the damage that occurs over time.[59,60] If the default network of APOE4 carriers is unable to efficiently transport KBs due to increased insulin sensitivity, they will not benefit from the metabolic efficiency imparted by KB metabolism. Therefore, APOE4 carriers will be chronically more dependent on glucose, and may generate more reactive oxygen species (ROS) relative to non-APOE4 carriers. Over time, the increased ROS will damage mitochondria and reduce energy generation, resulting in hypometabolism and the necessity of compensatory activation. The chronic hypometabolism may alter the processing of APP giving rise to the pathology associated with AD.[49] Such a model is consistent with analysis of the distribution of APOE4 across ethnic groups, which suggests that high carbohydrate diets are discordant with APOE4. A broader implication of this model is that current dietary practices may contribute directly to the development of AD and that dietary modification may prevent the disease.[8]

14.14 Conclusion

Without intervention, the incidence of AD is predicted to increase dramatically in the next forty years. Current treatments have failed to demonstrate significant disease modification, and new therapies are desperately needed. A tremendous research effort over the last 10 years has greatly increased the understanding of AD. Many exciting therapies are in development, particularly in the areas of Aβ clearance, *e.g.* vaccines and passive immunity, or Aβ prevention, *e.g.* secretase inhibitors (see other chapters in this book). Encouragingly, many of these therapies have entered clinic trials and, with successful results, hopefully will be available to the general patient population in the near future.

Addressing hypometabolism in AD with ketone bodies is a relatively new area of research. Yet, it is based on several key observations in the disease process. Hypometabolism is a well-characterised phenomenon in AD and maps to discrete brain regions. These regions converge on the brains default network. The default network may be conducive to attack by Alzheimer's disease due to its continuous activity and high rates of glucose metabolism. Hypometabolism occurs very early in the disease process, and can be detected in at-risk individuals decades before any clinical symptoms of AD are present. Ketone bodies are an attractive means to address hypometabolism. Ketone bodies are naturally produced by the body in times of low glucose availability and have been shown to be readily metabolised by the brain, even under normal blood glucose concentrations.[99] Ketone bodies are a very efficient fuel and may impart improved mitochondrial efficiency in the brain. In addition, metabolism of ketone bodies may reduce oxidative damage in the mitochondria in the brain.

Importantly, preliminary studies with the use of ketogenic MCTs gave encouraging results.[90,91] Future research will explore the utility of this approach and may lead to new therapeutic avenues into the treatment and prevention of AD.

Disclosure and Acknowledgements

The author of this paper is an employee of Accera Inc., a company developing ketone-body-based therapies for Alzheimer's disease. The author is the sole inventor of one issued patent (US 6835750) entitled: Use of medium chain triglycerides for the treatment and prevention of Alzheimer's disease and other diseases resulting from reduced neuronal metabolism II. The author and Accera Inc. have other published pending patent applications in this area: US 2002/0006959 A1, US 2003/0059824 A1, US 2006/0122270 A1, US 2008/0009467 A1, US 2006/0252775 A1, US 2007/0135376 A1, US 2008/0287372 A1, US 2007/0179197 A1.

References

1. S. Salloway and S. Correia, *Cleve. Clin. J. Med.*, 2009, **76**, 49–58.
2. D. J. Selkoe, *Ann. Intern. Med.*, 2004, **140**, 627–638.
3. E. M. Reiman, K. Chen, G. E. Alexander, R. J. Caselli, D. Bandy, D. Osborne, A. M. Saunders and J. Hardy, *Proc. Natl. Acad. Sci. USA.*, 2004, **101**, 284–289.
4. J. P. Blass and A. Zemcov, *Neurochem. Pathol.*, 1984, **2**, 103–114.
5. R. A. Velliquette, T. O'Connor and R. Vassar, *J. Neurosci.*, 2005, **25**, 10874–10883.
6. L. Bertram and R. E. Tanzi, *Nature Rev. Neurosci.*, 2008, **9**, 768–778.
7. W. J. Strittmatter and A. D. Roses, *Annu. Rev. Neurosci.*, 1996, **19**, 53–77.
8. S. T. Henderson, *Med. Hypotheses*, 2004, **62**, 689–700.
9. R. W. Mahley, K. H. Weisgraber and Y. Huang, *Proc. Natl. Acad. Sci. USA.*, 2006, **103**, 5644–5651.
10. R. N. Kalaria, G. E. Maestre, R. Arizaga, R. P. Friedland, D. Galasko, K. Hall, J. A. Luchsinger, A. Ogunniyi, E. K. Perry, F. Potocnik, M. Prince, R. Stewart, A. Wimo, Z. X. Zhang and P. Antuono, *Lancet Neurol.*, 2008, **7**, 812–826.
11. D. D. Clarke and L. Sokoloff, in *Basic Neurochemistry,* ed. G. J. Siegel, B. W. Agranoff, R. W. Albers and P. B. Molinoff, Raven Press, New York, 1994, pp. 645–680.
12. M. E. Raichle and M. A. Mintun, *Annu. Rev. Neurosci.*, 2006, **29**, 449–476.
13. N. R. Sibson, A. Dhankhar, G. F. Mason, K. L. Behar, D. L. Rothman and R. G. Shulman, *Proc. Natl. Acad. Sci. USA.*, 1997, **94**, 2699–2704.
14. R. G. Shulman, D. L. Rothman, K. L. Behar and F. Hyder, *Trends Neurosci.*, 2004, **27**, 489–495.

15. N. R. Sibson, A. Dhankhar, G. F. Mason, D. L. Rothman, K. L. Behar and R. G. Shulman, *Proc. Natl. Acad. Sci. USA.*, 1998, **95**, 316–321.

16. D. Attwell and S. B. Laughlin, *J. Cereb. Blood Flow Metab.*, 2001, **21**, 1133–1145.

17. L. Pellerin and P. J. Magistretti, *Proc. Natl. Acad. Sci. USA.*, 1994, **91**, 10625–10629.

18. F. Hyder, A. B. Patel, A. Gjedde, D. L. Rothman, K. L. Behar and R. G. Shulman, *J. Cereb. Blood Flow Metab.*, 2006, **26**, 865–877.

19. D. S. Dwyer, S. J. Vannucci and I. A. Simpson, in *Glucose Metabolism in the Brain,* ed. D. S. Dwyer, Academic Press, London, Editon edn, 2002, **vol. 51**, pp. 159–188.

20. D. Wang, J. M. Pascual, H. Yang, K. Engelstad, S. Jhung, R. P. Sun and D. C. De Vivo, *Ann. Neurol.*, 2005, **57**, 111–118.

21. G. Seidner, M. G. Alvarez, J. I. Yeh, K. R. O'Driscoll, J. Klepper, T. S. Stump, D. Wang, N. B. Spinner, M. J. Birnbaum and D. C. De Vivo, *Nature Genet.*, 1998, **18**, 188–191.

22. M. J. de Leon, S. H. Ferris, A. E. George, D. R. Christman, J. S. Fowler, C. Gentes, B. Reisberg, B. Gee, M. Emmerich, Y. Yonekura, J. Brodie, Kricheff II and A. P. Wolf, *AJNR. Am. J. Neuroradiol.*, 1983, **4**, 568–571.

23. S. Minoshima, B. Giordani, S. Berent, K. A. Frey, N. L. Foster and D. E. Kuhl, *Ann. Neurol.*, 1997, **42**, 85–94.

24. R. P. Friedland, T. F. Budinger, E. Ganz, Y. Yano, C. A. Mathis, B. Koss, B. A. Ober, R. H. Huesman and S. E. Derenzo, *J. Comput. Assist. Tomogr.*, 1983, **7**, 590–598.

25. E. M. Reiman, R. J. Caselli, L. S. Yun, K. Chen, D. Bandy, S. Minoshima, S. N. Thibodeau and D. Osborne, *N. Engl. J. Med.*, 1996, **334**, 752–758.

26. G. W. Small, L. M. Ercoli, D. H. Silverman, S. C. Huang, S. Komo, S. Y. Bookheimer, H. Lavretsky, K. Miller, P. Siddarth, N. L. Rasgon, J. C. Mazziotta, S. Saxena, H. M. Wu, M. S. Mega, J. L. Cummings, A. M. Saunders, M. A. Pericak-Vance, A. D. Roses, J. R. Barrio and M. E. Phelps, *Proc. Natl. Acad. Sci. USA.*, 2000, **97**, 6037–6042.

27. S. Gauthier, B. Reisberg, M. Zaudig, R. C. Petersen, K. Ritchie, K. Broich, S. Belleville, H. Brodaty, D. Bennett, H. Chertkow, J. L. Cummings, M. de Leon, H. Feldman, M. Ganguli, H. Hampel, P. Scheltens, M. C. Tierney, P. Whitehouse and B. Winblad, *Lancet*, 2006, **367**, 1262–1270.

28. A. Drzezga, N. Lautenschlager, H. Siebner, M. Riemenschneider, F. Willoch, S. Minoshima, M. Schwaiger and A. Kurz, *Eur. J. Nucl. Med. Mol. Imaging*, 2003, **30**, 1104–1113.

29. L. Mosconi, R. Mistur, R. Switalski, W. H. Tsui, L. Glodzik, Y. Li, E. Pirraglia, S. De Santi, B. Reisberg, T. Wisniewski and M. J. de Leon, *Eur. J. Nucl. Med. Mol. Imaging*, 2009, **36**, 811–822.

30. E. H. Corder, V. Jelic, H. Basun, L. Lannfelt, S. Valind, B. Winblad and A. Nordberg, *Arch. Neurol.*, 1997, **54**, 273–277.

31. L. Mosconi, B. Nacmias, S. Sorbi, M. T. De Cristofaro, M. Fayazz, A. Tedde, L. Bracco, K. Herholz and A. Pupi, *J. Neurol. Neurosurg. Psychiatry*, 2004, **75**, 370–376.
32. M. Piert, R. A. Koeppe, B. Giordani, S. Berent and D. E. Kuhl, *J. Nucl. Med.*, 1996, **37**, 201–208.
33. J. T. Becker, M. A. Mintun, K. Aleva, M. B. Wiseman, T. Nichols and S. T. DeKosky, *Neurology*, 1996, **46**, 692–700.
34. R. A. Sperling, J. F. Bates, E. F. Chua, A. J. Cocchiarella, D. M. Rentz, B. R. Rosen, D. L. Schacter and M. S. Albert, *J. Neurol. Neurosurg. Psychiatry*, 2003, **74**, 44–50.
35. C. L. Grady, A. R. McIntosh, S. Beig, M. L. Keightley, H. Burian and S. E. Black, *J. Neurosci.*, 2003, **23**, 986–993.
36. S. Y. Bookheimer, M. H. Strojwas, M. S. Cohen, A. M. Saunders, M. A. Pericak-Vance, J. C. Mazziotta and G. W. Small, *N. Engl. J. Med.*, 2000, **343**, 450–456.
37. M. W. Bondi, W. S. Houston, L. T. Eyler and G. G. Brown, *Neurology*, 2005, **64**, 501–508.
38. N. Filippini, B. J. MacIntosh, M. G. Hough, G. M. Goodwin, G. B. Frisoni, S. M. Smith, P. M. Matthews, C. F. Beckmann and C. E. Mackay, *Proc. Natl. Acad. Sci. USA.*, 2009, **106**, 7209–7214.
39. S. Chang, T. ran Ma, R. D. Miranda, M. E. Balestra, R. W. Mahley and Y. Huang, *Proc. Natl. Acad. Sci. USA.*, 2005, **102**, 18694–18699.
40. R. W. Mahley and Y. Huang, *Acta. Neurol. Scand. Suppl.*, 2006, **185**, 8–14.
41. H. Atamna and W. H. Frey 2nd, *Mitochondrion*, 2007, **7**, 297–310.
42. R. L. Veech, B. Chance, Y. Kashiwaya, H. A. Lardy and G. F. Cahill Jr, *IUBMB Life*, 2001, **51**, 241–247.
43. L. Devi, B. M. Prabhu, D. F. Galati, N. G. Avadhani and H. K. Anandatheerthavarada, *J. Neurosci.*, 2006, **26**, 9057–9068.
44. X. Wang, B. Su, G. Perry, M. A. Smith and X. Zhu, *Free. Radic. Biol. Med.*, 2007, **43**, 1569–1573.
45. P. H. Reddy and M. F. Beal, *Trends Mol. Med.*, 2008, **14**, 45–53.
46. X. Wang, B. Su, L. Zheng, G. Perry, M. A. Smith and X. Zhu, *J. Neurochem.*, 2009, **109**(Suppl 1), 153–159.
47. W. S. Liang, E. M. Reiman, J. Valla, T. Dunckley, T. G. Beach, A. Grover, T. L. Niedzielko, L. E. Schneider, D. Mastroeni, R. Caselli, W. Kukull, J. C. Morris, C. M. Hulette, D. Schmechel, J. Rogers and D. A. Stephan, *Proc. Natl. Acad. Sci. USA.*, 2008, **105**, 4441–4446.
48. D. Gabuzda, J. Busciglio, L. B. Chen, P. Matsudaira and B. A. Yankner, *J. Biol. Chem.*, 1994, **269**, 13623–13628.
49. T. O'Connor, K. R. Sadleir, E. Maus, R. A. Velliquette, J. Zhao, S. L. Cole, W. A. Eimer, B. Hitt, L. A. Bembinster, S. Lammich, S. F. Lichtenthaler, S. S. Hebert, B. De Strooper, C. Haass, D. A. Bennett and R. Vassar, *Neuron*, 2008, **60**, 988–1009.
50. M. E. Raichle, A. M. MacLeod, A. Z. Snyder, W. J. Powers, D. A. Gusnard and G. L. Shulman, *Proc. Natl. Acad. Sci. USA.*, 2001, **98**, 676–682.

51. R. L. Buckner, J. R. Andrews-Hanna and D. L. Schacter, *Ann. N Y Acad. Sci.*, 2008, **1124**, 1–38.
52. M. D. Greicius, B. Krasnow, A. L. Reiss and V. Menon, *Proc. Natl. Acad. Sci. USA.*, 2003, **100**, 253–258.
53. R. L. Buckner, A. Z. Snyder, B. J. Shannon, G. LaRossa, R. Sachs, A. F. Fotenos, Y. I. Sheline, W. E. Klunk, C. A. Mathis, J. C. Morris and M. A. Mintun, *J. Neurosci.*, 2005, **25**, 7709–7717.
54. R. M. Corbo and R. Scacchi, *Ann. Hum. Genet.*, 1999, **63**(Pt 4), 301–310.
55. J. V. Neel, *Am. J. Hum. Genet.*, 1962, **14**, 353–352.
56. J. V. Neel, *Nutr. Rev.*, 1999, **57**, S2–9.
57. J. Gowdy, *Limited Wants, Unlimited Means: A Reader On Hunter-Gatherer Economics And The Environment*, Island Press, Washington, DC, 1998.
58. L. Cordain, M. R. Eades and M. D. Eades, *Comp. Biochem. Physiol. A Mol. Integr. Physiol.*, 2003, **136**, 95–112.
59. S. T. Henderson, *Neurotherapeutics.*, 2008, **5**, 470–480.
60. M. Maalouf, J. M. Rho and M. P. Mattson, *Brain. Res. Rev.*, 2008.
61. M. L. Prins, *J. Cereb. Blood Flow Metab.*, 2008, **28**, 1–16.
62. J. Freeman, P. Veggiotti, G. Lanzi, A. Tagliabue and E. Perucca, *Epilepsy Res.*, 2006, **68**, 145–180.
63. J. Klepper, H. Scheffer, B. Leiendecker, E. Gertsen, S. Binder, M. Leferink, C. Hertzberg, A. Nake, T. Voit and M. A. Willemsen, *Neuropediatrics*, 2005, **36**, 302–308.
64. O. E. Owen, A. P. Morgan, H. G. Kemp, J. M. Sullivan, M. G. Herrera and G. F. Cahill Jr, *J. Clin. Invest.*, 1967, **46**, 1589–1595.
65. A. K. Taggart, J. Kero, X. Gan, T. Q. Cai, K. Cheng, M. Ippolito, N. Ren, R. Kaplan, K. Wu, T. J. Wu, L. Jin, C. Liaw, R. Chen, J. Richman, D. Connolly, S. Offermanns, S. D. Wright and M. G. Waters, *J. Biol. Chem.*, 2005, **280**, 26649–26652.
66. I. Van der Auwera, S. Wera, F. Van Leuven and S. T. Henderson, *Nutr. Metab. (Lond)*, 2005, **2**, 28.
67. Z. Zhao, D. J. Lange, A. Voustianiouk, D. MacGrogan, L. Ho, J. Suh, N. Humala, M. Thiyagarajan, J. Wang and G. M. Pasinetti, *BMC Neurosci.*, 2006, **7**, 29.
68. M. L. Prins, L. S. Fujima and D. A. Hovda, *J. Neurosci. Res.*, 2005, **82**, 413–420.
69. T. B. Vanitallie, C. Nonas, A. Di Rocco, K. Boyar, K. Hyams and S. B. Heymsfield, *Neurology*, 2005, **64**, 728–730.
70. P. G. Sullivan, N. A. Rippy, K. Dorenbos, R. C. Concepcion, A. K. Agarwal and J. M. Rho, *Ann. Neurol.*, 2004, **55**, 576–580.
71. H. S. Noh, Y. S. Hah, R. Nilufar, J. Han, J. H. Bong, S. S. Kang, G. J. Cho and W. S. Choi, *J. Neurosci. Res.*, 2006, **83**, 702–709.
72. M. Suzuki, M. Suzuki, Y. Kitamura, S. Mori, K. Sato, S. Dohi, T. Sato, A. Matsuura and A. Hiraide, *Jpn. J. Pharmacol.*, 2002, **89**, 36–43.

73. K. Tieu, C. Perier, C. Caspersen, P. Teismann, D. C. Wu, S. D. Yan, A. Naini, M. Vila, V. Jackson-Lewis, R. Ramasamy and S. Przedborski, *J. Clin. Invest.*, 2003, **112**, 892–901.

74. H. A. Lardy, R. G. Hansen and P. H. Phillips, *Arch. Biochem.*, 1945, **6**, 41–51.

75. H. A. Lardy and P. H. Phillips, *Arch. Biochem.*, 1945, **6**, 53–61.

76. K. Sato, K. Yoshihiro, C. A. Keon, N. Tsuchiya, M. T. King, G. K. Radda, B. Chance, K. Clarke and R. L. Veech, *Faseb. J.*, 1995, **9**, 651–658.

77. A. C. Bach and V. K. Babayan, *Am. J. Clin. Nutr.*, 1982, **36**, 950–962.

78. C. M. Studzinski, W. A. Mackay, T. L. Beckett, S. T. Henderson, M. P. Murphy, P. G. Sullivan and W. M. Burnham, *Brain Res.*, 2008, **1226**, 209–217.

79. M. Gasior, M. A. Rogawski and A. L. Hartman, *Behav. Pharmacol.*, 2006, **17**, 431–439.

80. P. F. Finn and J. F. Dice, *J. Biol. Chem.*, 2005.

81. M. Martinez-Vicente and A. M. Cuervo, *Lancet Neurol.*, 2007, **6**, 352–361.

82. G. B. Stokin and L. S. Goldstein, *Annu. Rev. Biochem.*, 2006, **75**, 607–627.

83. Y. Kashiwaya, T. Takeshima, N. Mori, K. Nakashima, K. Clarke and R. L. Veech, *Proc. Natl. Acad. Sci. USA.*, 2000, **97**, 5440–5444.

84. D. Moechars, I. Dewachter, K. Lorent, D. Reverse, V. Baekelandt, A. Naidu, I. Tesseur, K. Spittaels, C. V. Haute, F. Checler, E. Godaux, B. Cordell and F. Van Leuven, *J. Biol. Chem.*, 1999, **274**, 6483–6492.

85. L. Ho, W. Qin, P. N. Pompl, Z. Xiang, J. Wang, Z. Zhao, Y. Peng, G. Cambareri, A. Rocher, C. V. Mobbs, P. R. Hof and G. M. Pasinetti, *Faseb. J.*, 2004, **18**, 902–904.

86. J. A. Levin-Allerhand, C. E. Lominska and J. D. Smith, *J. Nutr. Health Aging*, 2002, **6**, 315–319.

87. R. D. Feinman, *Nutr. Metab. (Lond)*, 2005, **2**, 27.

88. D. Mungas, J. K. Cooper, P. G. Weiler, D. Gietzen, C. Franzi and C. Bernick, *J. Am. Geriatr. Soc.*, 1990, **38**, 999–1007.

89. J. M. Keene and T. Hope, *Appetite*, 1997, **28**, 167–175.

90. M. A. Reger, S. T. Henderson, C. Hale, B. Cholerton, L. D. Baker, G. S. Watson, K. Hyde, D. Chapman and S. Craft, *Neurobiol Aging*, 2004, **25**, 311–314.

91. S. T. Henderson, J. L. Vogel, L. J. Barr, F. Garvin, J. J. Jones and L. C. Costantini, *Nutr. Metab. (Lond)*, 2009, **6**, 31.

92. S. Craft, S. Asthana, G. Schellenberg, L. Baker, M. Cherrier, A. A. Boyt, R. N. Martins, M. Raskind, E. Peskind and S. Plymate, *Ann. N Y. Acad. Sci.*, 2000, **903**, 222–228.

93. M. A. Reger, G. S. Watson, W. H. Frey 2nd, L. D. Baker, B. Cholerton, M. L. Keeling, D. A. Belongia, M. A. Fishel, S. R. Plymate, G. D. Schellenberg, M. M. Cherrier and S. Craft, *Neurobiol. Aging*, 2006, **27**, 451–458.

94. M. E. Risner, A. M. Saunders, J. F. Altman, G. C. Ormandy, S. Craft, I. M. Foley, M. E. Zvartau-Hind, D. A. Hosford and A. D. Roses, *Pharmacogenomics J.*, 2006, **6**, 246–254.
95. S. Craft, E. Peskind, M. W. Schwartz, G. D. Schellenberg, M. Raskind and D. Porte Jr, *Neurology*, 1998, **50**, 164–168.
96. A. P. Halestrap and N. T. Price, *Biochem. J.*, 1999, **343**(Pt 2), 281–299.
97. M. K. Froberg, D. Z. Gerhart, B. E. Enerson, C. Manivel, M. Guzman-Paz, N. Seacotte and L. R. Drewes, *Neuroreport.*, 2001, **12**, 761–765.
98. A. M. Robinson and D. H. Williamson, *Physiol. Rev.*, 1980, **60**, 143–187.
99. S. G. Hasselbalch, P. L. Madsen, L. P. Hageman, K. S. Olsen, N. Justesen, S. Holm and O. B. Paulson, *Am., J. Physiol.*, 1996, **270**, E746–751.
100. H. S. Noh, Y. S. Kim, H. P. Lee, K. M. Chung, D. W. Kim, S. S. Kang, G. J. Cho and W. S. Choi, *Epilepsy Res.*, 2003, **53**, 119–128.
101. J. Klepper and B. Leiendecker, *Dev. Med. Child. Neurol.*, 2007, **49**, 707–716.
102. K. K. Tai, N. Nguyen, L. Pham and D. D. Truong, *J. Neural Transm.*, 2008, **115**, 1011–1017.
103. L. Massieu, M. L. Haces, T. Montiel and K. Hernandez-Fonseca, *Neuroscience*, 2003, **120**, 365–378.
104. J. Mejia-Toiber, T. Montiel and L. Massieu, *Neurochem. Res.*, 2006, **31**, 1399–1408.
105. M. Maalouf, P. G. Sullivan, L. Davis, D. Y. Kim and J. M. Rho, *Neuroscience*, 2007, **145**, 256–264.
106. R. Masuda, J. W. Monahan and Y. Kashiwaya, *J. Neurosci. Res.*, 2005, **80**, 501–509.
107. M. Suzuki, M. Suzuki, K. Sato, S. Dohi, T. Sato, A. Matsuura and A. Hiraide, *Jpn. J. Pharmacol.*, 2001, **87**, 143–150.
108. M. L. Prins, S. M. Lee, L. S. Fujima and D. A. Hovda, *J. Neurochem.*, 2004, **90**, 666–672.
109. K. Imamura, T. Takeshima, Y. Kashiwaya, K. Nakaso and K. Nakashima, *J. Neurosci. Res.*, 2006, **84**, 1376–1384.

Subject Index

Page references for Vol. 1 are given in Roman type. Page references for Vol. 2 are given in bold type.